BIOAEROSOLS

-

THE PROCEEDINGS OF THE 6TH INTERNATIONAL SCIENTIFIC CONFERENCE ON BIOAEROSOLS, FUNGI, BACTERIA, MYCOTOXINS IN INDOOR AND OUTDOOR ENVIRONMENTS AND HUMAN HEALTH.

SEPTEMBER 6 - 9, 2011

SARATOGA SPRINGS, NEW YORK, USA

Edited by

Dr. med Eckardt Johanning, M.D., M. Sc.

Philip R. Morey, Ph.D. CIH

Pierre Auger, M.D. M. Sc.

Published by the Fungal Research Group Foundation, Inc., Albany, New York

MOUNT SINAI SCHOOL OF MEDICINE

The conference meeting was organized by the Fungal Research Group Foundation, Inc., Albany, New York, USA in cooperation with U.S. ENVIRONMENTAL PROTECTION AGENCY and jointly sponsored with the Mount Sinai School of Medicine.

This activity was planned and implemented in accordance with the Essential Areas and policies of the Accreditation Council for Continuing Medical Education through the joint sponsorship of the Mount Sinai School of Medicine and the Fungal Research Group Foundation, Inc. The Mount Sinai School of Medicine is accredited by the ACCME to provide continuing medical education for physicians.

The production of this publication was made possible with funds from the U. S. Environmental Protection Agency (EPA), Office of Radiation and Indoor Air, Washington, D.C. (http://www.epa.gov/mold/index.html)

Publisher: Fungal Research Group Foundation, Inc., Albany, New York, N.Y.

www.fungalresearchgroup.com

Print: Boyd Printing, 595 New Loudon Road, #117, Latham NY 12110

info@boydprinting.com

Library of Congress Control Number: 2012953872

ISBN: 978-0-9709915-0-8

TABLE OF CONTENT:

PART III:

Prevention / Public Health

INDEX

INTRODUCTION AND OVERVIEW

Since our last meeting in 2003, the body of knowledge regarding indoor and occupational microbial exposure (fungi, bacteria and their allergenic, irritant and toxic by-products) and related diseases with important public health implications has grown significantly. Diseases such as allergy, asthma, inflammatory lung diseases, infections, and cases of mycotoxicosis and neurological or vascular disorders continue to be associated with exposure to bioaerosols. Complex reactions and interactions that result in adverse human health reactions pose great challenges to investigators, clinicians and public health officials.

Large scale natural disasters caused be storms and flooding that led to significant water damage and microbial contamination of homes and buildings have occurred in many countries throughout the world: USA, Australia, Pakistan, India, China, England, Poland, Germany, France, Belgium, and elsewhere. In addition to the human tragedy, the costs and technical challenges for cleaning and restoration are tremendous. How do we effectively protect the clean-up workers and the building occupants from harmful microbial contaminants (bacteria, mold, bio-toxins) and exposures?

We have learned that poor building and ventilation designs or maintenance, can contribute to increased microbial indoor exposure. Furthermore, the use of water-sensitive materials in areas with hot and humid climates, inadequate resources, as well as deficient renovation of existing architectural designs are factors that can lead to indoor mold growth. This has raised international concern about the impact of bioaerosols on the building occupants and workers' health.

Based on a focus group meeting at the Healthy Buildings 2009 meeting in Syracuse, N.Y., this 2011 Bioaerosols conference addressed the state of art research and practical experience to improve the understanding of microbials (bacteria, mold, bio- and mycotoxins), determine important agents and diagnosis of adverse human health effects, as well as explore new treatment approaches, and the control and prevention of such exposure. Scientific advances and knowledge gaps were discussed. Future research priorities were developed.

Precious artwork, books, paper documents and furniture contaminated with biologicals (mold, etc.) often contribute to occupant's exposure and patient complaints. Chemicals (called biocides) intended to kill bacteria and mold also are often harmful to humans and the environment. Little attention has been paid to these issues in hygiene practices and the systematic approach varies in the unregulated cleaning and restoration industry. Some say "a building keeps a memory" of a contamination, even after careful clean up! Some even suggest that such buildings or items should be condemned and destroyed. We would like to explore the scientific basis for safe materials and practice for the cleaning workers, users and building occupants.

Clearance criteria for re-occupancy and re-use shall be critically reviewed and any minimum consensus was explored

This meeting connected internationally-recognized researchers and leading investigators with “front line” practitioners and consultants addressing “real world” problems. We explored the scientific basis for what we do and recommend.

CONFERENCE RATIONALE, GOALS AND LEARNING OBJECTIVE

The 6th International Conference was a forum for the presentation and discussion scientific papers in the field of bioaerosols in order to enhance the knowledge of professionals in the field. .

There is a need for enhanced knowledge for health care practitioners and other professionals regarding the proper diagnosis, pathology and treatment of adverse health effects from bioaerosols exposures encountered in the environment and work place, in particular in indoor environments or workplaces that involve handling biological waste products (composting), wet and damp buildings, allergenic and toxic biological by-products from mold and bacteria.

The now dated National Academy of Science/ Institute of Medicine scientific committee formulated in 2004 that there is a public health interest in the topic of Indoor Dampness and Health. However, it also identified that there is further research required to learn about the causal connections of certain exposures, biological agents and adverse health effects. The following research gaps were identified (Institute of Medicine (U.S.), 2004):

- "Given the present state of the literature, the committee identified several kinds of research needs. Standard definitions of dampness, metrics, and associated dampness-assessment protocols need to be developed to characterize the nature, severity, and spatial extent of dampness... Any efforts to establish common definitions must be international in scope because excessive indoor dampness is a worldwide problem and research cooperation promoted the generation and dissemination of knowledge.
- Research is also needed to better characterize the dampness-related emissions of fungal spores, bacteria, and other particles of biologic origin and their role in human health outcomes; the microbial ecology of buildings, that is, the link between dampness, different building materials, microbial growth, and microbial interactions; and dampness-related chemical emissions from building materials and furnishings, and their role in human health outcomes.
- Studies should be conducted to evaluate the effect of the duration of moisture damage of materials and its possible influence on occupant health and to evaluate the effectiveness of various changes in building designs, construction methods, operation, and maintenance in reducing dampness problems…."
- "Indoor environments subject occupants to multiple exposures that may interact physically or chemically with one another and with the other characteristics of the environment, such as humidity, temperature, and ventilation rate. Few studies to date have considered whether there are additive or synergistic interacti-

ons among these factors. The committee encourages researchers to collect and analyze data on a broad range of exposures and factors characterizing indoor environments in order to inform these questions and possibly point the way toward more effective and efficient intervention strategies."

- "The committee encourages the CDC to pursue surveillance and additional research on acute pulmonary hemorrhage or hemosiderosis in infants to resolve questions regarding this serious health outcome. Epidemiologic and case studies should take a broad-based approach to gathering and evaluating information on exposures and other factors that would help to elucidate the etiology of acute pulmonary hemorrhage or hemosiderosis in infants, including dampness and agents associated with damp indoor environments; environmental tobacco smoke (ETS) and other potentially adverse exposures; and social and cultural circumstances, race/ethnicity, housing conditions, and other determinants of study subjects' health.
- Concentrations of organic dust consistent with the development of organic dust toxic syndrome are very unlikely to be found in homes or public buildings. However, clinicians should consider the syndrome as a possible explanation of symptoms experienced by some occupants of highly contaminated indoor environments.
- Greater research attention to the possible role of damp indoor environments and the agents associated with them in less well understood disease entities is needed to address gaps in scientific knowledge and concerns among the public."

The WHO Regional Office for Europe commissioned a study and concluded in its 2008 review that the most important health effects of mold and dampness exposures are increased prevalences of respiratory symptoms, allergies and asthma as well as perturbation of the immunological system. (WHO, 2009) The document also summarized the available information on the conditions that determine the presence of mold and measures to control their growth indoors. The guidelines were intended to protect public health under various environmental, social and economic conditions, and to support the achievement of optimal indoor air quality. However, while the guidelines provided objectives for indoor air quality management, they did not provide specific guidelines and strategies for achieving those objectives. The WHO-EU guidelines recommended formulating policy targets, and that governments should consider their local circumstances and select actions that will ensure achievement of their health objectives most effectively. This requires learning the latest research results, risk analysis and communication, team and interdisciplinary work. Physicians, industrial hygienist and air quality specialists and consultants need to better understand the technical exposure assessment methods, language, and successful intervention and control strategies.

In 2010 a 'New York State Toxic Mold Task Force' made up of politically appointed academic and non-academic members issued a report to the Governor and Legislature of New York State regarding the public health status, needs and research gaps. It was concluded that several information and data gaps exist regarding the timely recognition, assessment and control of environmental toxic and non-toxic biologicals (i.e., mold or fungi) in areas of indoor environments, public health and prevention (New York State Department of Health, 2010). The task force states that it focused their analysis on newly-emerging scientific information and on identifying areas where significant knowledge gaps still exist that appear to "substantially hinder decision making". Although some criticize that the committee lacked specific inside expertise and apparently failed to consider newer scientific papers and knowledge since the NAS analysis in the early 2000s and should have involved a broader spectrum of experts in the committee, the following conclusion and uncertainties were never the less summarized in their report:

- Exposure to building dampness and dampness-related agents including mold has been recognized nationally and at the state and local level as a potential public health problem.
- Asthma and other allergic respiratory diseases that can be exacerbated by mold exposures are common in NYS. This means many people are at risk for exacerbation of their respiratory conditions by exposure to mold conditions in buildings.
- Evidence for associations between non-respiratory effects and mold exposures in buildings is much more limited and generally does not allow clear conclusions to be drawn one way or the other.
- Molds, along with other organisms such as bacteria, mites and insects that proliferate in damp buildings, produce volatile compounds, spores and other minute particles that can cause irritant and allergic responses that range from annoying to serious depending on the amount of exposure and the immune system of the individual. Although some molds produce toxins, their contribution to adverse health effects in damp buildings, based on existing scientific information, is uncertain.

The 6th Annual International Scientific Conference on Bioaerosols, Fungi, Bacteria, Mycotoxins in Indoor and Outdoor Environments and Human Health addressed key areas of these identified knowledge gaps and provided scientific research, data, didactic materials and learning opportunities, that shall target change in knowledge, attitude, confidence and beliefs, practice-based clinical skills of health care providers with different professional background and specialty expertise. At the completion of the scientific meeting, the physician, industrial hygienist, health and safety specialist as well as public health officials and other participants gained a working knowledge of practical definitions, science based evidence to apply in their professional practice.

REFERENCES

1. Institute of Medicine (U.S.).Committee on Damp Indoor Spaces and Health. Damp Indoor Spaces and Health. Washington, DC: THE NATIONAL ACADEMIES PRESS; 2004.

2. WHO guidelines for indoor air quality: dampness and mould. World Health Organization WHO Regional Office for Europe. 2009.

3. New York State Department of Health. New York State Toxic Mold Task Force Final Report to the Governor and Legislature. New York State Department of Health - New York State Department of State. 2010.

Conference Chair and Director, Proceedings Editor:

Dr. med. Eckardt Johanning, MD, MSc

Co-Editors:

Philip R. Morey, Ph.D., CIH. - Gettysburg, Pennsylvania, USA

Pierre L. Auger M.D., M.Sc., FRCPC - Montréal, Québec, Canada

Conference Organization

Fungal Research Group Foundation, Inc.

4 Executive Park Drive, Albany, N.Y. 12203, USA

www.bioaersol.org info@bioaersol.org

About the Conference Organization:

The Fungal Research Group Foundation, Inc. and Dr. Eckardt Johanning organized the first international scientific meeting on " Bioaerosols, Fungi and Bacteria, Mycotoxins in Saratoga Springs, New York in 1994, and following in 1996, 1998, 2000 in Helsinki, Finland (together with Healthy Buildings 2000), and in 2003 which brought together a wide range of leading international researchers and health specialists devoted to public health and prevention. Proceeding books of the scientific presentations were published for the 1994, 1998 and 2003 and were made available for generally distribution.

Conference Chair and Director

Eckardt Johanning, M.D., M.Sc. - Fungal Research Group Foundation, Inc., Albany, New York, USA, Johanning MD PC.

E. Neil Schachter, M.D., Prof., Mount Sinai School of Medicine, New York, N.Y., (CME Course Director).

Scientific Committee

- Harriet M. Ammann, Ph.D., D.A.B.T. - Olympia, Washington, USA
- Pierre L. Auger M.D., M.Sc., FRCPC - Montréal, Québec, Canada
- Jürgen Bünger, M.D., Prof. - Bochum, Germany
- Denis Charpin, M.D., Prof. - Marseille, France
- Jean Cox-Ganser, Ph.D. (NIOSH) - Morgantown, West Virginia, USA
- Andrew Cutz, B.Sc., CIH - Markham, Ontario, Canada
- Nceba Gqaleni, Ph.D. - Durban, South Africa
- Manfred Gareis, D.V.M., Prof. - Kulmbach, Germany
- Michael R. Harbut, M.D., MPH - Royal Oak, Michigan, U.S.A.
- Ed Horn, Ph.D.,(NYSDOH,retired) - Albany, New York, USA
- Laura Kolb, Director, (EPA) - Washington, D.C., USA
- Kay Kreiss, M.D.(NIOSH) - Morgantown, West Virginia, USA
- Claude Mainville, P.E. - Montreal, Quebec, Canada
- Erwin Martlebauer, D.V.M., Prof. - Munich, Germany
- Robert K. McLellan, MD, MPH - Dartmouth, New Hampshire, USA
- Mark J. Mendell, Ph.D. - Berkely, California, USA
- Philip R. Morey, Ph.D., CIH - Gettysburg, Pennsylvania, USA
- Aino Nevalainen, Ph.D., Prof. - Kuopio, Finland
- Ed Olmsted, CIH - Garrison, New York, USA
- Jens Ponikau, M.D., Prof. - Buffalo, New York, USA
- Tuula Putus, M.D., Prof. - Turku, Finland
- Thomas G. Rand, Ph.D, Prof. - Halifax, Nova Scotia, Canada
- Rob A Samson, Ph.D., Prof. - Utrecht , The Netherlands
- E. Neil Schachter, M.D., Prof. - New York, New York, USA
- Magdalena Twaruzek, Ph.D. - Bydgoszcz, Poland
- Jan Grajewski, Ph.D., Prof. - Bydgoszcz, Poland
- Chin S. Yang, Ph.D. - Voorhees, New Jersey, USA
- Timothy E. Wallace, RS, CEHP, (FDOH) - Tallahassee, Florida, USA
- Donald Weekes, CIH, CSP - Ottawa, Ontario, Canada

Scientific conference coordination

Pierre L. Auger M.D., M.Sc., FRCPC - Montréal, Québec, Canada

Manfred Gareis, D.V.M., Prof., - Kulmbach, Germany

Laura Kolb, Director, (EPA) - Washington, D.C., USA

Claude Mainville, P.E., - Montreal, Quebec, Canada

Philip R. Morey, Ph.D., CIH - Gettysburg, Pennsylvania, USA

E. Neil Schachter, M.D., Prof. - MSSM, New York, New York, USA (CME Course Director)

Endorsement & Support

U.S. ENVIRONMENTAL PROTECTION AGENCY

Support – to present and discuss the latest research results, public health policy and state of the art technical education regarding indoor air and environmental health and prevention

The American Board of Industrial Hygiene awarded CM points to Certified Industrial Hygienists certified by the ABIH.

University of Natal (South Africa) - Centre for Occupational and Environmental Health

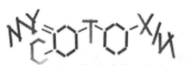

Gesellschaft für Mykotoxin Forschung (MycotoxinResearch Association, Germany)

CLIMATE CHANGE, THE INDOOR ENVIRONMENT, AND HEALTH - STUDY SUMMARY

BY

David A. Butler

Amid the considerable research on how climate change may affect public health, one subject has received relatively little attention- the impact of climate change on indoor environments and thereby on the health of people who live, work, study, or play in them. No government or private body has lead responsibility for investigating this question, and the lack of leadership is hindering action on identifying potential hazards, formulating solutions, and setting research and policy priorities.

Against this backdrop, the U.S. Environmental Protection Agency (EPA) asked the Institute of Medicine to convene an expert committee to summarize the current state of scientific understanding of the effects of climate change on indoor air and public health and to offer priorities for action.

Health Problems Here-and May Worsen

The committee's report, Climate Change, the Indoor Environment, and Health, points to extensive research on how climate change affects the outdoor environment, how the outdoor environment affects indoor environments under different climate conditions, and how indoor environments affect occupant health, among other related topics. But facing a dearth of research specifically directed at how these factors interact, the committee analyzed and synthesized information from the various independent lines of research.

The committee concludes that climate change influences indoor environmental quality, warranting attention and action. The committee based its conclusion on three key findings:

1. Poor indoor environmental quality is creating health problems today and impairs the ability of occupants to work and learn. By one estimate, poor indoor conditions cost the nation's economy tens of billions of dollars a year in exacerbation of illnesses and allergenic symptoms and in lost productivity.

2. Climate change may worsen existing indoor environmental problems and introduce new problems.

3. There are opportunities to improve public health while mitigating or adapting to alterations in indoor environmental quality induced by climate change.

Problematic Exposures Identified

To help in targeting research, the committee identified five major types of climate-induced indoor environmental problems.

Indoor air quality. Indoor environments can be contaminated by chemical, organic, and particulate pollutants that migrate from outdoors or that result from gas stoves and other indoor emission sources, such as building materials, radon, and environmental tobacco smoke. Climate change can affect these factors in various ways. For example, changes in the outdoor concentrations of a pollutant due to alterations in atmospheric chemistry or atmospheric circulation will affect indoor concentrations. Measures to reduce energy use in buildings, such as lowering ventilation rates may cause higher exposures to pollutants emitted from indoor sources. The expected increased use of air conditioning, if accompanied by reduced ventilation, could increase the concentrations of pollutants emitted from indoor sources. Additionally, power outages-caused by heat waves or other extreme weather events-could lead to the use of portable electricity generators that burn fossil fuels and emit poisonous carbon monoxide.

Dampness, moisture, and flooding. Extreme weather conditions associated with climate change may lead to more frequent breakdowns in building envelopes-the physical barrier between outdoor and indoor spaces-followed by infiltration of water into indoor spaces. Dampness and water intrusion create conditions that encourage the growth of fungi and bacteria and may cause building materials to decay or corrode, leading in turn to chemical emissions. Poorly designed or maintained heating, ventilation, and airconditioning systems may introduce moisture and create condensation on indoor surfaces. Humid conditions can, however, be improved by welldesigned and properly operating systems. Moldgrowth prevention and remediation activities also may introduce fungicides and other agents into the indoor environment.

Infectious agents and pests. Weather fluctuations and climate variability influence the incidence of many infectious diseases. Climate change may affect the evolution and emergence of infectious diseases, for example, by affecting the geographic range of disease vectors. The ecologic niches for pests will change in response to climate change, leading to changed patterns of exposure and, possibly, increased use of pesticides in some locations.

Thermal stress. Extreme heat and cold have several well-documented adverse health effects. High relative humidity exacerbates these effects in hot conditions. An increased frequency of extreme weather events may result in more frequent power outages that expose persons to potentially dangerous conditions indoors. The elderly, those in poor health, the poor, and those who live in cities are more vulnerable to both exposure to temperature extremes and the effects of exposure. Those populations experience excessive temperatures almost exclusively in indoor environments.

Building ventilation, weatherization, and energy use. Leaky buildings are common and cause energy loss, moisture problems, and migration of contaminants.

Poor ventilation is associated with occupant health problems or lower productivity. Climate change may make ventilation problems more common or more severe by prompting the implementation of energy efficiency (weatherization) measures that limit the exchange of indoor air with outdoor air. The introduction of new materials and weatherization techniques also may lead to unexpected exposures and health risks.

Priority Issues for Action and Recommendations

In formulating recommendations for ways to reduce the health effects caused by climateinduced indoor environmental conditions, the committee adopted a public health approach founded on three guiding principles. The overall effort, it said, should

- prioritize consideration of health effects into research, policy, programs, and regulatory agendas that address climate change and buildings;
- make prevention of adverse exposures a primary goal in designing and implementing strategies to address health effects; and
- include collection of data to be used in making better-informed decisions in the future.

The committee made a number of specific recommendations for actions to be taken by the EPA, in cooperation with other government agencies and with private-sector organizations where appropriate. These actions include:

- Initiating or expanding programs to identify populations at risk for health problems resulting from alterations in indoor environmental quality induced by climate change and implementing measures to prevent or lessen the problems.
- Developing or refining protocols and testing standards for evaluating emissions from materials, furnishings, and appliances used in buildings and promoting their use by standards-setting organizations and in the marketplace.
- Facilitating research to identify circumstances in which climate change mitigation and adaptation measures may cause or exacerbate adverse exposures.
- Facilitating the revision and adoption of building codes that are regionally appropriate with respect to climate-change projections and that promote the health and productivity of occupants.
- Developing model standards for ventilation in residential buildings and fostering updated standards for commercial buildings and schools, based on health-related criteria and aimed at providing a healthful environment under all design and operation conditions.
- Implementing a public health surveillance system that expands current ongoing surveys to gather information on how outdoor conditions, building characteristics, and indoor environmental conditions are affecting occupant health.

• Educating the public on issues of climate change, the indoor environment, and health.

• Evaluating actions taken in response to climate change-induced alterations in the indoor environment to determine whether they are enhancing occupant health and productivity in a cost-effective manner.

• Spearheading an effort across the federal government to make indoor environment and health issues an integral consideration in climate change research and action plans and, more broadly, coordinating work on the indoor environment and health.

CONCLUSION

The committee's observations and recommendations are based on scientific evidence that clearly shows that adverse indoor environmental quality is harming people's health. Altered climatic conditions will not necessarily introduce new risks for building occupants but may make existing indoor environmental problems more widespread and more severe and thus increase the urgency with which prevention and interventions must be pursued. Buildings that were designed to operate under the "old" climatic conditions may not function well under the "new." Considering the consequences of climate change adaptation and mitigation actions before they play out and thereby avoiding problems that can be anticipated will yield benefits in health and in avoiding costs of medical care, remediation, and lost productivity.

Committee on the Effect of Climate Change on Indoor Air Quality and Public Health

John D. Spengler (Chair) Akira Yamaguchi Professor of Environmental Health and Human Habitation, Department of Environmental Health, Harvard School of Public Health, Boston, Massachusetts

John L. Adgate Professor and Chair, Department of Environmental and Occupational Health, Colorado School of Public Health, University of Colorado, Aurora, Colorado

Antonio J. Busalacchi, Jr. Director and Professor. Earth System Science lnterdisciplin·ary Center, University of Maryland, College Park, Maryland

Ginger L. Chew Epidemiologist, Division of Emergency and Environmental Health Services, National Center for Environmental Health, Centers for Disease Control and Prevention, Atlanta, Georgia

Andrew Haines Director, London School of Hygiene and Tropical Medicine, London, UK

Steven M. Holland Chief, Laboratory of Clinical Infectious Diseases; Chief. lmmunopathogenesis Section, LCID; Tenured Investigator, lmmunopathogenesis

Section, National Institute of Allergy and Infectious Diseases, National Institutes of Health, Bethesda, Maryland

Study Sponsor

The U.S. Environmental Protection Agency

This study and key findings were presented by Dr. Butler at the conference. Summary above provided by Dr. Butler. For more information contact:

INSTITUTE OF MEDICINE Of THE NATIONAL ACADEMIES 500 Fifth Street, NW Washington, DC 20001 TEL 202.334.2352 FAX 202.334.1412 www.iom.edu

see: http://www.iom.edu/~/media/Files/Report%20Files/2011/Climate-Change-the-Indoor-Environment-and-Health/Climate%20Change%202011%20Report%20Brief.pdf 8/2012

PART I:

HEALTH EFFECTS – EPIDEMIOLOGICAL RESEARCH

NIOSH FIELD STUDIES ON DAMPNESS AND MOLD AND RELATED HEALTH EFFECTS

Jean M. Cox-Ganser

ABSTRACT

Over the past decade studies by our indoor environmental quality research team at NIOSH (Morgantown) have indicated increase in asthma onset from working in damp buildings. In addition, we found consistent associations between markers of exposure to dampness and mold and upper and lower respiratory symptoms that improve when away from the building (building-related), as well as with asthma onset after damp building occupancy. We have documented poorer quality of life and increased sick leave use in occupants with building-related respiratory illness, and found evidence for a nonallergic mechanism of building-related asthma associated with water damage. Our work in a large office building indicated that despite large remediation and repair efforts, there was no overall improvement in respiratory health indices over a three-year interval for employees for whom we had paired pulmonary function and questionnaire data. Employees classified as respiratory cases who were relocated within the building had decreases in medication use and sick leave.

A scoring system based on a standardized observational assessment of dampness and mold, as well as measures of hydrophilic fungi and ergosterol (a component of fungal cell walls) from floor dust have proven useful as indicators of adverse health effects. Endotoxin (a component of gram negative bacteria cell walls) exposure may change the effect of fungal exposure on respiratory health or vice versa.

We have found associations between the observational score and environmental measurements such as total culturable fungi, total culturable bacteria, β-D-glucan (a component of fungal cell walls), ergosterol, and the moisture content of walls and flooring. We are presently developing software for a practical assessment tool for dampness and mold and are working with a number of school districts to implement this as a way to help prioritize expenditures and respond promptly to water incursions and damage.

Visual contrast sensitivity testing has been used in studies of exposure to neurotoxins in situations other than damp indoor environments. Currently, its usefulness in investigations of mold-exposed populations in indoor environments is not well understood. A recent health hazard evaluation report by NIOSH investigators from Cincinnati described lower visual contrast sensitivity in staff at a severely water-damaged school in comparison to staff from a school without significant water damage. The investigators described limitations relating to the suitability of the comparison school. We are currently undertaking a similar investigation where a suitable comparison school is available.

Overview

We at NIOSH (Morgantown) have had a research initiative on indoor environmental quality with an emphasis on dampness and mold over the past decade. Over this time period we have conducted a number of field studies to investigate associations between building-related symptoms and conditions and indices of environmental exposure in damp buildings. Our studies have led us to believe the following: 1) building-related lung disease exists; 2) cases are sentinels for co-worker risk; 3) physicians should explore patient history for exposure to indoor damp environments; 4) there are consistent associations between health effects and markers of dampness and mold; 5) public health actions might best rely on signs of dampness; and 6) there is reason to prevent adverse health effects by early remediation of dampness. Individual findings from a number of our field studies are summarized below. NIOSH HSRB approved this research and all participants provided informed consent.

Community College Study

In 2000, we pioneered an observational check sheet for dampness and mold in a community college in which we compared 7 buildings with a history of recurrent water incursion with 6 buildings with very little history of water incursion (Park et al., 2004). The health concerns among employees were asthma, sinusitis, and hypersensitivy pneumonitis. There was a self-administered questionnaire survey offered to 554 fulltime employees from the 13 buildings, and the participation was 71% (393/554).

We developed an observational-based score sheet for our industrial hygienists, who rated 721 rooms in the 13 selected buildings. The crux of this instrument was rating for water stains, visible mold, mold odor, and moisture. We calculated an environmental exposure index for each participant based on the portion of time each spent in particular rooms.

The ratings for water stains, visible mold, mold odor, and moisture were used to calculate an environmental exposure index for each of the participants based on the portion of time each spent in particular rooms. We modeled work-related respiratory symptoms against the exposure index, adjusting for age, gender, smoking, job status, year of hire, allergies, and use of latex gloves. Positive exposure -response relationships existed between quartiles of individual exposure indices and the odds of having building-related wheeze, shortness of breath, nasal symptoms, and throat symptoms (Figure 1). These findings motivated us to use these inexpensive environmental observations in our epidemiologic field work, as we continued to explore damp buildings. Furthermore we concluded that visual and olfactory observation of water stains, visible mold, mold odors and dampness justifies action to control water damage.

Figure 1. Odds ratios and 95% confidence intervals for building-related respiratory symptoms in relation to quartiles of observational scoring for dampness and mold.

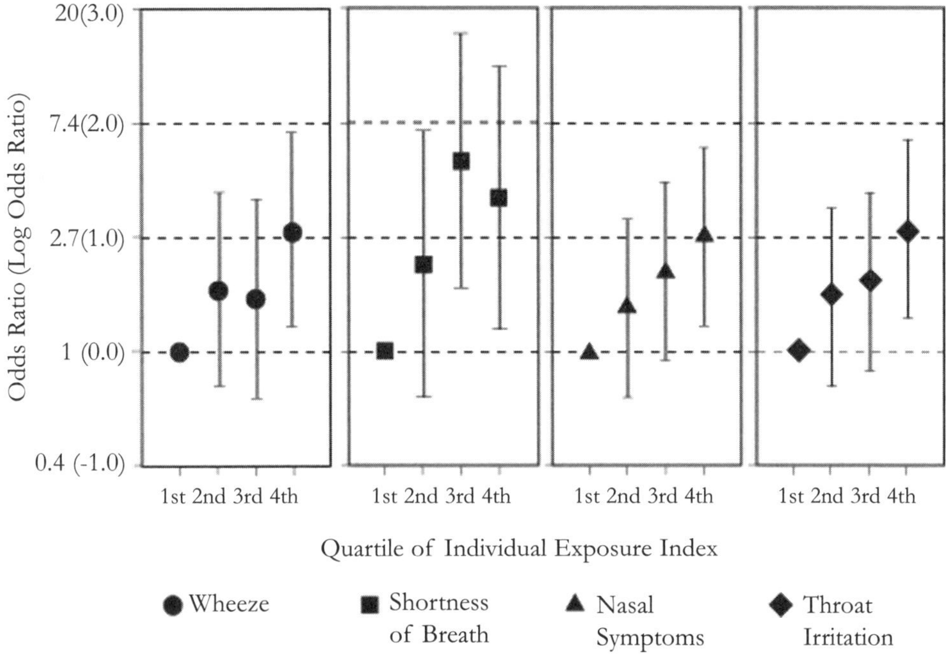

Hospital Study

Another major effort was a study of two hospitals (Rao et al., 2005; Cox-Ganser et al., 2009), one of which had had significant water damage from roof and window leaks, particularly on the top floor. The occupational physician for the hospital had documented a cluster of 6 asthma cases among nurses on that floor over a two-year period, most of whom had work-related peak flow abnormalities and all of whom had no evidence of latex asthma.

When we walked through the implicated facility during an initial survey of the top two floors, there was fungal contamination inside the walls and on ceilings. We conducted a questionnaire survey of current employees from both hospitals and had 64% (1171/1834) participation. The environmental survey included both observational assessment for dampness and mold and air, chair dust, and floor dust samples.

Using models adjusted for age, gender, smoking status, reported mold or dampness at home, and hospital worked in, we examined the association of dampness scores (as tertile values) with post-hire onset asthma, and work-related asthma and chest symptoms and found positive, exposure dependent relationships. There were no significant associations between the dampness score and work-related upper respiratory symptoms or the non-work-related upper or lower respiratory symptoms, or pre-hire diagnosed asthma.

We also found that work-related respiratory symptoms were related to a number of biological exposures in dusts, including ergosterol (a marker for fungal biomass) (Fig 2), extra-cellular polysaccharides specific for *Penicillium/Aspergillus*, and cat allergen.

Figure 2. Odds ratios and 95% confidence intervals for work-related lower and upper respiratory symptoms in relation to quartiles of ergosteral in floor dust.

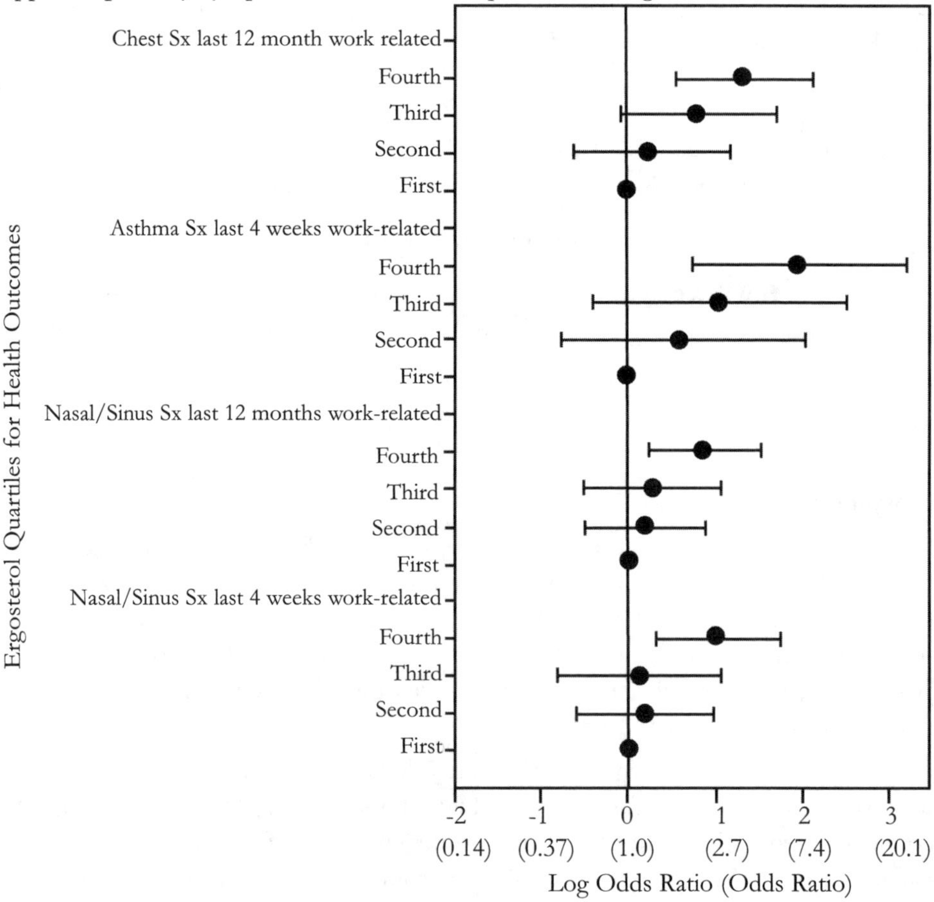

Study In Three Schools

We carried out observational assessments, moisture meter readings from walls and floors, as well as floor dust sampling in three schools in one district in the Northeastern US. There were associations between the observational assessment and results of environmental sampling of floor dusts and the results of moisture meter assessments. Rooms with scores above the median had significantly higher floor dust levels of total culturable fungi, total culturable bacteria, β-D-Glucan, and ergosterol. Furthermore the moisture content of walls and flooring was higher in rooms with observational assessment scores above the median.

Current Public Health Practice Project

Our findings that observational assessment for dampness and mold was associated with health outcomes as well as with objective measures of microbial contamination led us to a goal of working towards the practical application of a dampness/mold tool in schools. To this end we have developed a simplified version of the assessment sheet for dampness and mold, and are developing software for use of computer tablets for data entry and producing summary reports.

The aims for this tool are: to identify and record areas of dampness and mold throughout buildings; to trigger early repair and remediation to avoid potential health effects and more costly repair and remediation; to create awareness of potential problem areas; and to track past and present problem areas by repeating the use of this tool.

Office Building Study

We conducted a study of occupants of a large office building in Connecticut from 2001 to 2007 (Cox-Ganser et al., 2005; Park et al., 2006; Park et al., 2008; Iossifova et al., 2011). Historically, the balconies, windows, and roof of this building leaked, and there had been extensive water damage, particularly on the upper floors. Office workers had concerns about work-related symptoms and many new cases of asthma, hypersensitivity pneumonitis, and sarcoidosis. We carried out repeated cross-sectional surveys as the building underwent remediation for water damage through the unlined brick façade, the balconies on the upper floors, through window flashing, and through the roof.

In 2001 we offered a questionnaire survey to all 1327 employees working in building and had 888 participants (67%). Results indicated that employees had 7.5 times the rate of new onset asthma after starting working in the building compared to their incidence of asthma before building occupancy. There were also 8 hypersensitivity pneumonitis cases and 6 sarcoidosis cases reported. We compared our findings for this building population with national data and found the prevalence ratio for ever being diagnosed with asthma was 2.2 (1.9-2.6) and for current asthma was 2.4

(2.0-3.0). Compared to the state adult population prevalence ratios were 1.4 (95% CI 1.2-1.6) for lifetime asthma, and 1.5 (95% CI 1.3-1.9) for current asthma.

In 2002 we followed up with a case-comparison group study and found that two-thirds of our cases had either objective abnormalities on lung function tests or were on respiratory medications. Respiratory cases had poorer quality of life and increased use of sick leave related to respiratory problems. We also evaluated skin test reactivity to common aeroallergens in three groups - those with post-occupancy asthma those with pre-occupancy asthma and those with no asthma. As expected, pre-existing asthmatics had greater skin test reactivity to all aeroallergens: grasses, trees, ragweed, roach, weeds, dust mites, and cat allergen, with the last 4 being statistically significant, in comparison to the other two groups. In contrast, those with post-occupancy asthma were slightly less likely than nonasthmatics to be atopic. We did not use mold sensitivity in our definition of atopy. Interestingly, those with post-occupancy asthma were least likely to be mold-sensitive by prick test, although the difference was not statistically significant.

From our work comparing microbial levels in dusts to health outcomes we found that total fungal colony counts in floor dust were associated in an exposure dependent way with lower respiratory symptoms. Models indicated a synergistic effect between endotoxin and fungal exposures in association with health outcomes. We also found linear associations between respiratory health and the levels of hydrophilic fungi in floor and chair dust.

Some of our recent longitudinal analyses documented that persons with building-related respiratory disease generally did not improve despite remediation efforts, even though new cases became less common. Paired pulmonary function and questionnaire data from 2002 and 2005 for 97 employees found no overall improvement in respiratory symptom scores, medication use, lung function or use of sick leave. Respiratory cases who had been relocated in the building had a decrease in medication use and sick leave in 2005.

In summary: this water-damaged building was associated with new-onset asthma; there had been considerable personal, social, and economic burden on both employees and employers; assessing exposure to both fungi and endotoxin appears important; ergosterol and hydrophilic fungi appear useful markers of health risk in damp buildings; and evidence suggests that adverse health outcomes may not reverse after remediation (or reverse slowly);

Visual Contrast Sensitivity

Visual contrast sensitivity testing has been used in studies of exposure to neurotoxins; e.g. mercury, organic solvents. Currently, its usefulness in investigations of mold-exposed populations in indoor environments is not well understood. Investigators from NIOSH (Cincinnati) conducted a Health Hazard Evaluation in a water-damaged school in New Orleans and a comparison school in Ohio. Staff at the water-damaged school had higher symptoms prevalences and lower visual contrast sensitivity across all spatial frequencies (NIOSH, 2010). The investigators described limitations relating to the suitability of the comparison school. We have ongoing work in a water-damaged school and a comparison school in New England where both are in a high socio-economic suburban district, about 1 mile apart. We have conducted questionnaire surveys, lung function testing, visual acuity testing, and visual contrast sensitivity testing. Data analysis and report writing is presently being carried out.

DISCLAIMER: The findings and Conclusions in this report are those of the author and do not necessarily represent the views of the National Institute for Occupational Safety and Health.

REFERENCES

- Cox-Ganser JM, Rao CY, Park J-H, Schumpert JC, Kreiss K. Asthma and respiratory symptoms in hospital workers related to dampness and biological contaminants. *Indoor Air.* 2009:19(4), 280-290.
- Cox-Ganser JM, White SK, Jones R, Hilsbos K, Storey E, Enright PL, Rao CY, Kreiss K. Respiratory morbidity in office workers in a water-damaged building. *Environmental Health Perspectives* 2005:113:485-490.
- Iossifova YY, Cox-Ganser JM, Park JH, White SK, Kreiss K. Lack of respiratory improvement following remediation of a water-damaged office building. *Am J Ind Med. Apr,* 2011:54(4):269-277.
- NIOSH 2010. Health hazard evaluation report: comparison of mold exposures, work-related symptoms, and visual contrast sensitivity between employees at a severely water-damaged school and employees at a school without significant water damage, New Orleans, LA. By Thomas G, Burton NC, Mueller C, Page E. Cincinnati, OH: U.S. Department of Health and Human Services, Centers for Disease Control and Prevention, National Institute for Occupational Safety and Health, *NIOSH HETA* No. 2005-0135-3116.
- Park J-H, Cox-Ganser JM, Kreiss K, White SK, Rao CY. Hydrophilic Fungi and Ergosterol Associated with Respiratory Illness in a Water-damaged Building. *Environmental Health Perspectives.* 2008:116(1):45-50.
- Park J-H, Cox-Ganser J, Rao C, Kreiss K. Fungal and endotoxin measurements in dust associated with respiratory symptoms in a water-damaged office building. *Indoor Air,* 2006:16: 192-203.
- Park J, Schleiff PL, Attfield MD, Cox-Ganser J, Kreiss K. Building-related respiratory symptoms can be predicted with semi-quantitative mold exposure index. *Indoor Air;* 2004:14:425-433.
- Rao CY, Cox-Ganser JM, Chew GL, Doekes G, White S. Use of Surrogate Markers of Biological Agents in Air and Settled Dust Samples to Evaluate a Water-Damaged Hospital. *Indoor Air* 2005:15(Suppl 9):89-97.

HOW TO FIND, HOW TO MANAGE? DEVELOPING GUIDELINES FOR MOLD PROBLEMS THROUGH RESEARCH.

Aino Nevalainen, Martin Täubel, Helena Rintala and Anne Hyvärinen

BACKGROUND

The indoor air quality and health problems associated with building dampness and mold were originally brought up by practical experiences of people and less by means of learning through science. This paper summarizes the process how dampness, moisture and mold problems have been managed as a public health issue in Finland, and gives an overview of the present research in the field.

The phenomenon was originally documented with extensive epidemiological population studies e.g. in Canada and USA (Dales et al., 1991; Brunekreef et al., 1989) and the findings were also verified in Finland (Pirhonen et al., 1996). The results of those studies clearly showed an association between building moisture problems and adverse effects on human health, although the causal relationship between exposing agents and symptoms were far from clear. This hazard identification can be considered the first step of a risk assessment process, which is still going on. After this association between observations of damp, moisture and mold in indoor premises and ill health of the occupants was repeatedly shown in a number of studies in different countries, climates and conditions around the world, more systematic documentation on different aspects of the problem and its causal links was gradually increasing in the scientific literature. Today, the documentation of the phenomenon has been thoroughly evaluated by several expert committees (e.g., IOM 2004; WHO 2009) but we are still underway to reveal the characteristics of microbial exposures that are harmful to health and that should be controlled in buildings.

Guideline development is problematic for building mold. For one thing, the control of the indoor air quality is practically in the hands of the owner of each individual building, and cannot be approached in a similar, region or society-scale approach than, say, outdoor air quality. Furthermore, since the causal links between exposure and health effects are not known, no health-based threshold limit values can be set to prevent the exposure of the occupants. With these limitations in mind, it is however necessary to develop meaningful guidance and instructions for the public how to manage the problem. A pragmatic approach can be used as the evidence-based understanding concerning the roots of the problem is relatively clear. The microbial growth on building surfaces and structures, "mold", is the critical source of exposure, and this growth is only possible with the presence of excess moisture, i.e., water. As microbial exposures are complex to measure, the detection of signs of moisture can be used as a surrogate of the exposure. While the causal connection between an individual's illness and mold exposure may be hard to verify, it is beneficial for both the occupants' health and for the building itself to repair any dampness or moisture problems and remove any mold from the building surfaces.

The characterization of the moldy building issue as a public health problem in Finland was started with a cross sectional study on prevalence of moisture damage observations in the Finnish housing stock (Nevalainen et al., 1998). A random sample was drawn from the building registry to obtain a representative sample of single family houses built in each decade between 1960ies and 90ies. The houses were inspected by trained civil engineers to register all the signs of moisture, e.g., damp or moldy patches, discolored wood or peeling paint. As a result, it was observed that, when even the slightest signs were recorded, 80% of houses had some signs of present or previous dampness or moisture, and 50% of homes needed at least some minor repair or more thorough investigation. As it was already known that such dampness or mold is a risk factor for respiratory and other symptoms, the efforts to find tools to identify problematic buildings and understand the whole phenomenon were started. The guideline development has been strictly based on both national and international research.

Development of guidelines for microbial measurements

The very simple prerequisites of microbial growth on any surfaces are good to keep in mind. Environmental microbes, present everywhere and hugely diverse in their environmental adaptability, usually find enough nutrients and acceptable temperatures anywhere, and therefore the limiting factor for their growth is the availability of water. Buildings and their surfaces should be dry – this is one of the goals of building's physical design – and thus any excess moistening of the assemblies other than bathroom or kitchen surfaces means a failure in that goal. Wherever there is water available, fungi and bacteria start to grow.

Environmental microbes such as fungi and bacteria are ubiquitously present. Fungal spores and bacteria, whether as spores or as vegetative cells, can be found on any indoor surfaces and in indoor air. Normal sources of microbes are soil, vegetation and outdoor air, and among the major intramural normal sources are house plants, pets, vegetables, fruit, firewood and other materials of organic origin. A majority of indoor bacteria originates from human skin (Täubel et al., 2009). These normal microbes can be found on any surfaces and in indoor air in concentrations that are characteristic to the climatic and geographical conditions. The normal species seem to be relatively stable around the world (Shelton et al., 2002). Any microbial growth which results from a dampness or moisture problem, means the presence of an "abnormal" source and it alters the concentrations and composition of the indoor mycobiota and bacterial flora. This can be utilized in practical investigations of indoor spaces as such changes can be detected with microbial sampling and analyses.

The concentration and species in a sample must be seen in a right context since both depend on the method used. According to a number of studies carried out in Finland, the airborne concentrations of fungi and their variation appeared to behave in a logical way when analyzed with cultural methods (Nevalainen et al., 1991; Reponen et al., 1992; Lehtonen et al., 1993; Pasanen et al., 1992; Hyvärinen et al., 2001; Meklin et al., 2005). The results of these studies were the basis of the conclusions that (1) the wintertime concentrations of normal homes were usually less than 100 cfu/m^3, (2) the presence of dampness/moisture damage increased the indoor air concentration of fungi, (3) the mycobiota of indoor air in moisture damaged indoor environments differed from normal and moisture indicators could be observed, (4) the concentrations in school buildings were remarkably lower than in homes, and (5) after a thorough remediation, the airborne concentrations of fungi decreased to normal levels and the species were normal. These aspects have been the core of the guidelines given by the Ministry of Health and Welfare (Asumisterveysohje, 2003). The main users are municipal health inspectors but the guidelines are in wide use among indoor environmental consultants and other actors.

During the recent years, efforts have been put for development of more sophisticated methods for indoor microbial investigations. While culture methods – in spite of their many limitations – could be utilized for discrimination between moisture damaged and reference buildings, sequence analyses of the indoor microbial communities mainly failed to show such differences. Effect of remediation was not readily evident when molecular profiling was applied (Pitkäranta et al., 2008; Rintala et al., 2010; Pitkäranta et al., 2011). It may be speculated that what we see in cultural analyses are fresh emissions of existing sources of fungi, while such emissions are only a minute amount of microbial material that actually exists in the indoor environment. These microbial communities may be long-time residents of the indoor space and their variation less dependent on local sources such as a moisture damaged, moldy spot of limited area.

The comparison of results from different studies is difficult if the sampling and analysis have been made in a different way. However, many kinds of methods can be used to analyze whether the microbial situation in one building or room differs from another, and whether remediation measures have been efficient to eliminate undesired fungal sources from the building. These are usually the goals of microbial sampling in building investigations. At present, quantitative PCR methods are being validated for the purposes of healthy housing guidance. As with any methodology, enough background reference material must be available before the interpretation of the results can be done in a meaningful way.

Moisture, mold and health

There is not yet enough documentation or understanding on how the exposure in indoor environments to harmful microbial emissions actually takes place, how much and what kind of microbial material must be inhaled in order to develop symptoms in question, and via which pathophysiological mechanisms the symptoms develop. Another story are the individual factors which also contribute to the development of symptoms or disease on individual level.

In practice, the identification of cases where dampness, moisture and mold might be connected to symptoms that occupants experience, may be tricky. The Finnish Health Protection Law, which is the legislation covering the prevention of environmental factors that threat human health, includes the possibility to control the conditions that are known to pose a risk for human health. It means that even the indoor conditions that may be harmful to health, should be remediated. According to the principle of the legislation, this should be done whether or not there are disease cases that could strictly be connected to such indoor exposures.

It is, however, necessary to study also the health effects of mold exposure, the exposing agents that may be critical, and the pathophysiological pathways that lead to disease. The task is complex and no simple causal links can be expected to be found. In the following, a concise summary of ongoing health-related studies is presented.

HITEA, "Health Effects of Indoor Pollutants: Integrating microbial, toxicological and epidemiological approaches" is a project aiming for understanding the health effects of microbial exposures indoors. Funded mainly by European Commission FP7, it has eight partners from seven European countries and is coordinated by THL, (Institute for Health and Welfare), Finland.

The aim of HITEA is to identify the role of indoor biological agents that lead to long term respiratory, inflammatory and allergic health impacts among children and adults. The focus is on microbial exposures due to dampness problems of buildings; and the role of allergens, chemicals and poor ventilation is also studied. Microbiological, toxicological, analytical and immunological techniques are applied to elucidate the exposures, their short term and long term health impacts, and the mechanisms behind them. The field work and health assessment phases have been completed, and at present, HITEA –project is in an intensive phase of laboratory analyses, statistical analyses and reporting.

The HOTES study is an intervention study on effects of mold remediation on the microbial exposure and health of occupant of single family houses. Severe cases of moisture and mold problematic homes are selected to the study by trained experts from the Finnish Society for the pulmonary disabled (Heli). The selected families get help and support for the remediation of their home and their exposure and health status are followed throughout the process. The field phase is still underway in this study which is funded by THL, Heli and the Ministry of the Environment.

In the series of studies called Indoor MoBi, qPCR methods are validated for the purposes of indoor microbial investigations. The aim is to find more accurate and faster methods to characterize the microbial conditions in indoor environments. QPCR concentrations of fungi in mold-infested materials were usually about two orders of magnitude higher than those obtained with culture methods. The results correlated well so that the conclusion of whether the material was more or less infested was mainly the same with both qPCR and culture (Pietarinen et al., 2008).

QPCR concentrations of fungi in house dust vs. observed moisture damage were analyzed in the study by Lignell et al. (2008). In this study, the following microbes were associated with moisture damage:

Cladosporium sphaerospermum, Eurotium amstelodami / chevalieri / herbariorum / rubrum / repens, Penicillium brevicompactum / stoloniferum, Penicillium / Aspergillus / Paecilomyces variotii, Trichoderma viride / atroviride / koningii, Wallemia sebi.

Microbes that did not associate with moisture damage were:

Acremonium strictum, Aspergillus fumigatus, A. penicillioides, A. versicolor, Aureobasidium pullulans, Cladosporium cladosporioides, Penicillium chrysogenum, P. variabile, Stachybotrys chartarum, Streptomyces spp.

These findings raised an intriguing question on which microbes could actually be considered indicators of moisture damage. It is evident that more studies are needed for these conclusions.

It is becoming more and more evident that the role of microbial toxins as exposing agents in moldy indoor environments should be revealed. At present, we have preliminary data available about the occurrence of fungal and bacterial toxins in house dust (Täubel et al., 2011). Among the bacterial toxins that were detected were valinomycin, a potent ionophore in mitochondrial membranes which is produced by *Streptomyces spp.*, and monactin with similar activity, also produced by *Streptomyces spp.* Several field studies are ongoing and the results will give an overview on the occurrence and types of toxic metabolites in indoor environments. Whether their occurrence is associated with moisture damage and whether they seem to link with health effects, is to be learned through these ongoing studies.

REFERENCES

- Asumisterveysohje (Guidance for healthy housing). Sosiaali- ja terveysministeriö (Ministry for Social Affairs and Health), Helsinki (in Finnish). 2003.
- Brunekreef B, Dockery DW, Speizer FE, Ware JH, Spengler JD, Ferris BG. Home dampness and respiratory morbidity in children. *Am Rev Respir Dis.* 1989:140:1363-1367
- Dales RE, Burnett R, Zwanenburg H. Adverse health effects among adults exposed to home dampness and molds. *Am Rev Respir Dis.* 1991:143:505-509

- Hyvärinen A, Reponen T, Husman T, Ruuskanen J, Nevalainen A. Characterizing mold problem buildings – concentrations and flora of viable fungi. *Indoor Air.*1993:3:337-343
- Koskinen OM, Husman TM, Meklin TM, Nevalainen AI. The relationship between moisture or mold observations in houses and the state of health of their occupants. *Eur Resp J.* 1999:14:1363-1367
- Lehtonen M, Reponen T, Nevalainen A. Everyday activities and variation of fungal spore concentrations in indoor air. *Int Biodeterior Biodegrad.* 1993:31:25-39
- Lignell U, Meklin T, Rintala H, Hyvärinen A, Vepsäläinen A, Pekkanen J, Nevalainen A. Evaluation of quantitative PCR and culture methods for detection of house dust fungi and streptomycetes in relation to moisture damage of the house. *Lett Appl Microbiol.* 2008:47:303-308
- Nevalainen A, Pasanen AL, Niininen M, Reponen T, Kalliokoski P, Jantunen M.The indoor air quality in Finnish homes with mold problems. *Environ Int.* 1991:17:299-302
- Nevalainen A, Partanen P, Jääskeläinen E, Hyvärinen A, Koskinen O, Meklin T, Vahteristo M, Koivisto J, Husman T. Prevalence of moisture problems in Finnish houses. *Indoor Air.* 1998:(Suppl.)4:45-49
- Pasanen AL, Niininen M, Kalliokoski P, Nevalainen A, Jantunen M. Airborne Cladosporium and other fungi in damp versus reference residences. *Atm Environ.* 1992: 26B:121-124
- Pietarinen VM, Rintala H, Hyvärinen A, Lignell U, Nevalainen A. Quantitative PCR analysis for fungi and bacteria in building materials and comparison to culture-based analysis. *J Environ Monit.* 2008:10(5):655-663
- Pirhonen I, Nevalainen A, Husman T, Pekkanen J. Home dampness, moulds and their influence on respiratory infections and symptoms in adults in Finland. *Eur Resp J.* 1996:9:2618-2612
- Pitkäranta M, Meklin T, Hyvärinen A, Paulin L, Auvinen P, Nevalainen A, Rintala H. Analysis of fungal flora in indoor dust by ribosomal DNA sequence analysis, quantitative PCR, and culture. *Appl Environ Microbiol.* 2008:74:233-244
- Pitkäranta M, Meklin T, Hyvärinen A, Nevalainen A, Paulin L, Auvinen P, Lignell U, Rintala H. Molecular profiling of fungal communities in moisture-damaged buildings before and after remediation – comparison of culture-dependent and –independent methods. *BMC Microbiology.* 2011:11:235
- Reponen T, Nevalainen A, Jantunen M, Pellikka M, Kalliokoski P. Normal range criteria for indoor air bacteria and fungal spores in a subarctic climate. *Indoor Air.* 1992:2:26-31
- Rintala H, Pitkäranta M, Toivola M, Paulin L, Nevalainen A. Diversity and seasonal dynamics of bacterial community in indoor environment. *BMC Microbiology.* 2008:8:56
- Shelton BG, Kirkland KH, Flanders WD, Morris GK. Profiles of airborne fungi in buildings and outdoor environments in the United States. *Appl Environ Microbiol.* 2002:68(4):1743-1753
- Täubel M, Rintala H, Pitkäranta M, Paulin L, Laitinen S, Pekkanen J, Hyvärinen A, Nevalainen A. The occupant as a source of house dust bacteria. *J Allergy Clin Immunol.* 2009:124:834-840

- Täubel M, Sulyok M, Vishwanath V, Bloom E, Turunen M, Järvi K, Kauhanen E, Krska R, Hyvärinen A, Larsson L, Nevalainen A. Co-occurrence of toxic bacterial and fungal secondary metabolites in moisture-damaged indoor environments. *Indoor Air.* 2011:21(5):368-75

HYPERSENSITIVITY PNEUMONITIS

E. Neil Schachter

Diseases of sifters and measurers of grain. Ramazzini

"I have often wondered how so noxious a dust can come from grain as wholesome as wheat... All kinds of grain and in particular wheat... have mixed in with them a very fine dust... hence, whenever it is necessary to sift wheat or other kinds of grain to be ground in the mill, or to measure it when corn merchants convey it hither and thither, the men who sift and measure are so plagued by this kind of dust that when the work is finished they heap a thousand curses on their calling.

The throat, lungs and eyes are keenly aware of serious damage... almost all who make a living by sifting or measuring grain are short of breath and cachectic and rarely reach old age". (Ramazzini, 1940)

INTRODUCTION:

Hypersensitivity Pneumonitis (aka Extrinsic Allergic Alveolitis) is an immunologically mediated inflammation of the pulmonary parenchyma, involving alveolar walls and terminal airways secondary to the repeated inhalation of sensitizing agents including micro-organisms, organic dusts and simple chemicals.

Histopathologically it is characterized by a granulomatous interstitial pneumonitis with occasional bronchiolitis obliterans

Diagnosis is usually based on clinical presentation, exposure history and radiologic appearance. Bronchoscopic and histologic features are helpful in characterizing the disease.

Removal of the affected individual from the toxic environment, particularly early on in the course of the disease, can arrest and reverse the progression of hypersensitivity pneumonitis to an irreversible stage. Anti-inflammatory treatment may reduce symptomatic manifestations of the disease.

Historical perspective:

Ramazzini, as with so many inhalational diseases, appears to be the first to describe what would subsequently be the clinical picture of hypersensitivity pneumonitis.

The earliest description of this syndrome in contemporary literature is attributed to the report of JM Campbell who in 1932 described: "Acute symptoms following work with hay". In this report he described five English farmers with progressive symptoms following work with moldy hay. These farm workers experienced dyspnea which progressed over a period of several weeks (Campbell, 1932). The syndrome was later studied in detail with surgical biopsies by Rankin et al. (1962)

studying initially 39 agricultural workers and later 34 additional workers. The exposures consisted of moldy fodder, hay, and grain in 31; moldy silage in 24; shredding and threshing dusts in 8; various organic dusts in 8 and moldy tobacco in 2. The major symptoms described by these individuals included dyspnea, cough, fever, chills, weight loss and hemoptysis. Chest X-rays varied from clear to frank interstitial disease. On biopsy they describe a granulomatous pneumonitis.

The importance of serologic findings was subsequently stressed by investigators in both Britain and the United States. This characterization led to a broader understanding of the disease, to include pulmonary responses to a wide variety of antigens representing numerous environmental and occupational exposures. Gregory et al identified the antigens of Thermopolyspora polyspora and Micropolyspora vulgaris in moldy but not good hay (Gregory et al., 1963; Gregory, Lacy, 1963). Work by investigators such Pepys led to the characterization of precipitins against an extract of moldy hay in the sera of patients with farmer's lung disease (Pepys et al., 1962). The importance of these antigens in causing disease was subsequently demonstrated by inhalation challenge leading to the reproduction of acute attacks characteristic of the disease (Williams, 1963). Similar findings were described for a wide variety of patients exposed to different antigens including: bird fancier's lung, pituitary snuff disease bronchopulmonary aspergillosis, and detergent enzyme lung (Pepys et al.,1962; Williams, 1963; Faux et al., 1971; Pepys et al., 1966).

Antigens associated with hypersensitivity pneumonitis:

Hundreds of different sources of different antigens have been associated with the syndrome of hypersensitivity pneumonitis. Colorful names associated with the sources of the antigens have been adopted to describe the context in which the illness occurs. Table 1 lists some of the clinical presentations. These antigens can be grouped under three major headings:

1. Microbial agents: These are the most commonly recognized cause of HP. (eg Bacteria and Fungi)
2. Animal Proteins: Particulates from a wide variety of animal sources (eg Birds, Insects, Small mammals)
3. Chemicals: eg Di-isocyanates (TDI, MDI) detergent enzymes, pesticides (Pyrethrum).

The effectiveness of antigens in producing disease is felt to relate to the nature of the inhaled dust (particle size, solubility, antigenicity) and the intensity and frequency of the exposure.

Host risk factors:

Although many people are exposed to environmental antigens associated with HP only some develop antibodies and fewer still, manifest disease. Host susceptibility undoubtedly plays a major role since many workers and exposed individuals do not manifest illness. Of interest are the following observations:

1. Smoking is protective
2. Pregnancy may trigger HP
3. Although rare HP can occur in children

Clinical presentation:

Features of the clinical illness are usually similar regardless of the nature of the inhaled antigen. The illness is usually divided into 2 stages:

Acute syndrome:

1. Illness begins 4 to 12 hours after exposure
2. Symptoms may persist for up to 18 hours following the onset and recovery is usually spontaneous.
3. The attacks usually recur each time the individual is exposed to the offending dust.
4. The acute attack may mimic acute viral illnesses such as influenza, bronchitis or pneumonia
5. Respiratory and constitutional symptoms (cough, chest tightness, fever, chills, malaise and myalgias).
6. Physical findings (tachypnea, tachycardia, rales).
7. Labs (leukocytosis, neutrophilia, lymphopenia)
8. Additional symptoms may include anorexia, weight loss, and progressive dyspnea (particularly for severe attacks)

Subacute and chronic syndromes:

Some patients with hypersensitivity pneumonitis with prolonged exposure may develop irreversible damage. It also happens that patients bypass the acute stage of the illness and present with the features of the chronic disease.

1. Dyspnea, sputum production, fatigue, anorexia, and weight loss
2. Physical findings: rales, wheezing, cyanosis, right sided heart failure, clubbing
3. Chest X-ray pulmonary fibrosis, honeycombing

Favorable prognostic factors:

* Early age of onset
* Short duration of exposure
* The clinical presentation of the disease, acute being more favorable.

Laboratory findings:

1. Precipitating antibodies: The serum in suspected individuals can be assayed for many potential antigens including: molds, grain dust, fungi, blood or other fluids from animal sources. In general these indicate exposure sufficient to cause disease but are not diagnostic.

   Precipitins are present in 3 to 30% of asymptomatic farmers and 50% of asymptomatic pigeon breeders

   The absence of precipitins does not rule out the presence of hypersensitivity pneumonitis.

   Skin testing with these antigens is not felt to be helpful in diagnosing the disease.

2. ESR, CRP and immunoglobulins are elevated. ANA and other auto antibodies are not elevated.

3. Inhalation challenge by exposure of the patient to the antigen either in the natural setting or in the controlled environment of an environmental chamber can recreate the findings of the acute disease thereby supporting the diagnosis. The response to antigen challenge takes two forms:

   a. Delayed response: The onset of fever, malaise, headache, rales on examination, an elevation of the white blood count (Neutrophils), and a decrease in spirometric parameters 8-12 hours after the challenge. Hypoxemia and abnormal radiographic findings may also be documented.

   b. A biphasic response can occur particularly in atopic individuals with an immediate asthma-like response occurring immediately after challenge and subsiding rapidly, but followed by the above delayed response.

4. Chest X-rays may be within normal limits during the acute response. More commonly patchy, ill-defined infiltrates are seen in all lung fields. As the disease progresses sharply demarcated reticulo-nodular findings are noted. In the end-stage chronic phase diffuse fibrosis, lung contraction and honeycombing may be seen.

5. CT scans in the acute stage reveal ground glass infiltrates, but as with the plain X-ray may be negative. In the subacute stage diffuse micronodules are noted and there may be air-trapping and mild fibrosis. In the chronic phase the radiographic appearance is frequently similar to that of pulmonary fibrosis.

6. Pulmonary Function: Spirometry demonstrates a restrictive pattern in the acute phase with decreased DLCO and hypoxemia. Resolution may take four to six weeks in the acute situation.

   Both restrictive and obstructive findings are seen in the chronic disease. DLCO and hypoxemia are marked. Many patients exhibit an abnormal methacholine tests.

7. Bronchoalveolar Lavage (BAL): The disease is characterized by a marked BAL lymphocytosis, but this finding also occurs in asymptomatic exposed individuals. Normally the percentage of lymphocytes population of BAL fluid is in the single digits. In Hypersensitivity pneumonitis it can range from 20 to 50 %. It may persist despite clearing of the clinical picture. The presence of mast cells and plasma cells is usually a sign of active disease.
8. Lung Biopsy: If the clinical and laboratory findings are not specific enough a lung biopsy, frequently open (as opposed to bronchoscopic), may be required. Pathologic findings in the disease include:
    a. Noncaseating granulomas located in or around smaller airways.
    b. An alveolitis characterized by lymphocytes, plasma cells and histiocytes.
    c. Fibrosis around the smaller airways

Epidemiology:

The epidemiology of hypersensitivity pneumonitis is poorly characterized. Some isolated studies indicate the wide variability in the prevalence of exposed individuals.

1. Prevalence among farmworkers in Britain 7 % (Grant et al., 1970)
2. Prevalence among farmers in the United States 0.5 to 3% (Gruchow et al., 1981; Madsen et al., 1976)
3. Prevalence among pigeon breeder's 16 to 21% (Christensen et al., 1975)
4. Incidence among office workers exposed to a contaminated air conditioning system 52% (Ganier et al., 1980)
5. Incidence among lifeguards exposed to a public swimming pool 37% (Rose et al., 1998)

Indoor environments are host to many common species of fungi including: *Cladosporium, Alternaria, Epicoccum, Fusarium, Penicillium, Aspergillus, Geotrichum Rhodotorula,* and *Chaetomium.* These fungi in general invade the indoor environment from the outdoor environment. However because of favorable environments, particularly in water damaged homes they can proliferate indoors to levels which are independent of their numbers in the outdoor environment. Clinical disease in these settings can result from host susceptibility (eg immunocompromise) or host sensitivity (the host develops an abnormal response to the fungal antigens) as in the case of hypersensitivity pneumonitis. Fungus contaminated showers, air conditioning systems, and humidifiers have been reported as sources (Hogan et al., 1996; Banaszak et al., 1970; Volpe et al., 1991).

Diagnosis:

Early diagnosis is important because the disease is potentially reversible early in its course.

Diagnostic Criteria include:

1. Exposure to antigen by history, environmental measurements or by the presence of antibodies
2. a. Symptoms compatible with the clinical picture

   b. A characteristic chest X-ray/CT scan

   c. Characteristic physiologic findings
3. BAL with lymphocytosis (usually low CD4/CD8)
4. Positive inhalation challenge
5. Histopathology showing compatible changes

The diagnosis is established by one of the following 4 combinations:

a. 1, 2, 3

b. 1, 2, 4

c. 1, 2a, 3 and 5

d. 2, 3, 5

A probable diagnosis is established with 1, 2a and 3. (Schuyler, Cormier, 1997)

Prevention and treatment:

Prevention requires recognition and elimination of the causative agent. This can be accomplished by four approaches:

1. Reduction or elimination of the antigen in the environment. Examples of this strategy include the use of cleaning and abatement strategies in the environment, eliminating the antigen source (eg replacing or boarding the bird or the animal) treating contaminated surfaces with antimicrobial products to limit or eliminate the growth of organisms.
2. Better facility design by reduction of moisture, improvement of air conditioning and ventilation. Removal of water damaged materials, carpeting and other furnishings that collect dust and encourage the growth of organisms.
3. Preventive maintenance with inspections and replacement of damaged materials in the home or work environment.
4. Wearing protective devices and providing air filtration in the home or work environment.

Treatment consists of avoidance of the antigenic material which may require extensive reconditioning of the environment or relocation of home or change of job.

Prevention and avoidance are frequently successful in the early acute stages of the disease, however if these strategies fail treatment with corticosteroids may reduce symptoms particularly in the acute severely ill patient. The onset of chronic disease characterized by pulmonary fibrosis and severe lung impairment limit the expected recovery.

Corticosteroid therapy (in conjunction with antigen avoidance) is usually continued for one to three months, with initial doses of 30-60 mg of prednisone and tapering following the initial phase for about one month (Monkare, 1983; Kokkarinen et al., 1992; Patel et al., 2001).

Table 1:

Some characteristic presentations of hypersensitivity pneumonitis.

Disease	Environmental source	Antigen
Farmer's Lung	Moldy hay, grain, silage	Thermophylic antinomycetes
Bagassosis	Sugarcane	Thermoactinomyces saccharii
Malt worker's lung	Moldy malt	*Aspergillus* species
Suberosis	Moldy cork	*Penicillium* frequentans
Mushroom worker's lung	Mushroom compost	Thermoactinomyces viridis
Wine maker's lung	Mold on grapes	*Botrytis* cincrea
Woodworker's lung	Moldy wood dust	*Alternaria* species
Tobacco grower's lung	Tobacco plants	*Aspergillus* species
Wheat weevil disease	Weevil infested flour	Sitophus granarius
Laboratory technician lung	Rodent urine	Rodent urinary protein
Air conditioner pneumonitis	Contaminated system	Thermoactinomyces vulgaris, Acanthamoeba polyphaga, Naegleria gruberi
Hot tub lung	Mists, molds on walls and tub	*Cladosporum* species
Bird fancier's lung	Parakeets, budgerigars, pigeons	Droppings, feathers, proteins

REFERENCES

- Banaszak E, Thiede W, Fink J. Hypersensitivity pneumonitis due to contamination of an air conditioner. *N Engl J Med* 1970:283:271
- Campbell JM: Acute symptoms following work with hay. *Br. Med. J.* 1932:2:1143-44.
- Christensen L, Schmidt C, Robbins L. Pigeon breeder's Disease: A prevalence study and review. *Clin Allergy*. 1975:5:417
- Faux JA, Wells ID, Pepys J. Specificity of Avian serum proteins in tests against the sera of bird fanciers. *Clin Allergy* 1971:1:159-70.
- Ganier M, Lieberman P, Fink J et al. Humidifier lung. An outbreak in office workers. *Chest* 1980:77:183
- Grant I, Blyth W, Wardrop VE et al. Prevalence of farmer's lung in Scotland: A pilot survey. *Br Med J.* 1970:1:530.
- Gregory P, Festenstein G, Skinner F. Microbial and biochemical changes during the moulding of hay. *J Gen Microbiol.* 1963:33:147
- Gregory P, Lacey M. Mycological examination of the dust from mouldy hay associated with farmer's lung disease. *J Gen Microbiol.* 1963:30:75.
- Gruchow H, Hoffman R, Marx J et al. Precipitating antibodies to farmer's lung antigens in a Wisconsin farming population. 1981:124:411
- Hinson KFW, Moon AJ, Plummer NS. Broncho-Pulmonary Aspergillosis. *Thorax* 1952:7:317-33.
- Hogan M, Patterson R, Pore R et al. Basement shower hypersensitivity pneumonitis secondary to *Epicoccum nigrum*. *Chest.* 1996:110:854
- Kokkarinen J, Tukolainen H, Terno E. Effect of corticosteroid treatment on the recovery of pulmonary function in farmer's lung. *Amer Rev Resp Dis.* 1992:145:3
- Madsen D, Klock L, Wenzel F et al. The prevalence of farmer's lung in an agricultural population. *Am Rev Resp Dis.* 1976:113:171
- Monkare S. Influence of corticosteroid treatment on the course of farmer's lung. *Eur J Resp Dis.* 1983:64:283
- Patel A, Ryu J, Reed C. Hypersensitivity Pneumonitis: Current concepts and future questions. *J Allerg Clin Immunol* 2001:108:661.
- Pepys J, Jenkins PA, Lachmann PJ. An Iatrogenic antibody: immunological responses to "Pituitary Snuff" in patients with Diabetes Insipidus. *Clin Exp Immunol* 1966:1:377-89.
- Pepys J, Longbottom JL, Hargreave FE et al. Allergic reactions of the lungs to enzymes of Bacillus subtilis. *Lancet* 1969:1; 1181-4.
- Pepys J, Riddell RW, Citron KM et al. Precipitins against extracts of hay moulds in the serum of patients with farmer's lung, aspergillosis, asthma and sarcoidosis. *Thorax* 1962:17:366-74.
- Ramazzini B. De Morbus Artificum. 1713, Chicago 1940, University of Chicago Press.
- Rankin J, Jaeschke WH, Callies QC et al. Physiopathologic Features of the acute interstitial granulomatous pneumonitis of agricultural workers. *Ann. Int. Med.* 1962:57:606-6

- Rose C, Matyny J, Newman L et al. Lifeguard lung: Endemic granulomatous pneumonitis in an indoor swimming pool. *Am J Pub Health* 1998:88:1795
- Schuyler M, Cormier Y. The diagnosis of hypersensitivity pneumonitis. *Chest.* 1997:111:534.
- Volpe B, Sulavik S, Tran P et al. Hypersensitivity pneumonitis associated with a home humidifier. *Conn Med.* 1991:55:571
- Williams JV. Inhalation and skin tests with extracts of hay and fungi in patients with farmer's lung. *Thorax* 1963:18:182.

HEALTH COMPLAINTS, LUNG FUNCTION, AND IMMUNO LOGIC EFFECTS IN GERMAN COMPOST WORKERS FROM LONG-TERM EXPOSURES TO BIOAEROSOLS

Jürgen Bünger, Anja Deckert, Frank Hoffmeyer, Dirk Taeger, Thomas Brüning, Monika Raulf-Heimsoth, Vera van Kampen

KEYWORDS: organic dust, composting, chronic bronchitis, lung function, molds, specific antibodies

ABSTRACT

Employees at workplaces in composting facilities are exposed to high bioaerosol concentrations containing airborne actinomycetes and molds (organic dust). We investigated work related symptoms and diseases of 190 currently exposed compost workers and 38 non-exposed control subjects in a cross sectional study at 31 composting plants situated in Northern Germany.

Participants of the study were asked for symptoms and diseases, exposures to bioaerosols, atopic diseases, and smoking habits using a standardized protocol. They underwent a physical examination by an occupational physician and a lung function measurement according to ATS criteria. Contents of specific IgE to environmental allergens and molds were measured as possible markers of allergic occupational asthma. Specific IgG to molds and actinomycetes was investigated since they may serve as biomarkers of exposure.

Compost workers suffered more often from chronic cough and mucous membrane irritation syndrome (MMIS) than the controls. Work-related influenza-like symptoms were reported by 11 compost workers (5.8%) but none control subject. Lung function parameters of compost workers were within the reference ranges. Nevertheless, the values for the forced vital capacity (FVC% predicted) of compost workers were significantly lower than for control subjects ($p = 0.01$). The significance was even higher when only non-smoking study participants were compared ($p = 0.001$). Specific IgE values to environmental allergens and molds were positive in 25.3% and 7.4% of currently exposed compost workers. There were no significant differences in specific IgE and IgG concentrations between the two groups.

Compost workers suffered significantly more often from MMIS which is likely to be associated with occupational exposures to organic dust. The elevated incidence of chronic cough indicates cases of chronic bronchitis according to WHO-criteria. Also the reduced FVC% predicted may be caused by this exposure. There is no higher frequency of mold sensitization in the group of compost workers probably due to a healthy worker effect. Reports of work-related influenza-like symptoms may indicate cases of hypersensitivity pneumonitis or organic dust toxic syndrome in the compost workers.

INTRODUCTION

Employees at workplaces in composting facilities are exposed to high bioaerosol concentrations containing airborne actinomycetes and molds (organic dust). In general, these microorganisms can show infectious, allergenic, and toxic properties. Diseases like infections, chronic bronchitis, occupational asthma, and hypersensitivity pneumonitis were reported in workers exposed to similar bioaerosols in other industries and agriculture. The review of twenty epidemiological studies resulted in an excess of upper airway and eye irritation - so called mucous membrane irritation syndrome (MMIS) - in bioaerosol exposed workers (Schlosser et al., 2009). The evaluation of case reports revealed cases of hypersensitivity pneumonitis, organic dust toxic syndrome (ODTS), and allergic bronchopulmonary aspergillosis (ABPA) (Schlosser et al., 2009). Especially allergens of molds can trigger type I allergies, such as bronchial asthma and allergic rhinitis (Bünger et al., 2007). Also skin and gastrointestinal problems have been reported in biowaste collectors (Bünger et al., 2000, Ivens et al., 1999). The aim of this study was monitoring of the health status of bioaerosol-exposed workers in German composting plants.

METHODS

At 31 composting plants in North-western Germany, 190 currently exposed compost workers were examined in 2009. Also 38 office employees who never had been occupationally exposed to organic dust were studied with identical protocols. An occupational physician asked for work-related symptoms, exposures to bioaerosols at their workplaces, atopic diseases, and smoking habits using a questionnaire. In addition, all study participants underwent a physical examination by the physician. Forced expiratory volume in 1 second (FEV_1) and forced vital capacity (FVC) were measured with a portable spirometer (MasterScope, Viasys Health Care, Hoechberg, Germany) according to the guidelines of the American Thoracic Society (ATS 1995). Reference values were chosen according to Quanjer et al., (1993). Specific IgE antibodies to a mixture of ubiquitous allergens (sx1; atopy screen) and a mold mixture (mx1) and total IgE were measured in the sera of the participants by ImmunoCAP (Phadia, Uppsala, Sweden) according to the manufacturer's recommendations. Specific IgG antibodies to *Aspergillus fumigatus* (Gm3), *Penicillium* spp. (Gm27) *Thermoactinomyces vulgaris* (Gm23), and *Sacharopolyspora rectivirgula* (Gm22) were measured with IgG-ImmunoCAP (Phadia).

The study design and the protocol were reviewed and approved by the ethics committee of the Ruhr-University Bochum in accordance with the Declaration of Helsinki. All study participants gave written informed consent to the study protocol.

RESULTS

Male participants were predominant in both groups (Table 1). The control subjects were older than the compost workers. Because lung function decreases with age, the measured values were calculated according to age-adjusted reference values (as %predicted).

Table 1: Characteristics of study participants.

	Compost workers (n = 190)		Controls (n = 38)	
Male gender: n (%)	181	(95.3%)	37	(97.4%)
Female gender: n (%)	9	(4.7%)	1	(2.6%)
Age: yrs [mean ± SD]	45.4 ± 9.3		58.0 ± 6.4	
BMI: [kg/m²]	27.4 ± 4.3		27.1 ± 4.1	
Smoking status:				
Smokers: n (%)	59	(31.0%)	5	(13.2%)
Ex-smokers: n (%)	71	(37.4%)	14	(36.8%)
Non-smokers: n (%)	60	(31.6%)	19	(50.0%)

For bioaerosol-exposed workers, the mean duration of employment in the composting facilities was 12.0 ± 7.8 years.

Compared to the control subjects a higher number of compost workers suffered from chronic cough and irritation of the eyes, although significant differences were observed only for the latter. Additionally, up to 18% of compost workers reported nose symptoms and up to 11% skin irritations. However, this was the same in the control group. Influenza-like symptoms were only reported by compost workers (n=11, 5.8%).

The mean lung function values of both study groups were within the reference ranges. The reproducibility of the spirometric measurements within 200 mL was comparable between groups (FEV_1: compost 94.2%, controls 94.7%; FVC: compost 91.1%, controls 87.0%), therefore differences in participants' cooperation could be excluded. While values for FEV_1 %predicted were slightly lower, values for FVC %predicted were significantly lower in compost workers compared to controls (p=0.010). The significance was even increased when only non-smoking study participants were taken into account (p=0.001, figure 1).

Figure 1: Lung function parameters of non-smoking compost workers and control subjects. Testing was done according to ATS criteria (1995) and reference values were chosen from Quanjer et al. (1993).

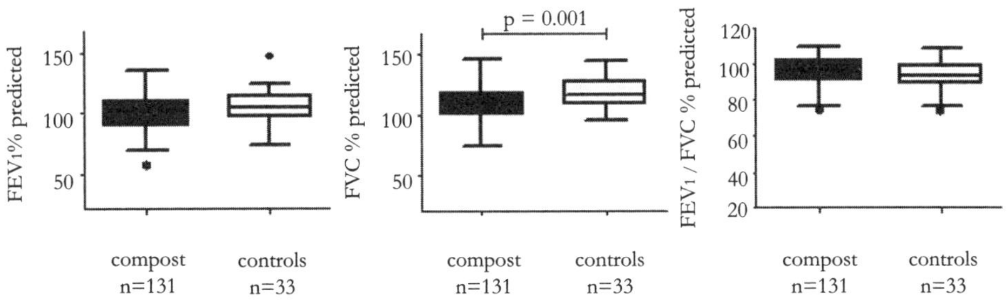

There were no significant differences in the number of positive results or in the concentrations of total and specific IgE antibodies between compost workers and control subjects. However, high specific IgE values to molds (>CAP-class 2 (>3.5 kU/L)) were found only in the group of compost workers. Also in the case of total IgE and specific IgE to ubiquitous allergens (sx1), the highest values were observed in compost workers. Concerning specific IgG antibodies neither the number of positive tests nor the antibody concentrations were significantly different between compost workers and the control group. However, for all antigens the maximum IgG concentrations in the group of compost workers were higher than in the control subjects.

Female gender and age were significantly associated with decreased lung function parameters FEV_1 and FVC in the multivariate analysis. While occupation as a compost worker, cough, dyspnoea, and smoking showed also a negative association, there was no relationship between lung function and duration of employment. Neither the concentrations of specific IgE antibodies to ubiquitous allergens and molds nor those of specific IgG antibodies to molds and actinomycetes showed a significant association with lung function parameters.

DISCUSSION

Symptoms of MMIS due to occupational exposure to organic dust have been reported previously (Schlosser et al., 2009; Domingo and Nadal, 2009). The results of this study add further evidence to the causal association between bioaerosol exposure and MMIS of the upper airways and eyes.

The higher prevalence of chronic cough in the compost workers compared to the control group indicates induction of chronic bronchitis according to the WHO-criteria. This observation was not reported in earlier publications and should be confirmed by longitudinal cohort studies.

In contrast to other authors (Schlosser et al., 2009), MMIS of the nose was not reported more often by compost workers than by control subjects in the present study. However, already Bünger et al. (2007) supposed that watering and itching eyes seem to be closely associated with bioaerosol exposure, while irritations of the nose also may be provoked by other circumstances, e.g. dry air due to central heating or air conditioning. Wearing respiratory protection masks in areas with high exposure to organic dust could also be an explanation why eye, but not nose irritation has been reported more frequently by compost workers.

In an earlier cross sectional study, work related health complaints and diseases of 58 German compost workers were compared with those of 40 control subjects. Compost workers had significantly more symptoms and diseases of the airways and the skin. Additionally, they showed significantly increased IgG antibody concentrations to *Aspergillus fumigatus*, *Saccharopolyspora rectivirgula*, and *Streptomyces thermovulgaris* that were associated with the duration of employment (Bünger et al., 2000). Unlike these results, in our study compost workers did not show significantly higher concentrations of specific IgG than the control group. However, while Bünger et al. (2000) used IgG measurements based on antigens that have been prepared from pure cultures of molds and actinomycetes which were directly identified in the air at composting plants, in our study commercial ImmunoCAPs were used.

Gastrointestinal complaints which had been reported in biowaste collectors by Bünger et al. (2000) and Ivens et al. (1999) were not observed in the regularly higher to bioaerosols exposed compost workers.

A healthy worker survivor effect, the phenomenon that sensitive individuals leave the workplace due to health problems, is suggested by several authors to explain the low rates of atopic sensitization in compost workers (Schlosser et al., 2009). This effect can also not be excluded in our study, since there was no higher frequency of sensitization to ubiquitous allergens and molds in the group of compost workers, although they were certainly exposed to a higher degree than the control subjects.

In a prospective cohort study in German composting plants, 218 workers reported a significantly higher prevalence of MMIS than 66 control subjects. Lung function measurements showed a significant decline of FVC %predicted during the observation period of five years in non-smoking compost workers (Bünger et al., 2007). In accordance with this, in our study compost workers showed slightly lower values for FEV_1 %predicted and significantly lower ones for FVC %predicted than the controls. Relatively few groups have measured lung function parameters in occupational settings in the compost industry. In an Austrian study, no statistically significant impairment of the lung function could be found in 137 employees from two composting facilities and three waste sorting plants, but most of the workers had a very short duration of employment (Marth et al., 1997).

Results of the present study indicate that MMIS and chronic cough is frequent in bioaerosol-exposed compost workers. Furthermore an impairment of lung function was observed, indicating the need of improved risk assessment and state-of-the-art protective measures in composting plants.

ACKNOWLEDGMENTS

The authors thank Nina Rosenkranz, Marita Kaßen, and Ulla Meurer for their excellent technical assistance. The study was supported by a grant of the German Federal Institute for Occupational Safety and Health (BAuA, F 2063).

REFERENCES

- American Thoracic Society (ATS). Standardization of spirometry (1994 update). *Am. J. Respir. Crit. Care Med.* 1995 Sep; 152 (3): 1107-36.
- Bünger J, Antlauf-Lammers M, Schulz TG, Westphal GA, Müller MM, Ruhnau P, Hallier E. Health complaints and immunological markers of exposure to bioaerosols among biowaste collectors and compost workers. *Occup. Environ. Med.* 2000 Jul; 57(7): 458-64.
- Bünger J, Schappler-Scheele B, Hilgers R, Hallier E. A 5-year follow-up study on respiratory disorders and lung function in workers exposed to organic dust from composting plants. I*nt. Arch. Occup. Environ. Health* 2007 Feb; 80 (4): 306-12.
- Domingo JL, Nadal M. Domestic waste composting facilities: a review of human health risks. *Environ. Int.* 2009 Feb; 35 (2): 382-9.
- Ivens UI, Breum NO, Ebbehøj N, Nielsen BH, Poulsen OM, Würtz H. Exposure-response relationship between gastrointestinal problems among waste collectors and bioaerosol exposure. *Scand. J. Work Environ. Health.* 1999 Jun; 25 (3): 238-45.
- Marth E, Reinthaler FF, Schaffler K, Jelovcan S, Haselbacher S, Eibel U, Kleinhappl B. 1997. Occupational health risks to employees of waste treatment facilities. *Ann. Agric. Environ. Med.* 1997; 4.1: 143-7.
- Quanjer PH, Tammeling GJ, Cotes JE, Pedersen OF, Peslin R, Yernault JC. Lung volumes and forced ventilatory flows. Report working party standardization of lung function tests European community for steel and coal. *Eur. Respir. J.* 6 Suppl. 1993 Mar; 16: 5-40.
- Schlosser O, Huyard A, Cartnick K, Yañez A, Catalán V, Quang ZD. Bioaerosol in composting facilities: occupational health risk assessment. *Water Environ. Res.* 2009 Sep-Oct; 81 (9): 866-77.

AIRBORNE WORKPLACE EXPOSURE TO MICROBIAL METABOLITES IN WASTE RECYCLING PLANTS

Stefan Mayer, Vinay Vishwanath, Michael Sulyok

ABSTRACT

In waste recycling plants employees are exposed to a complex mixture of airborne bioaerosols. To which degree the exposure to airborne mould in waste recycling plants is associated with a toxic health risk emanating from microbial metabolites is largely unknown.

Settled dust samples were taken at waste recycling plants for municipal waste and for waste paper. The settled dust samples were analyzed for 186 fungal and bacterial metabolites using liquid chromatography / tandem mass spectrometry (LC-MS/MS). Additionally the concentration of airborne dust was measured. Based on these data the concentrations of airborne microbial metabolites and further on the airborne exposure was calculated for regular and worst case conditions.

In total 38 microbial metabolites were detected in the settled dust samples consisting of 33 fungal and 5 bacterial metabolites. Samples from municipal waste recycling plants contained a higher number of microbial metabolites which occurred in higher concentrations compared to recycling plants for waste paper.

Apart from outliers the inhalable airborne dust concentration is usually below the German occupational threshold limit value of 10 mg/m^3.

The calculated amount of inhaled microbial metabolites under regular working conditions during an 8 hours work shift averaged to 37 and 8 ng/70 kg bw for municipal waste and waste paper recycling respectively. Under worst case conditions the corresponding amounts would be 5573 and 4841 ng/70 kg bw. Different approaches to assess these exposure rates are discussed.

BACKGROUND / INTRODUCTION

Waste recycling is an industrial sector of growing importance. In Germany about 85 % of the municipal waste is recycled. The growing relevance is accompanied by an increase of employment. Beside the separated sampling of wastes like waste paper, compostable wastes and glass waste the efficient sorting is a key process in gaining the recyclable fractions. Although a lot of sorting is automated, there are still employees engaged in manual waste sorting as well as in control-, maintenance- and inspection tasks.

Exposure to municipal waste is usually accompanied by an exposure to a complex mixture of bioaerosols. The European directive 2000/54/EG on the protection of workers from risks related to exposure to biological agents at work forces employers, in case of any activity likely to involve a risk of exposure to biological agents, to determine the nature, degree and duration of workers' exposure in order to assess any risk to the workers' health or safety and to lay down the measures to be taken. The health risks emanating from biological agents are infections, sensitization and toxicological threats.

Workplace exposure measurements usually show elevated mould concentrations ranging from 10^4 to 10^7 cfu/m^3 (Rabe et al., 2003). Moulds and some bacteria, especially actinomyces are classified as sensitizing agents. Some of them can also exhibit infectious potential, e.g. Aspergillus fumigatus. Toxic effects of moulds and bacteria are also possible due to cell fragments like endotoxins (Rylander 2006) and microbial metabolites like mycotoxins or bacterial toxins (Mayer et al., 2008).

While information about infectious and sensitizing properties are easily available; information about occurrence and exposure to microbial metabolites in waste sorting plants are scarce. Most investigations cover bioaerosol exposure in composting plants (Bünger et al., 2000). Concerning waste handlers Heldal et al. (2003) observed an association between exposure to fungal spores and ß(1-3)-D-glucans and nasal swelling and neutrophil influx. Degen et al. (2003) observed elevated concentrations of OTA in blood samples of waste handlers. This could be regarded as evidence for direct exposure to microbial metabolites in such work places. However, apart from this part of bioaerosols other microbial metabolites may occur.

The aim of our preliminary study was to derive information about the broad spectrum of microbial metabolites occurring in dust samples from different waste recycling plants and to estimate the airborne exposure and inhalative intake.

MATERIALS AND METHODS

Methods being appropriate to determine airborne mycotoxins or bacterial toxins directly are not available. In order to get information about the occurrence of microbial metabolites settled dust samples were taken in 16 waste recycling plants by sweeping settled dust from floor and surfaces of technical equipment; 8 samples were taken in recycling plants for municipal waste and 8 samples in recycling plants for waste paper.

Additionally airborne dust concentrations were determined in several occupational environments in waste recycling plants. Air samplers with a throughput of 10 l/min were used in combination with membrane filters. The amounts of sampled dust were determined gravimetrically. The sampling system collects only the inhalable dust fraction (particle size $\leq$100 µm). 23 airborne dust samples were taken in 8 recycling plants for waste paper and 7 samples were taken from 5 recycling plants for municipal waste. Sampling took place either personally or stationary depending on workers compliance. Duration of measurements was adapted to cover a representative period to calculate the time-weighted average shift concentration and varied between 1 and 4.5 hours. Peaks of dust concentrations were not measured separately.

The settled dust samples were analyzed for 159 nonvolatile secondary fungal and 27 bacterial metabolites. In brief, the dust samples were extracted in acetonitrile/water/acetic acid (79:20:1, v/v/v), raw extracts were diluted and subsequently analyzed without further clean-up. Detection and quantification was done with a QTrap 4000 LC-MS/MS (Applied Biosystems, Foster City, CA) equipped with a TurboIonSpray electrospray ionization source and a 1100 Series HPLC System (Agilent, Waldbron, Germany). Sample preparation, instrumental parameters, method performance characteristics (coefficients of variation, limits of detection, recoveries) and standards used in this multi-target method have been described in detail by Vishwanath et al. (2009).

Table 1. Parameters used to calculate the inhaled amounts of microbial metabolites

Parameter	Regular case	Worst case
Airborne dust concentration WP[1] [mg/m^3]	2.9[a]	33.1[b]
Airborne dust concentration MW[2] [mg/m^3]	1.9[a]	22.9[b]
Breathing volume of exposed workers [l/h]	600	3000

[1] Waste paper, [2] municipal waste, [a] median concentrations, [b] maximum concentrations

The airborne exposure was calculated using the airborne dust concentration and the sum of the concentrations of microbial metabolites in settled dust samples from recycling plants for municipal waste and waste paper respectively.

The calculation of the inhaled amounts of microbial metabolites has been performed for regular and worst case conditions using the parameters summarized in table 1 and the median and maximum concentrations of airborne microbial metabolites in recycling plants for municipal waste and waste paper respectively. It was assumed that the settled dust has been airborne and the distribution of microbial metabolites in the settled dust and the airborne dust is comparable.

RESULTS

Airborne dust concentrations were lower when municipal waste was handled compared to waste paper (table 1). This may be due to usually lower moisture in waste paper compared to municipal waste. However, dust exposure can vary according to the type of waste the composition and the moisture content of the waste.

In the settled dust samples in total 38 nonvolatile secondary microbial metabolites were determined, consisting of 33 fungal and 5 bacterial metabolites (table 2). The data revealed different exposure situations between recycling plants for municipal waste and paper waste. Samples from recycling plants dealing with municipal waste contained an average number of 26 microbial metabolites compared to 14 microbial metabolites in samples from waste paper recycling plants.

The differences between recycling plants for municipal waste and waste paper concerned mainly the fungal metabolites malformin C, griseofulvin and myriocin. Chlamydosporol, dechlorogriseofulvin, deoxybrevianamide E and patulin also occurred exclusively in municipal waste samples but only with a low frequency. The 5 bacterial metabolites were found in settled dust samples from both municipal wastes as well as from paper waste recycling plants with the only exception of puromycin which hasn't been detected in settled dust samples from waste paper plants. The bacterial metabolites comprised at least 3 metabolites with antibiotic properties like chloramphenicol. Not only the number of metabolites but also the concentrations of microbial metabolites were higher in recycling plants for municipal waste.

One of the most frequent metabolite in both types of recycling plants was enniatin B. However, apart from two exceptions the most frequent metabolites were not identical with the metabolites occurring in elevated concentrations and vice versa.

Aflatoxins were not detected, despite the fact that mold species like Aspergillus flavus were detected in elevated concentrations of 10^4 - 10^5 cfu/m^3. However, occurrence of aflatoxins or other metabolites not yet observed cannot be excluded.

Standard deviation indicated at least in some cases a broad variation of concentrations. This can be due to variations in the composition and microbial contamination of waste fractions. Nonetheless the data, in particular the composition of the metabolites, can be taken as representative. The settled dust acts as a kind of long-term storage for substances which occurred in the workplace. The settled dust sampled for this study was at least one week old.

The calculated amounts of inhaled microbial metabolites in municipal waste plants averaged to 37 ng/70kg bw under regular working conditions and 5573 ng/70 kg bw under worst case conditions. The respective amounts in waste paper recycling plants were 8 and 4841 ng/70 kg bw.

The concentrations in the settled dust and the further calculation of exposure should be judged carefully and taken as a first impression. First because of the low number of samples and second because settled dust is an extremely challenging matrix: Matrix effects and incomplete extraction can hamper quantitative analysis (Vishwanath et al., 2009).

Table 2: Concentration and frequency of occurrence of microbial metabolites in settled dust samples from recycling plants for municipal waste and waste paper

Fungal metabolites	Municipal waste (n = 8)			Waste paper (n = 8)		
	Median [ng/g]	StaDev	No. pos. samples	Median [ng/g]	StaDev	No. pos. samples
3-Methylviridicatin	8,6	1,8	5	8,8	0,3	2
Alamethicin F30	25,0	12,7	5	14,0	1,2	3
AME	25,7	13,1	6	16,3	11,2	4
Apicidin	1,3	0,2	2	0,7	0,2	2
Beauvericin	10,5	7,4	8	0,9	3,8	7
Chaetoglobosin A	215,0	78,0	4	209,0	47,0	2
Chanoclavine	2,1	1,6	7	1,0	0,2	4
Chlamydosporol	37,5	30,1	2	-	-	0
Cyclopenin	49,2	98,8	8	46,5	52,0	6
Cyclopeptine	30,0	17,1	8	7,7	3,2	4
Cyclosapeptide A	16,7	9,9	3	11,2	0,6	2
Cytochalasin D	-	-	0	136,0	120,0	2
Dechlorogriseofulvin	230,0	-	1	-	-	0
Deoxybrevianamid E	221,0	-	1	-	-	0
Emodin	157,0	123,0	8	57,1	19,0	4
Enniatin A	13,8	4,6	6	3,8	1,2	2
Enniatin A1	15,7	9,5	8	4,6	4,0	6
Enniatin B	9,9	9,0	8	1,8	2,7	8
Enniatin B1	23,5	18,6	8	6,6	6,3	7
Equisetin	91,0	154,0	8	86,3	59,6	4
Fumigaclavine	61,9	30,1	8	10,6	2,3	3
Griseofulvin	100,0	616,0	6	-	-	0
Malformin C	61,7	31,3	4	-	-	0

Fungal metabolites	Municipal waste (n = 8)			Waste paper (n = 8)		
	Median [ng/g]	StaDev	No. pos. samples	Median [ng/g]	StaDev	No. pos. samples
Meleagrin	40,9	12,5	8	41,4	16,0	4
Mevinolin	1459,0	272,0	2	2512,0	1359,0	3
Myriocin	1159,0	794,0	4	-	-	0
Patulin	49144,0	-	1	-	-	0
Pentoxyfyllin	67,3	71,5	8	61,1	56,8	2
Physcion	836,0	269,0	4	380,0	543,0	5
Roquefortine C	160,0	85,0	7	25,5	10,3	2
Stachybotrylactam	70,6	23,8	4	153,0	40,0	3
Sterigmatocystine	16,1	12,6	8	20,5	10,1	5
Viridicatin	635,0	315,0	8	108,0	-	1
Bacterial metabolites						
Chloramphenicol	50,5	34,1	4	9,1	10,1	3
Monactin	27,6	42,3	7	1,7	0,7	5
Nonactin	10,7	17,0	8	0,8	0,3	4
Puromycin	88,0	42,9	3	-	-	0
Valinomycin	2,5	2,5	8	1,1	0,5	5

Threshold limit values for inhalatory intake of microbial metabolites which can be used for health hazard assessment are not available. For a selective number of mycotoxins health recommendations based on the tolerable daily intake (TDI) via food exists. In order to get a first estimation about the possible health risk the inhalative intake of microbial metabolites in recycling plants was compared with the lowest of these recommendations, which has been set for T-2/HT-2 toxins at 4.2 µg/70 kg bw (table 3).

Table 3: Comparison of inhaled amounts of microbial metabolites with the TDI value for T-2/HT-2 toxins

	Regular conditions [ng/70 kg bw]	Worst case conditions	TDI*
Municipal waste	37	5.573	4.200
Waste paper	8	4.841	4.200

* TDI for T-2/HT-2 toxins

Under regular conditions the inhalatory intake during recycling of municipal waste and waste paper respectively is at least 2 orders of magnitude lower than the lowest TDI value for mycotoxins. This may be interpreted that no actual health hazard exists during regular working conditions. Under worst case conditions the TDI can be exceeded. However, the worst conditions which have been covered in the present study may occur only a few days a month. But on the other hand exposure conditions are conceivable resulting in an even higher exposure. Especially when massively mould contaminated waste is handled a very high exposure to microbial metabolites can be supposed which may lead to acute toxic effects. In a next study such exposure conditions should be considered.

The comparison of the sum of microbial metabolites with the TDI value may underestimate the health risk. The TDI is calculated for a single mycotoxin. In recycling plants a complex mixture of microbial metabolites occur. It must be taken into account that interactions between the microbial metabolites can occur which may lead to an elevated health risk.

Further evidence for an underestimation results from a study of Amuzie et al. (2008) indicating that inhalatory (nasal) exposure to DON was more toxic to the mouse than oral exposure.

Risk estimates based on a summation of the concentrations of all detected microbial metabolites, ignore that the detected metabolites cover a broad range of health effects ranging from noxious (e.g. puromycin) to potential carcinogenic effects (e.g. sterigmatocystine). The hazardous potential of a single metabolite with a high cytotoxicity or other relevant health effect can be determinative for a whole mixture of metabolites. Thus, for a sound assessment of health risks the microbial metabolites have to be assessed separately.

However, also this approach is problematic. Apart from the fact that for most of the detected metabolites no sufficient data about possible adverse health effects exist, the approach to asses metabolites separately does not consider possible interactions. On this field the availability of data is even worse. This is in particular true for the inhalation route of exposure.

The problem of the lack of data for a sound assessment will still exist for the next years, if not for decades. Therefore new approaches and assessment criteria are necessary to assess health risks at workplaces like in the recycling industry.

One approach could be to define a background level, for instance by sampling and analyzing dust samples from areas without a known pollution. Mücke et al. (1999) measured up to 49 ng AfB1/g dust and up to 55 ng OTA/g dust in the upwind area of a composting plant. They calculated an airborne concentration of up to 1.3 pg AfB1/m^3 and up to 0.8 pg OTA/m^3 respectively. However, as wind can change it is questionable if the chosen sampling site represents the background concentration. To make this approach possible strategies have to be developed for sampling places and conditions.

However, exposure studies are only a first step. Additionally more medical investigations are needed to prove the possible link between exposure and health effects.

CONCLUSIONS

The high number of determined microbial metabolites and especially the calculated airborne concentrations for worst case conditions underline, that toxic effects due to microbial metabolites should be taken into account when assessing health risks in waste recycling plants.

Beside fungal metabolites also bacterial metabolites have to be considered.

Recycling of municipal waste seems to lead to a higher exposure compared to waste paper recycling.

The used method has shown to be appropriate for screening complex workplace samples for multiple metabolites and can expedite the knowledge about workplace risks in further workplace environments.

REFERENCES:

- Amuzie CA, Harkema JR, Pestka JJ. Tissue distribution and proinflammatory cytokine induction by the trichothecene deoxynivalenol in the mouse: Comparison of nasal vs. oral exposure. *Toxicology* 2008:248: 39-44
- Bünger J, Antlauf-Lammers M, Schulz TG, Westphal GA, Müller MM, Ruhnau P, Hallier E. Health complaints and immunological markers of exposure to bioaerosols among biowaste collectors and compost workers. *Occup Environ Med* 2000:57:458-464
- Degen GH, Blaskewicz M, Lektarau V, Grüner C. Ochratoxin a analyses of blood samples from workers at waste handling facilities. *Mycotoxin Research* 2003:19 (1): 3-7
- European commission. Opinion of the scientific committee on food on fusarium toxins: Part 5: T-2 toxin and HT-2 toxin. 2001
- Heldal KK, Halstensen AS, Thorn J, Djupesland P, Wouters I, Eduard W, Halstensen TS. Upper airway inflammation in waste handlers exposed to bioaerosols. *Occup. Environ. Med.* 2003:60: 444-450
- Mayer S, Engelhardt S, Kolk A, Blome H. The significance of mycotoxins in the framework of assessing workplace related risks. Mycotoxin Research 2008:24(3)151-164
- Mücke W (ed.) Keimemissionen aus Kompostierungs- und Vergärungsanlagen. Herbert Utz Verlag München, 1999 506 p.
- Rabe R, Schwirblat P, Felten C, Küppers M. Belastung durch luftgetragene Schimmelpilze an Arbeitsplätzen in der Abfallwirtschaft. Gefahrstoffe - Reinhaltung der Luft 2003:63: 25-34
- Rylander R. Endotoxin and occupational airway disease. Current Opinion in Allergy & Clinical Immunology: 2006:6(1): 62-66

- Vishwanath V, Sulyok M, Labuda R, Bicker W, Krska R. Co-occurrence of toxic bacterial and fungal secondary metabolites in moisture-damaged indoor environments. *Analytical and Bioanalytical Chemistry* 2009:395; 5:1355-1372

RESPIRATORY HEALTH AND FLOOD RESTORATION WORK IN THE POST-KATRINA ENVIRONMENT

Roy J. Rando, John J. Lefante, Laurie M. Freyder, Robert N. Jones

ABSTRACT

Post-Katrina flood restoration workers have been exposed to bioaerosols from microbial overgrowth of flooded materials and debris. As part of a 5-year longitudinal examination of flood restoration workers, the prevalence of post-Katrina respiratory illness and symptoms was assessed. Spirometry and interview were performed on 791 participants who worked for various public and private institutions, or were private residents of the New Orleans area, and most worked in the building construction and maintenance trades or custodial services. Administered questionnaire included information on respiratory health and symptoms, smoking history, and time spent performing post-Katrina restoration work including demolition, trash removal, landscape restoration, sewer/drain repair, and mold remediation. Prevalence of symptoms and percent predicted lung function parameters were examined statistically for correlation with time spent in restoration work. 74% of study participants reported time spent in restoration work since Katrina with an average of 1,646 hours. Since Katrina, 29% of study participants reported at least one episode of transient fever/cough, 48% reported sinus symptoms, and 4.5% developed new onset asthma. Prevalence rate ratios of fever/cough (PRR: 1.7) and sinus symptoms (PRR: 1.3) were statistically elevated for those who did any restoration work and increased with restoration work time; new onset asthma prevalence (PRR: 2.2) was elevated but not statistically significant. Lung function parameters were slightly depressed in the overall population, but were not significantly different between those with and without restoration work exposure. Analysis of symptoms and lung function in this cohort of post-Katrina New Orleans workers indicates moderate adverse respiratory effects, including ODTS and sinusitis, associated with restoration work.

INTRODUCTION/GOAL OF STUDY

In August of 2005, Hurricane Katrina devastated the New Orleans area with high wind, heavy rainfall, and a storm surge of about 7 m which caused the collapse of the surrounding levee system and flooded approximately 80% of the city for many weeks. In the aftermath of the flood event, the infrastructures of the city along with residences and commercial buildings were grossly contaminated with sediments deposited by the floodwaters and subsequently by microbial overgrowth supported by the residual moisture, high humidity, and elevated temperatures in the area. Indicators of microbial contamination in air, dust, and damaged building materials, including total and culturable mold spores, fungal fragments, mycotoxins,

1→3-β-D-glucan, and bacterial endotoxin were generally elevated, often extremely so, were related to the depth and duration of flooding, and indoor levels were typically higher than those in the surrounding outdoor environment.(Rao et al., 2007)

There has been extensive rebuilding in the New Orleans area after Katrina and residents performing repairs of their property, as well as those working in the construction and building maintenance trades, were at risk for inhalation exposures arising from demolition, removal, and repair of flood-damaged and contaminated infrastructure and building materials.(Adhikari et al., 2009) There were widespread reports of persistent non-productive cough, often with sore throat and rhinorrhea, in the population residing in New Orleans in the Fall of 2005. This symptom complex became known as the Katrina Cough.(Barbeaui et al., 2010) While Katrina Cough was generally ascribed to be an irritant phenomenon resulting from a combination of elevated levels of atmospheric particles, dry weather conditions, and the Fall allergy season, various upper and lower respiratory illnesses and adverse effects including rhinitis, hay fever, organic dust toxic syndrome (ODTS), hypersensitivity pneumonitis (HP), respiratory infections including pneumonia, and exacerbation or initiation of asthma, have been linked to exposures to microbial contaminants in the environment. The potential for respiratory illness arising from post-Katrina inhalation exposures was therefore of particular concern. As part of a 5-year longitudinal study of post Katrina work, exposure, and health, the prevalence of respiratory illness and symptoms associated with flood restoration work after Katrina is examined cross-sectionally.

METHODS AND MATERIALS

Beginning in 2007, 791 adults residing or working in the greater New Orleans metropolitan area were recruited to participate in the study. Study participants included 488 employees (mostly in facilities maintenance and housekeeping) of three large institutions in the City of New Orleans, 63 members of a local union hall for the skilled and unskilled building trades, 95 private building contractors and self-employed tradesmen, and 143 other residents of the New Orleans area, many of whom performed restoration work on their own properties. Overall, 54% of the study cohort reported a skilled or unskilled building or maintenance trade as their primary occupation, including carpentry, electrical, plumbing, paint/drywall, HVAC, grounds keeping, general construction, general maintenance, operating/building engineer, and mechanic/machinist. An additional 15% of study participants worked in custodial or janitorial services.

Testing was conducted in a mobile laboratory van outfitted with spirometry and interview work stations and ancillary equipment. Spirometry testing procedures and equipment complied with American Thoracic Society spirometric test criteria. All spirometric test results were quality assured and interpreted by senior study investigators.

Predicted lung function parameters for forced expiratory volume in one second (FEV1), forced vital capacity (FVC) and FEV1/FVC ratio, were computed from the predictive equations of Hankinson and were based on race, age, gender, and height. Those participants with chronic obstructive pulmonary disease (COPD) were identified according to the GOLD criteria, i.e., FEV1/FVC less than 70% and FEV1 % predicted less than 80%.

A questionnaire was administered to the study participants which accounted for a variety of putative and established risk factors and potential confounders for the development of airways disease including asthma, allergic disease, historical confounding exposures, serious childhood respiratory illness, cigarette smoking history, environmental tobacco smoke, and age, gender and race. Additional questions were designed to target specific symptoms that might be associated with the post Katrina environment including post-Katrina onset of asthma, sinus symptoms, pneumonia, and transient fever and cough absent infection, the latter an indicator of possible hypersensitive (HP) or toxic (ODTS) reaction.

Participants reported the number of hours spent in each of five specific restoration work activities (demolition and rip-out, trash removal, landscape restoration, sewer line repair and mold remediation) for each year since the hurricane, and the type and relative frequency of any respiratory protective equipment that may have been used during the work. For analysis, restoration of personal property was included in the total time spent in restoration work along with any from the subject's regular employment.

The unadjusted prevalence rate ratios for each symptom or condition for those doing any restoration work vs. those not doing any restoration work, were calculated within smoking categories based on 2 X 2 contingency tables. Interactions between age, smoking category, use of respiratory protective equipment, and total hours of restoration work were considered, and were not statistically significant. Multiple linear regression related %P FEV1, FVC and FEV1/FVC to restoration work hours, use of respiratory protection, gender, asthma classification and smoking category. All possible interactions were considered and were not statistically significant, and no significant exposure associations were detected.

The study protocol was approved by the authors' Institutional Review Board and all study participants provided written informed consent.

RESULTS

The majority of the study cohort was African American (70%) and male (75%). Current smokers comprised 28% of the cohort, whereas 18% were ex-smokers and 54% had never smoked. Almost 75% of the study participants performed some restoration work activity post Katrina. Demolition/rip out was the most commonly reported restoration work activity, followed by landscape restoration, trash/debris removal, mold remediation and sewer line repair. Many study subjects worked as much as 16 hours per day, seven days per week, for extended durations after Katrina and the majority reported time spent in more than one type of restoration work activity. For total combined hours spent in any of the specific restoration work activities, the mean and median values reported by 473 subjects with valid data were 1646 and 620 hours, respectively.

Upper and lower respiratory symptoms were prevalent in the cohort. Episodes of transient fever and cough occurring post-Katrina were reported by about 29% of the study cohort with multiple episodes being common. The median number of episodes was 3, and 10% of the group reported having 12 or more such occurrences. New onset sinus symptoms were reported by almost half the cohort. The prevalence rate ratios for transient fever and cough, and new onset sinusitis were significantly elevated ($P < 0.05$) for those reporting any restoration work (PRR: 1.7 and 1.3, respectively; Table 1). The results suggest a possible protective effect of smoking as the prevalence rate ratios were statistically significant for ex- and never-smokers but not for current smokers.

Table 1: Post-Katrina Restoration Work Exposure and Respiratory Symptoms, Conditions and Lung Function in a Cohort of New Orleans Workers

	Prevalence Rate Ratio: Exposed vs. Non-Exposed (95% CI)			% P, Mean ± S.E. Exposed (E) Non-Exposed (NE)		
	Fever & cough	New onset sinus	New onset asthma	FEV_1	FVC	FEV_1/FVC
current smokers	1.3	1.1	1.0	E 92.7±1.4	94.9±1.2	97.2±0.8
	(0.8 – 2.1)	(0.8 – 1.6)	(0.1 – 9.1)	NE 95.3±1.8	96.2±1.8	98.9±1.0
ex- and never-smokers	2.0*	1.4*	2.7	E 96.4±1.3	95.1±0.7	101.1±0.4
	(1.3 – 3.0)	(1.1 – 1.9)	(0.8 – 8.8)	NE 95.0±1.3	94.9±1.3	99.7±0.7
overall	1.7*	1.3*	2.2	E 95.5±0.7	95.1±0.6	100.0±0.4
	(1.3 – 2.4)	(1.1 – 1.7)	(0.8 – 6.2)	NE 95.2±1.1	95.4±1.0	99.4±0.6

*$P < 0.05$

Among those reporting never having had asthma prior to Katrina (*n* = 539), about 4.5% reported new onset asthma. The prevalence rate ratios for new onset asthma were elevated for those who performed restoration work overall (PRR: 2.2) and especially for ex- and never-smokers (PRR: 2.7), which was borderline significant (*P* = 0.09).

Figure 1: Unadjusted Prevalence of Respiratory Symptoms and Conditions with Quartiles of post-Katrina Restoration Work Time

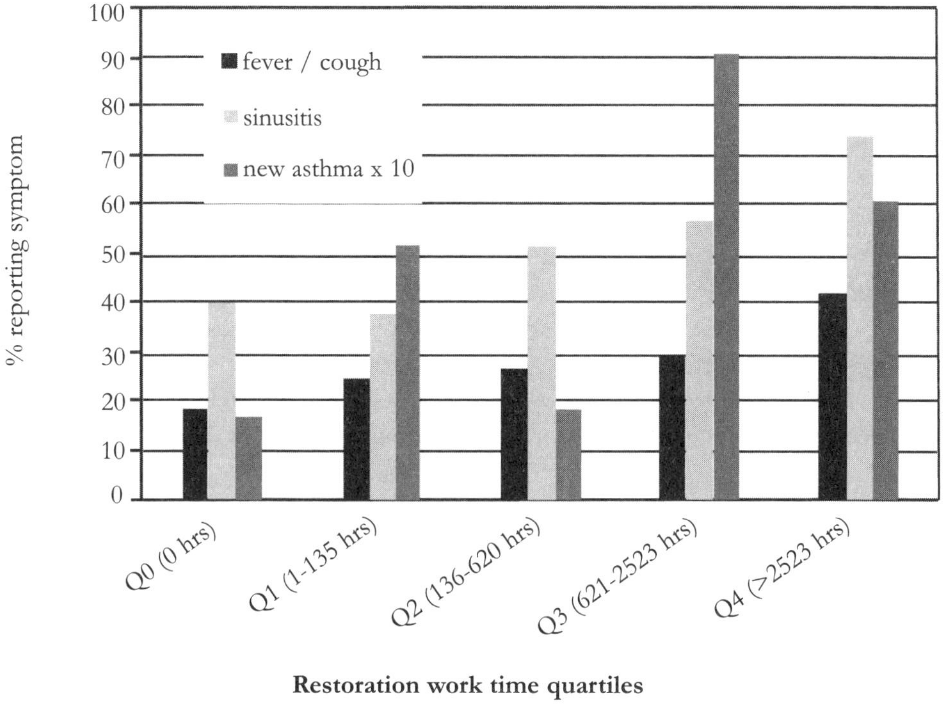

Figure 1 shows the unadjusted prevalence of new onset asthma, sinusitis, and fever & cough with quartiles of reported time spent in restoration work activities. The unadjusted proportions for each of these are higher for those with restoration work in comparison to those without and the prevalences generally increase with increasing restoration work time quartiles.

Significant elevations in prevalence rate ratios for those having done any restoration work were not observed for pneumonia, dyspnea, and COPD.

Overall, lung function parameters were somewhat depressed in the cohort (Table 1) with group averages for %P FEV1, FVC, and FEV1/FVC ratio being a few percent below 100%. Current smokers who did restoration work showed lower overall predicted lung function compared to smokers who did not; however, the differences did not reach the level of statistical significance.

CONCLUSIONS

There was an observable elevation in prevalence of new onset asthma, particularly among ex- and never smokers, with restoration work exposure. Prevalence rates for sinusitis among those who performed restoration work were significantly higher in comparison to those who did not. The prevalence of episodes of fever & cough in the present study population is clearly elevated for those who have done restoration work. Some of these cases may be relatable to Katrina Cough but ODTS is likely to be underlying many of these reports and appears to be a common adverse effect of restoration work exposures in the post Katrina environment. While this study did not identify any restoration work related decrements in lung functional parameters, annual excess declines of small magnitude are difficult to detect over short time period and are unlikely to result in a detectable group difference in function, measured cross-sectionally, after only a few years in the post Katrina environment.

This study provides further evidence that workers performing restoration work on flood-damaged structures are at risk of respiratory health impacts from exposure to microbial contaminated dust and debris. Appropriate exposure controls such as the use of respiratory protective equipment and dust suppression techniques should be employed to reduce exposure and manage the risk of acquiring exposure-related respiratory disease for workers performing flood restoration activities.

This study did not identify any restoration work related decrements in lung functional parameters. Annual excess declines of small magnitude are difficult to detect over short time period and are unlikely to result in a detectable group difference in function, measured cross-sectionally, after only a few years in the post Katrina environment.

Overall, this study suggests moderate adverse respiratory health effects among post-Katrina restoration workers including ODTS and sinusitis, which were commonly reported in the study cohort, and the prevalence of new onset asthma among restoration workers was noticeably elevated. While it is unclear from this cross-sectional analysis whether restoration work exposures have adversely affected pulmonary function in the population, the functional parameters overall are depressed in the cohort. Ongoing health surveillance of this study cohort, along with a quantitative exposure assessment, is examining whether there is increased risk for long term or irreversible effects on respiratory health and how the risks relate to the nature and magnitude of the exposures occurring during flood restoration work.

ACKNOWLEDGEMENT

This work was supported by Grant Number R01OH008938 from CDC/NIOSH. Its contents are solely the responsibility of the authors and do not necessarily represent the official views of CDC/NIOSH. We dedicate this work to the memory of our late friend and colleague, Dr. Henry W. Glindmeyer, who the memory of our late friend and colleague, Dr. Henry W. Glindmeyer, who initiated this study.

REFERENCES

- Adhikari A, Jung J, Reponen T, Lewis JS, Degrasse EJ, Grimsley LF, Chew GL, Grinshpun SA. Aerosolization of fungi, (1-3)-β-D-glucan and endotoxin from flood-affected materials collected in New Orleans homes. *Environ Res.* 2009;109(3):215-224.
- Barbeau DN, Grimsley LF, White LE, El-Dahr JM, Lichtveld M. Mold exposure and health effects following hurricanes Katrina and Rita. *Annual Rev Public Health.* 2010:31:165-178.
- Rao CY, Riggs MA, Chew GL, Muilenberg ML, Thorne PS. Characterization of molds, endotoxins, and glucans in homes in New Orleans after Hurricanes Katrina and Rita. *Appl Environ Microbiol.* 2007:73(5):1630-1634.

MOLD SPECIES IDENTIFIED IN FLOODED DWELLINGS

Denis Charpin

INTRODUCTION

Nowadays, heavy rains and flooding occur more frequently in some regions, probably due to climate change, and they are responsible for mold proliferation in dwellings (Diaz, 2007). Around 80 mold species are considered to live indoors (Kuhn, Ghannoun, 2003). In vitro, these mold species demonstrate different metabolic production, and have different allergenic properties (Nielsen, 2003). Thus, the health impact of mold exposure and the decontamination procedures which need to be implemented may depend on the mold species under consideration. The mold species developing in a given dwelling depend on environmental conditions (Nielsen, 2003). Thus, one may anticipate that mold species encountered in flooded dwellings may be different from those encountered in other unhealthy buildings. Surprisingly, very few data are available on this important issue. In this paper, we compared mold species identified in flooded dwellings compared to those found in unhealthy dwellings.

Materials and methods Selection of dwellings

- The group of flooded dwellings consisted of 185 dwellings flooded on 2–5 December 2003. Flooding followed very heavy rains, which were responsible for an inflow of the Rhone River. The northern part of the town of Arles, located in southeast France, was flooded. Water level in houses went up to 1.5 m and stayed in the houses for 10–15 days. Slow receding waters allowed mold to proliferate to a large extent. A team of architects visiting the flooded area identified the most severely damaged dwellings. Our organization was asked to evaluate damage in these houses in order to provide appropriate advice to tenants to minimize damage and to draw an inventory of damage for insurance compensation. The fieldwork began in April 2004 and ended in November 2004. All dwellings were at that time unremediated or partially remediated.
- Other unhealthy dwellings (n = 341) were those visited by our organization during its routine activity (Charpin-Kadouch et al., 2007), i.e., those of tenants whose health status was considered by the attending physician to be negatively influenced by the housing conditions.

The protocol included:

- A visual inspection of dwellings and the computation of the moldy area in each room. Molds were also investigated behind damaged wallpaper. The moldy surface was classified into 3 grades, according to the recommendations of the New York City Department of Health and Mental Hygiene (New York City Department of Health, Mental Hygiene 2005).

- A sampling of visible molds. Each macroscopically different area was sampled using the papergummed technique (Porto, 1953), which is recommended by the Centraal Bureau voor Schimmeltechnique, The Netherlands. The sampling kit consisted of a slide covered by a transparent papergum, which is removed from the slide (Menzel Glaser, pressure sensitive 3 M), gently pressed onto the moldy surface, then put back on the slide. After color straining using chloral biphenyl cotton blue, the slide was studied using an optical microscope.
- Other data on the building characteristics and indoor climatic parameters were also gathered, but will not be presented in this paper because they do not appear to be relevant to its objective.

RESULTS

About 329 samples were collected from the 185 flooded dwellings and 936 samples from the other unhealthy dwellings. Flooded dwellings were mainly located on the ground floors of private houses. The percentage of moldy dwellings, as well as the moldy surface, was comparable in flooded and unhealthy dwellings (Table 1).

Table 1 Some descriptive characteristics of both groups of studied dwellings

	Flooded dwelling n=185	Unhealthy dwelling n=341	P value
Private houses/apartments Distribution of moldy area per dwelling	166/19	137/204	0.001
No mold	78 (42.2%)	137 (40.1%)	
<1 m^2	25 (13.5%)	43 (12.6%)	0.001
1–3 m^2	38 (20.5%)	107 (31.4%)	
>3 m^2	44 (23.8%)	54 (15.9%)	

Figure 1 displays the distribution of mold species from both series. *Cladosporium* was the most prevalent mold species in both types of dwellings. *Alternaria* species was identified in 24% of samples taken in flooded dwellings and in 6.5% of samples taken from other unhealthy dwellings (P \ 0.0001). The percentage of samples including *Stachybotrys chartarum* was equal to 8.5 and 5.5%, respectively (P = 0.0007). In contrast, *Aspergillus* and *Penicillium* species were encountered more frequently in unhealthy dwellings (P = 0.000007 and P = 0.009, respectively).

Figure: 1 Percentage of mold species identified in flooded dwellings (gray bars) and in unhealthy dwellings (black bars)

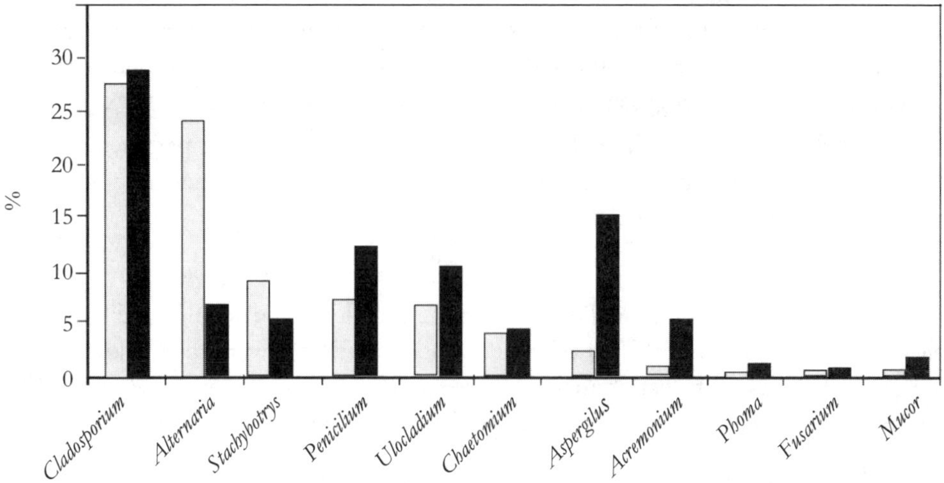

DISCUSSION

The main result of the study lies in the demonstration that two mold species, namely *Alternaria* and *Stachybotrys chartarum*, predominated in flooded dwellings. Surprisingly, there are very few data in the literature on this issue of increasing importance. Many websites provide data on basic mold cleaning and relevance of molds to human health (e.g., bt.cdc.gov/disasters/ mold/protect.asp; bt.cdc.gov/hurricanes/index.asp; dehs. umn.edu/iag/flood.html; fema.gov/news/-event.fema) but these sites do not provide any data on mold species identified in flooded dwellings. Papers dealing with this issue relate to buildings which have had accidental water damage, which is a completely different issue because water does not stay long in the building.

The other preliminary data available on development of molds following flooding were gathered following flooding in New Orleans, LA. It is note-worthy that initial mold sampling in flooded dwellings in New Orleans was performed a few days after the flooding, whereas it began 4 months later in our survey. In the New Orleans study, the predominant mold species were *Cladosporium* in both settings and *Penicillium/Aspergillus* (Rao et al., 2007; Solomon et al., 2006) in the group of unhealthy dwellings. The difference might be accounted for by this different timing. However, if, on a given wall surface, mold species may vary over time following varying environmental conditions, flooding was from the beginning responsible for the 100% humidity on surfaces. Besides, the higher percentage of *Stachybotrys chartarum* in our group of flooded dwellings may be related to the sampling method because this mold species is known to be underrepresented in airborne sampling which has been performed in the New Orleans study. Finally, the difference may also be related to different building materials used in those different geographic areas. For example, damp gypsum board is highly susceptible to *Stachybotrys chartarum* proliferation (www.gypsum.org).

Under experimental conditions, *Alternaria* and *Stachybotrys chartarum* species demonstrate the highest water activity, up to 0.94, whereas this parameter, for other mold species, ranged from 0.70 to 0.85 (Northolt and Bluesman, 1982; Boutin-Forzano et al., 2004). Thus, it is no surprise that they were overrepresented in flooded dwellings. When the wall surfaces have dried, the mold is still viable and may contain mycotoxins. From a medical viewpoint, *Alternaria* and *Stachybotrys chartarum* species deserve a special consideration because they are able to produce potent mycotoxins (Kuhn, Ghannoun, 2003; Nielsen, 2003) as secondary metabolites. These mycotoxins may have a health impact, especially during decontamination. Thus, use of protective personnel equipment and exposure mitigation techniques among residents in order to prevent respiratory morbidity should be encouraged.

In conclusion, two mold species, *Alternaria* and *Stachybotrys chartarum* were overrepresented in flooded dwellings. This finding is important to consider because those mold species may produce mycotoxins with health impact.

ACKNOWLEDGMENT I thank for financial support the Direction Départementale des Affaires Sanitaires et Sociales (DDASS) des Bouches-du-Rhône.

REFERENCES

- Boutin-Forzano S, Charpin-Kadouch C, Bennedjai N et al. Wall relative humidity: A simple and reliable index for predicting *Stachybotrys chartarum* infestation in dwellings. *Indoor Air*, 2004:14, 196–199. doi:10.1111/j.1600-0668.2004.00233.x.
- Charpin-Kadouch C, Mouche JM, Quéralt J et al. Housing and health counselling. *Environmental Research*, 2007:103, 149–153. doi:10.1016/j.envres.2006.06.006.
- Diaz JH. The influence of global warming on natural disasters and their public health outcomes. *American Journal of Disaster Medicine*, 2007:2, 33–42.
- Kuhn DM, Ghannoun MA. Indoor molds, toxigenic fungi, and *Stachybotrys chartarum*: Infectious disease perspective. *Clinical Microbiology Reviews,* 2003:16, 144–172. doi: 10.1 128/CMR.16.1 .144-172.2003.
- New York City Department of Health, Mental Hygiene. Guidelines on assessment and remediation of fungi in indoor environments. New York, NY: Department of Health and Mental Hygiene. 2005.

- Nielsen KF. Mycotoxin production by indoor molds. *Fungal Genetics and Biology*, 2003:39, 103–117. doi:10.1016/ S1087-1845(03)00026-4.
- Northolt D, Bluesman LB. Prevention of mold growth and toxic production through control of environmental conditions. *Journal of Food Protection*, 1982:45, 519– 526.
- Porto JM. The use of cellophane tape in the diagnostic of *Tinea versicolor*. The *Journal of Investigative Dermatology*, 1953:21, 229.
- Rao CY, Riggs MA, Chew GL et al. Characterization of airborne molds, endotoxins, and glucans in homes in New Orleans after Hurricanes Katrina and Rita. *Applied and Environmental Microbiology,* 2007:73, 1630– 1634. doi: 10.1 128/AEM.01973-06.
- Solomon GM, Hjelmroos-Koski M, Rotkin-Ellman M, Hammond SK. Airborne mold and endotoxin concentration in New Orleans, Louisiana, after flooding, October through November 2005. *Environmental Health Perspectives,* 2006:114, 1381–1386.

ATTRIBUTABLE FRACTIONS OF RISK FACTORS OF RESPIRATORY DISEASES AMONG CHILDREN IN MONTREAL, CANADA

Louis Jacques, Céline Plante, Sophie Goudreau, Leylâ Deger, Michel Fournier, Audrey Smargiassi, Stéphane Perron, Robert L. Thivierge

KEYWORDS: Molds, moisture, asthma, rhinitis, respiratory infections, attributable fractions

ABSTRACT

Goal of study

The objective of this cross-sectional study was to estimate the attributable fractions among the child population living in Montréal, in Canada, of asthma, respiratory infections and winter allergic rhinitis associated with known important risk factors, including both the indoor and the outdoor environment.

Methods

Families with at least a child aged 6 months to 12 years living on the Island of Montreal were selected from a random list provided by the Québec Health Insurance Board. Data collection was completed in summer 2006 by a mixed mode survey, either by phone or Internet. The final sample contained 7956 subjects. Log-binomial regression models were used to estimate adjusted prevalence-ratios (PRs) with the corresponding 95% confidence intervals for the association between exposure variables and disease. We calculated the attributable fractions among the population for major risk factors derived from our model, whilst isolating the modifiable factors.

Results

The main risk factors found to be associated with these respiratory diseases are 1) personal and family history of allergy, asthma or eczema, 2) perinatal factors (low birth weight, lack of breastfeeding and exposure to tobacco smoke) and 3) environmental factors related to indoor and outdoor air quality. Among the modifiable factors, exposure to environmental factors related to housing conditions and indoor air quality emerges as most important, including indicators of excessive moisture and molds, cockroaches, tobacco smoke and pollution from nearby highways or roads. Indicator of excessive moisture and molds has the highest attributable fractions, explaining respectively 17%, 26% and 14% of cases of active asthma, respiratory infections and winter allergic rhinitis.

Conclusion

Indicator of excessive moisture or molds appears to be the most important modifiable risk factor of main respiratory diseases of children in Montréal. Control of this factor would prevent many health outcomes among children and their families.

Background

Asthma, respiratory infections and allergic rhinitis are the most prevalent respiratory diseases in children in Canada and their occurrence is closely related to one another (Garner and Kohen 2008; Thomas 2010; ISAAC 1998; Bousquet et al., 2001). They lead to many medical consultations, emergency room visits, hospitalizations, and loss of work and school days (Krahn et al., 1996).

Previous data have indicated that hospital admissions and emergency room visits for pediatric asthma are unevenly distributed geographically throughout the Island of Montreal (total population: 2 millions), affecting particularly disadvantaged areas (Kosatsky et al., 2004; Lajoie et al., 2004), but few data exists as to the geographical distribution of respiratory infections and allergic rhinitis. Socioeconomic status, quality and organization of health care, access to health care services, lifestyle and environmental factors could explain the uneven distribution of asthma rates (Gold and Wright, 2005). Furthermore, respiratory infections and allergic rhinitis may share several genetic, social and environmental determinants and risk factors of childhood asthma. Determining the relative importance of these factors within a population is key to guide priority actions to reduce the prevalence of respiratory diseases among children.

The objective of this study was to estimate the attributable fractions among the child population living in Montréal, in Canada, of asthma, respiratory infections and winter allergic rhinitis associated with known important risk factors, including both the indoor and the outdoor environment.

Materials

The study design has been described in detail in a previous publication (Değer et al., 2010). The objectives of this cross-sectional study were 1) to estimate the geographic distribution of prevalence rates of asthma, respiratory infections and allergic rhinitis, among children aged 6 months to 12 years and living on the Island of Montreal, and 2) to identify the risk factors associated with the distribution of these diseases.

Study participants

The sample size calculation was based on the estimated prevalence of asthma among children (12%) and a capacity to detect a difference of 5% between the sub territories with a statistical power of 80%. Participants were recruited from a random list of Montreal households (N=17,697), with at least one child aged 6 months to 12 years old, provided by the Quebec Health Insurance Board (RAMQ). The survey targeted only one child per family. The eligibility criteria were: 1) the child had to reside on the Montreal Island and live at least 50% of the time with the responding mother, father or legal guardian; 2) the child's parent or legal guardian had to be a French- or English-speaking adult (≥18 years). Potential participants were initially approached by mail. Follow-up contacts were made by telephone (up to 10 per household) when a number could be paired to the sampled address (N=12,680 paired telephone numbers of which 77% were confirmed as accurate and within the Montreal Island), or by mail when no residential telephone number could be obtained. The overall response rate, estimated at 60%, was calculated with the number of completed surveys divided by the number of eligible households (e.g., with valid telephone and mail contact information). The final sample consisted of 7956 subjects. Data collection was completed in the summer 2006. A mixed mode survey was used, either by phone or Internet, the latter being chosen by 52% of respondents. The web and interview questionnaires were identical. Consent of the child's parent or legal guardian was obtained verbally or by means of the electronic questionnaire transmitted. The main study protocol was approved by the Montreal Public Health and Social Services Agency Research Ethics Board.

Variables

The outcomes studied included various degrees of severity of asthma, two groups of respiratory infections and winter allergic rhinitis, as defined in Table 1. The definition of active asthma was adapted from Health Canada's 1995-1996 Student Lung Health Survey (Health Canada, 1998). Asthma control was assessed on the basis of five specific symptom-based criteria outlined in the Canadian Pediatric Asthma Consensus Guidelines (Canadian Thoracic Society, 2004) which were adapted for our study questionnaire.

The main risk factors known to be associated with the occurrence or aggravation of these diseases in the scientific literature, with a probable or possible causal relation, were included in the study (Boulet et al., 1999; Jacques et al., 2005; Gold and Wright 2005; Subbarrao et al., 2009). Questions related to diseases were based on international (ISAAC, 1998; Burney et al., 1994) and national (Health and social survey of Quebec children and youth 1999) surveys. The following groups of variables were examined: current and past respiratory health, use of medication and health care services, history of parental and child atopy, home environmental exposures, lifestyles and socioeconomic status (SES). Exposure assessment was based on the questionnaire data, except for exposure to vehicle pollutants which are quantitative estimates derived from previous studies (Smargiassi et al., 2004). The definitions of the risk factors relevant for the following results are included in Table 1.

Table 1. Definitions of respiratory diseases and major risk factors, Respiratory health study among children aged 6 months to 12 years, Montréal, Canada, 2006

Disease or risk factor	Definition
Active Asthma	Asthma diagnosed by a physician and one or more of the following features in the last 12 months preceding the survey:
	1) wheezing or whistling in the chest
	2) a dry cough at night
	3) an asthma exacerbation
	4) use of asthma medication
Uncontrolled active asthma	Active asthma and one or more of the following features during the past 3 months:
	1) daytime symptoms, e.g., wheezing or whistling in the chest ≥ 3 times a week
	2) night time symptoms, e.g., awakened by dry cough or wheezing ≥ 1 night a week
	3) the need for >3 doses a week of short-term acting β2-agonists (SABAs; rescue medication)
	4) physical activity limitations
	5) absence from school or daycare due to asthma
Severe active asthma	Active asthma with emergency visit or hospitalization in the last 12 months preceding the survey

Table 1. continued	
Disease or risk factor	**Definition**
Respiratory infections	At least one infection of each following group in the last 12 months:
	1) otitis or sinusitis
	2) bronchitis, bronchiolitis or pneumonia
Winter allergic rhinitis	1) Sneezing, or runny nose or blocked nose when the child didn't have a cold or flu in the past 12 months AND
	2) symptoms more important during November to March
Maternal education level	Secondary and less vs post-secondary
Parental atopy	Reported history of asthma, allergic rhinitis or eczema in either one parent
Lack of breastfeeding	Out from hospital, the child was fed with milk formulae only
Exposure to tobacco smoke *in utero*	The mother smoked or has been exposed to tobacco smoke during pregnancy
Actual exposure to tobacco smoke	One or more occupant(s) currently smoke inside residence, every day or nearly everyday
Current or past exposure to tobacco smoke	Since birth, child has been exposed to tobacco smoke inside residence
Excessive moisture/mould indicator at home	At least one of the followings: visible moulds, mould odour, signs of water infiltration, history of past water damages, room or apartment in the basement
Vehicle and wood burning pollutants	One or more of the following features:
	1) Residence 200 m or less from a highway
	2) Residence located on a street with high traffic volume ($\geq$3160 vehicles estimated/morning period)
	3) Use of fire place or wood stove three times or more per week (winter)
Exposure to cockroaches	Presence of cockroaches noticed in the residence in the last 12 months

Validation of data

Three types of validation were conducted. 1) Data from the two survey modes were compared on some key variables and no important mode effect was observed (e.g. difference due to socioeconomic status, desirability; refer to Plante et al. (2012) for further information). The data from the two survey modes were combined for all analyses. 2) Parents' responses to questions on reported diagnosis by a physician and use of health services and medication were compared with the RAMQ database. This comparison showed generally good agreement (kappa = 0.63 for prevalence of lifetime asthma). Using RAMQ data does not change the prevalence rates obtained for the Island, nor the rank of local health and social services territories by prevalence rate (see Table 2). 3) The socio-economic status of survey participants was compared to the 2006 census data. The education level of mothers in the study is somewhat higher than census data, but a difference in the question wording may have influenced the distribution of responses. Moreover, the rank of the sub territories by level of education is very little modified by the use of census data.

Statistical analysis

Sample weights were used in all analyses to adjust for variability in the response rate by local territory and to match the distribution of children aged 6 months to 12 years by age and sex on the Island of Montreal of the 2006 Census Statistics Canada.

Geographic variation of prevalence rates were looked at local (N=29) and sub-regional (N=12) territories of health and social services. Log-binomial regression models were used to estimate adjusted prevalence-ratios (PRs) with the corresponding 95% confidence intervals for the association between exposure variables and disease outcomes. The log-binomial regression was favored to logistic regression because of high outcome prevalence (>10%) and to minimize potential bias in risk estimation (Behrens et al., 2004). The control group varies according to the type of asthma as follows: non asthmatics for active asthma, controlled asthmatics for uncontrolled asthma and non severe asthmatics for severe asthma. In a separate analysis, non asthmatics were used as controls for uncontrolled active asthma and severe active asthma. Potential confounders included in the multiple regressions are sex, age of child and parental history of atopy. The effect of maternal educational level as a proxy for socioeconomic status was assessed but not included in the final analyses because the results were not significantly modified when all other factors were taken into account. Children taking medication for asthma but with no asthma diagnosis were excluded from the regression analyses (N=325). Cases with missing data ("do not know" or "refusal") on any of the variables included in the multivariable regression models were excluded (9 % for severe asthma and 15 % for other outcomes). However, the comparison between the regression model subset excluding the missing data and the corresponding subset of missing data showed little differences in socioeconomic status, atopy, sex, age, and outcome prevalence. These differences were not statistically significant for outcomes and family income (Chi square test), and slight or non significant for the other factors tested depending on the regression model subset (discrepancies of 2% to 5 % and ≤two years for mean age). Among the various measures of exposure to tobacco smoke, the one with the largest effect was retained in the regression model of each disease.

We calculated the attributable fractions among the population for major risk factors derived from our model, whilst isolating the factors that can be prevented (modifiable factors) from those that cannot, using the method described in Eide and Gefeller (1998). Only strong and statistically significant factors were considered in the calculation of attributable fractions given the limited number of risk factors that can be analyzed altogether.

RESULTS

Prevalence rates

Prevalence rates for each disease are presented in Table 2. Prevalence rates according to local territories varied roughly by a factor of 3 for active asthma and winter allergic rhinitis and by a factor of 5 for respiratory infections (Table 2).

Prevalence rate of winter allergic rhinitis was much lower among non asthmatics (12%) and increased according to severity of asthma, namely 21% for active asthma, 28% for uncontrolled active asthma and 33% for severe active asthma.

Similarly, prevalence rates of risk factors according to local territories varied largely. Table 2 shows the range of prevalence rates per local territory within the Montreal Island for the main risk factors.

Table 2. Range of prevalence rates per territory of respiratory diseases and of exposure to major risk factors among children in Montreal, Canada, 2006

Disease or risk factor	Prevalence rate (%)		
	Montreal Island (95% CI)	Min	Max
Active asthma[1]	12.8 (±0.7)	7.3	21.9
Uncontrolled active asthma[2]	4.5 (±0.5)	2.8	6.2
Severe active asthma[2]	3.1 (±0.4)	1.6	4.8
Respiratory infections[1]	6.3 (±0.5)	2.7	14.3
Winter allergic rhinitis[1]	13.3 (±0.7)	6.3	18.0
Parental atopy[1]	36.2 (±1.1)	20.3	47.5
Exposure to tobacco smoke *in utero*[1]	13.4 (±0.8)	3.8	30.1
Low birth weight (<2.5kg)[1]	6.3 (±0.5)	3.7	10.6
Absence of breastfeeding[1]	21.8(±0.9)	41.9	7.1
Day care before school[1]	75.5 (±8.5)	65.0	86.5
Maternal education secondary or less[1]	24.5 (±1.0)	5.6	52.6
Actual exposure to tobacco smoke[1]	11.6 (±0.7)	4.7	26.9
Exposure to indicator of excessive moisture/mould at home[1]	36.3 (±1.0)	29.8	51.9
Exposure to cockroaches[1]	4.5 (±0.5)	~0	25.9
Residence 200 m or less from a highway[1]	6.5 (±0.5)	~0	27.2

[1] For active asthma, respiratory infections, winter allergic rhinitis and risk factors, the range of prevalence rates was assessed among 29 territories of local health and social services.

[2] For uncontrolled active asthma and severe active asthma, the range of prevalence rates was assessed among 12 territories of sub regional- health and social services

Multivariable analyses

The main risk factors found to be associated with these respiratory diseases were 1) personal and family history of allergy, asthma or eczema, 2) perinatal factors (low birth weight, lack of breastfeeding and exposure to tobacco smoke) and 3) environmental factors related to indoor and outdoor air quality (Table 3).

Table 3. Prevalence Ratio of risk factors of respiratory diseases estimated from multiple log binomial regression models, among children (6 months to 12 years old), Montreal, Canada.

Risk factor[1]	**Prevalence ratio (95% CI)**[2]				
	Active asthma	Uncontrolled active asthma[3]	Severe active asthma[3]	Respiratory infections	Winter allergic rhinitis
Parental atopy	2.18 (1.93-2.47)	1.34 (1.12-1.60)	1.16 (0.94-1.44)	1.95 (1.58-2.39)	1.22 (1.08-1.39)
Exposure to tobacco smoke *in utero*[3]	1.26 (1.10-1.44)				
Low birth weight (<2,5kg)	1.38 (1.13-1.70)				
Lack of breastfeeding	1.19 (1.04-1.36)				
Day care before school age	1.17 (1.02-1.38)			1.42 (1.09-1.86)	1.18 (1.02-1.37)
Actual exposure to tobacco smoke				1.53 (1.14-2.05)	
Current or past exposure to tobacco smoke					1.16 (1.00-1.34)
Excessive moisture/mould indicator	1.14 (1.01-1.29)	1.21 (1.01-1.52)	1.24 (1.01-1.53)	1.44 (1.17-1.77)	1.25 (1.11-1.42)
Exposure to cockroaches					1.26 (0.97-1.60)
Vehicle and wood burning pollutants		1.24 (1.02-1.42)			

[1] See Table 1 for definitions. [2] Adjusted for age and sex. [3] Among active asthmatic children

Among the modifiable factors (exposure to tobacco smoke, lack of breastfeeding, excessive moisture and mold indicator, vehicle and wood burning pollutants and cockroaches), exposure to environmental factors related to present or recent housing conditions and indoor air quality emerged as most important, including indicator of excessive moisture and molds, cockroaches, tobacco smoke and pollution from nearby highways or roads. Within the excessive moisture and mold indicator, location of the apartment or child room in the basement was the most strongly associated variable with the outcome.

When non asthmatics were used as controls for uncontrolled active asthma and severe active asthma, the prevalence ratios for the same risk factors were higher (data not presented). As to uncontrolled active asthma, taking into account the use of corticosteroids did not change substantially the values of the prevalence ratios. For all outcomes, the inclusion of the mother education level as a proxy for the socioeconomic status did not change significantly the values of the prevalence ratios when other significant risk factors were included in the model. This would indicate that, in our analysis, the association between a low SES and higher risk of respiratory disease in the primary analysis (data not shown) is mainly explained by a higher frequency of exposure to the direct risk factors identified, including those related to the environment.

Population attributable fractions

Population attributable fractions for each disease are presented at Figure 1. The white portion of each pie includes non modifiable risk factors, such as atopy and age, and the factors for which the fraction could not be calculated. This can include a number of factors for which the fraction was too small or not significant enough to be estimated.

Figure 1. Population attributable fractions of active, uncontrolled and severe asthma, respiratory infections and winter allergic rhinitis, Montreal children, Canada, 2006

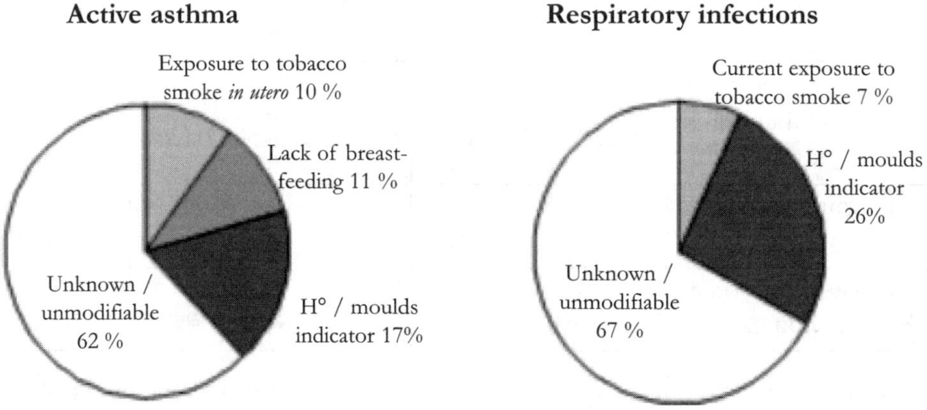

Uncontrolled asthma (among children with active asthma)

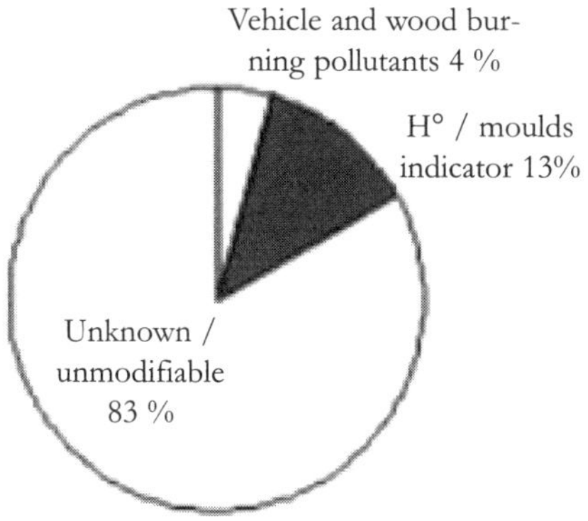

Winter allergic rhinitis

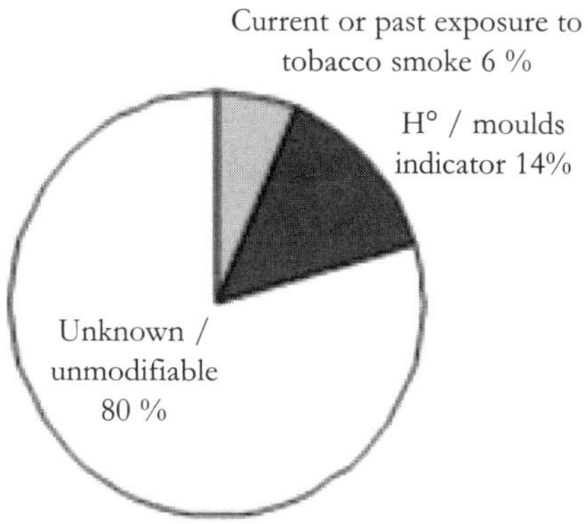

Severe asthma (among children with active asthma)

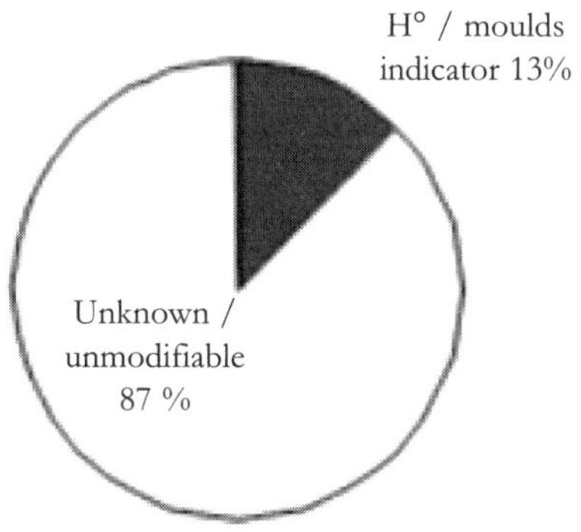

Excessive moisture and molds are the risk factors with the highest attributable fractions among the child population of the Montreal Island: 13% to 17% of the prevalence rates of active, uncontrolled and severe asthma, 26% of the prevalence rate of respiratory infections and 14% of the prevalence rate of winter allergic rhinitis. Exposure to second hand smoke is an important risk factor for active asthma, respiratory infections and winter allergic rhinitis with attributable fractions of 10%, 7%, and 6% respectively. In addition, we calculated the attributable fractions of lack of breastfeeding for active asthma (11 %), and exposure to vehicle and wood burning pollutants for uncontrolled asthma (4%). Due to its low prevalence in the sample, the attributable fraction for exposure to cockroaches could not be calculated.

DISCUSSION

This study looks at the determinants of the most prevalent respiratory diseases among children. These share many common risk factors and their occurrence is closely related to one another (ISAAC 1998; Bousquet et al., 2001). Indicator of excessive moisture and molds as well as other potential contaminants associated with excessive moisture (Trasher and Crawley, 2009) appear to be the most important modifiable risk factor of these respiratory diseases of children in Montreal, explaining respectively 17%, 26% and 14% of cases of active asthma, respiratory infections and winter allergic rhinitis. Excess moisture and molds are accepted risk factors for the diseases studied, as shown by high consistency between various study designs and systematic reviews (Mendell et al., 2011; Fisk et al., 2007; Bornehag et al., 2001, 2004; Antova et al., 2008). These fractions are close to those estimated for the USA by Mudarri and Fisk (2007), which are 21% (CI: 12%-29%) of asthma cases for the total population and 9-19% of respiratory infections among children, all attributable to excessive moisture and molds.

The estimates of prevalence ratios and population attributable fractions for excessive humidity and molds are probably underestimated as a result of unreported presence of mold in the home which are most often hidden in the walls (thus residents are unaware of such a problem) as found in our experience and that of others (Moularat et al., 2008), and because the history of water damage in the apartment and in the building is often unknown by the actual tenants, these two factors probably causing non-differential misclassification. Moreover, the poorest families who are most susceptible to live in unsanitary buildings tend to be underrepresented in any type of study requiring voluntary participation, which would also contribute to underestimate the effect.

Although tobacco smoke is an important risk factor of these respiratory diseases, it was not significant among children with uncontrolled asthma and severe asthma. This may be explained by the fact that smoking was reported by the parents to be one of the first changes made following the diagnosis of asthma in their child, while excessive moisture or molds was not mentioned at all, probably because it is generally not under the control of tenants or simply unknown.

The other risk factors for which significant population attributable fractions were estimated (air pollutants from traffic and woodburning, lack of breastfeeding) are also known risk factors for these diseases (HEI 2010; WHO 2003; Naeher et al., 2007), although results have been somewhat contradictory in the literature as to lack of breastfeeding (Fredriksson et al., 2007). In a separate analysis (data not shown), we have found that breastfeeding had a slightly greater protective effect among young children exposed to tobacco smoke, as observed in another study (Chulada et al., 2003). As to cockroaches and allergic rhinitis, population attributable fractions could be not estimated for a statistical reason, but was found to be an important risk factor in the regression model.

In term of strengths and limitations, these estimates are derived from a population-based sample rather than a health care organization clientele. We have analyzed a wide range of risk factors known from previous studies, including many important environmental factors. A cross-sectional study does not generally allow by itself to determine causality. However, this study builds on information accumulated from numerous studies of various designs. Excess humidity and molds indicator and the other significant risk factors have been found in the same order of magnitude as in previous studies, suggesting good validity (Mendell et al., 2011; Fisk et al., 2007; Bornehag et al., 2001; Antova et al., 2008).

Time sequence can be questionable in a prevalence study. In this regard, atopy, perinatal factors and SES are not an issue in this study. Concurrent or recent exposure to environmental factors were considered in association with the respiratory diseases, which is consistent with our knowledge that these factors can aggravate or even cause those diseases in a short term. Moreover, it is unlikely that the disease is the cause of the exposure rather than the reverse. For example, it is unlikely that the occurrence of uncontrolled asthma in a child has led parents to choose to live in an unsanitary dwelling, in a basement or near a highway.

The data are mostly derived from information reported by parents. The methods used in this survey are comparable with those of international (ISAAC, 1998; Burney et al., 1994) and national surveys of the same type already performed. Various types of analysis were conducted to assess the validity of reported data. For the majority of exposure factors though, we were not able to validate the information or not able to obtain objective data with the exception of outdoor pollution from road traffic. Regarding excessive moisture and molds, studies have shown that the information reported by the occupants tend to an underestimation of the risk (Fisk et al., 2007). Although based on qualitative reported data, the use of an indicator including five variables captures the main conditions associated with excessive moisture, molds and other associated contaminants (Mendell et al., 2011). Moreover, basements, which were found within this indicator to have the strongest association with the outcome, are most affected by water infiltrations, excess humidity and molds, as found in our experience of hundreds of investigations.

In conclusion, control of excessive moisture and molds, and other environmental allergens and irritants would prevent many health outcomes among all members of the families. Excessive moisture and molds is a very common risk factor to which the population is most often exposed unwittingly. Removal of exposure to this factor would be an effective and sustainable way to control the inflammation which causes the development and persistence of these diseases. Major public policies and financial investments are required to implement regular maintenance of buildings, in particular to prevent water infiltrations from the roof and other sources, to act promptly in case of water damages and to decontaminate and restore effectively the affected buildings.

DECLARATION OF INTEREST

This study was funded by the Direction de santé publique de l'Agence de la santé et des services sociaux de Montréal. The authors report no conflicts of interest. The authors alone are responsible for the content and writing of the paper.

REFERENCES

- Antova T, Pattenden S, Brunekreef B, Heinrich J, Rudnai, P, Forastiere F, Luttmass-Gibson H, Grize L, Katsnelson B, Moshammer H, Nikiforov B, Slachtova H, Slotova K, Zlokowska R, Fletcher T. Exposure to indoor mould and children's respiratory health in the PATY study. *J Epidemiol Community Health* 2008; 62: 708-714.
- Behrens T, Taeger D, Wellmann J, Keil U. Different methods to calculate effect estimates in cross-sectional ctudies. A Comparison between Prevalence Odds Ratio and Prevalence Ratio. *Methods Inf Med* 2004; 43: 505–509
- Bornehag C-G, Blomquist G, Gyntelberg F, Järvholm B, Malmberg P, Nordvall L, Nielsen A, Pershagen G, Sundell J. Dampness in buildings and health. *Indoor Air* 2001; 11: 72-86.
- Bornehag C-G, Sundell J, Bonini S, Custovic A, Malmberg P, Skerfving S, Sigsgaard T, Verhoeff A. Dampness in buildings as a risk factor for health effects, EUROEXPO: a multidisciplinary review of the literature (1998-2000) on dampness and mite exposure in buildings and health effects. *Indoor Air* 2004; 14: 243-257.
- Boulet LP, Becker A, Berube D, Beveridge R, Ernst P. Provocative factors in asthma. *CMAJ* 1999; 161:8S-14S.
- Bousquet J. Cauwenberge PV, Khaltaev N, in collaboration with WHO. Allergic rhinitis and its impact on asthma (ARIA). *J Allergy Clin Immunol* 2001; 108 (5 Suppl): 147-334.
- Burney PG, Luczynska C, Chinn S, Jarvis D. The European Community Respiratory Health Survey (ECRHS). *Eur. Respir J* 1994; 7: 954-960
- Canadian Thoracic Society. Canadian Pediatric Asthma Consensus Guidelines 2003. http://www.respiratoryguidelines.ca/asthma-pediatric-guidelines, Dec. 2004.
- Chulada PC, Arbes SJ, Dunson D, Zeldin DC. Breast-feeding and the prevalence of asthma and wheeze in children: Analyses from the Third National Health and Nutrition Examination Survey, 1988-1994. *J Allergy Clin Immunol* 2003; 111: 328-36
- Değer L, Plante C, Goudreau S, Smargiassi A, Perron S, Thivierge RL, Jacques L. Home environmental factors associated with poor asthma control in Montreal children: a population-based study. *J. Asthma* 2010; 47(5): 513-20.
- Eide GE, Gefeller O. Sequential and average attributable fractions as aids in the selection of preventive strategies. *J. Clin. Epidemiol* 1998; 48: 645-655.
- Fisk WJ, Lei-Gomez Q, Mendell MJ. Meta-analyses of the associations of respiratory health effects with dampness and mold in homes. *Indoor Air* 2007; 17: 284–296.

- Fredriksson P, Jaakkola N, Jaakkola JK. Breastfeeding and childhood asthma: a six-year population-based cohort study. *BMC Pediatrics* 2007; 7: 39
- Garner R. Kohen D. Changes in the prevalence of asthma among Canadian children. Statistics Canada, Health Reports (Catalogue 82-003) 2008; 19(2): 1-6.
- Gold DR, Wright R. Population disparities in asthma. *Annu. Rev. Public Health* 2005; 26: 89-113
- HEI. Panel on the Health Effects of Traffic-Related Air Pollution: A Critical Review of the Literature on Emissions, Exposure, and Health Effects. Health Effect Institute 2010. HEI special report 17, Boston, MA.2010.
- Health and social survey of Québec children and youth. 1999. http://www.stat.gouv.qc.ca/publications/sante/enfant-ado_an.htm
- Health Canada. Childhood Asthma in Sentinel Health Units. Findings of the Student Lung Health Survey 1995-1996. http://www.phac-aspc.gc.ca/publicat/ashu-auss/index-eng.php
- International Study of Asthma and Allergies in Childhood (ISAAC). Worldwide variation in prevalence of symptoms of asthma, allergic rhinoconjunctivitis and atopic eczema. The International Study of Asthma and Allergies in Childhood (ISAAC) Steering Committee. *Lancet* 1998; 351: 1225-1232.
- Jacques L, L'Heureux F, Kosatsky T, Fortier I, Drouin L, King N. Portrait de l'asthme et de la rhinite allergique chez les jeunes montréalais de 6 mois à 12 ans. Direction de santé publique de l'Agence de la santé et des services sociaux de Montréal 2005; 27 p.
- Krahn MD, Berka C, Langlois P, Detsky AS. Direct and indirect costs of asthma in Canada, 1990. *CMAJ* 1996; 154: 821-31.
- Kosatsky T, Smargiassi A, Boivin M-C, Drouin L, Fortier I. Évaluation de l'excès de maladies respiratoires dans les secteurs de Pointe-aux-Trembles — Montréal-Est et Mercier-Est — Anjou. Une analyse des données sanitaires et environnementales (1995-2000), Direction de santé publique, Agence de développement des réseaux locaux de services de santé et de services sociaux de Montréal. 2004.
- Lajoie P, Laberge A, Lebel G, Boulet LP, Demers M, Mercier P, Gagnon MF. Cartography of emergency department visits for asthma - targeting high-morbidity populations. *Can Respir J* 2004; 11: 427-433.
- Mendell M J, Mirer A G, Cheung K, Tong M, Douwes J. Respiratory and Allergic Health Effects of dampness, mold, and dampness-related agents: A Review of the Epidemiologic Evidence. *Environ Health Perspect* 2011; 119: 748-756.
- Moularat S, Robine E, Draghi M, Derbez M, Kirchner S, Ramalho O. Moisissures dans les environnements intérieurs: état des connaissances et détermination de la contamination fongique des logements français par un indice chimique. *Pollution atmosphérique* 2008; 197:33-46
- Mudarri D, Fisk W J. Public health and economic impact of dampness and mold. *Indoor Air* 2007; 17: 226-235.
- Naeher LP, Brauer M, Lipsett M, Zelikoff JT, Simpson CD, Koenig JQ, Smith KR Woodsmoke health effects: a review. *Inhalation Toxicology* 2007; 19: 67–106.

- Plante C, Jacques L, Chevalier S, Fournier M. Comparability of data between Internet and telephone in a survey on the respiratory health of children. *Can Respir J* 2012; 19(1): 13-18.
- Smargiassi A, Berrada K, Fortier I, Kosatsky T. Traffic intensity, dwelling value, and hospital admissions for respiratory disease among the elderly in Montreal (Canada): a case-control analysis. *J Epidemiol Community Health* 2006; 60(6): 507-512.
- Subbarrao P, Becker A, Brook JR, Daley D, Mandhane PJ, Miller GE, Turvey SE, Sears MR. Epidemiology of asthma: risk factors for development. *Expert Rev. Clin. Immunol* 2009; 5(1): 77-95.
- Thomas EM. Recent trends in upper respiratory infections, ear infections and asthma among young Canadian children. Statistics Canada, Catalogue no. 82-003-XPE. Health Reports 2010; 21(4): 47-52.
- Trasher JD, Crawley S. The biocontaminants and complexity of damp indoor spaces: more than what meets the eyes. *Toxicology and Industrial Health* 2009; 25 (9-10): 583-615.
- WHO. Health Aspects of Air Pollution with Particulate Matter, Ozone and Nitrogen Dioxide. Report on a WHO working group, Bonn, Germany. 2003.

FUNGI AND CHRONIC RHINOSINUSITIS (CRS): CAUSE AND EFFECT

E B Kern, J U Ponikau, D A Sherris and H Kita

ABSTRACT

Fungi are present in the airway mucus of chronic rhinosinusitis (CRS) patients and in normal healthy controls. Fungi (especially *Alternaria*) induce the production of cytokines (IL-13 and IL-5) crucial for the eosinophilic inflammation. This immune response occurred only in CRS patients but not in healthy controls. Fungi induce an eosinophilic tissue airway inflammation in mammals (mice), which is in contrast to a neutrophilic response to bacteria. Fungi can induce an eosinophilic airway inflammation and congestion in patients. Eosinophils, in vivo, target fungi in the mucus with CRS and nasal polyps. Fungal antigens with a molecular weight of 61 kilo Daltons (kDa) cause activation and degranulation of human eosinophils via the beta-2 integrin on the CD11b receptor. Clinically, antifungal drugs can reduce nasal polyps, improve computed tomography (CT) scans, and decrease levels of interleukin-5 (IL-5) and markers of eosinophilic inflammation. However, using different antifungal applications and different outcome measures can produce conflicting results. While fungi are present in the nasal and sinus airway mucus in both CRS patients and in normal controls, the evidence suggests that it is the fungi that initiates the cytokine cascade that produces CRS. In other words, evidence points to CRS as an immunologic disease triggered by inert fungi in the airway mucus and mediated by the activated and degranulating eosinophil.

BACKGROUND:

Every patient who has nasal polyps also exhibits evidence of chronic inflammation of the upper airway and has chronic rhinosinusitis (CRS). This chronic inflammation is associated with spotty (heterogeneous) damage of the nasal and sinus airway respiratory epithelium, and this is identical to the epithelial damage seen in asthma (Ponikau et al., 2003; Wladislavosky-Waserman et al., 1984). Nasal polyps are the end-stage of this chronic inflammatory reaction of the upper airway. Notable exceptions include both antral-choanal polyps and the polyps of cystic fibrosis.

The predominant inflammatory cell seen in patients with CRS is the eosinophil not the neutrophil. The evidence supporting the contention that fungi located in the airway mucus alone (not in the tissue of the nose or sinuses) can induce this eosinophilic inflammatory response will be presented. This inflammation is not a fungal infection (since an infection requires microorganisms to enter tissue) and is differentiated from other forms of fungal involvement.

Overview: Fungi is everywhere

To find fungi in the nasal airway mucus requires specific culture techniques especially harvesting of adequate amounts of airway mucus and the chemical breaking of the disulfide bridges contained in the mucin to allow the entrapped fungi to be released (Braun et al., 2003; Ponikau et al., 1999). Utilizing these culture methods, fungi are found in the mucus of 96% (n > 202) and 91.3% (n > 92) of unselected, consecutive patients with chronic eosinophilic inflammation of the nose and paranasal sinuses, respectively (Braun et al., 2003; Ponikau et al., 1999). In addition, fungi are present in almost every healthy control. Fungi are ubiquitous. In fact, fungi are present in the mucus after birth: 20% of newborns, 72% of infants at 2 months, and 94% at 4 months (Lackner et al., 2005). Careful collection and proper staining are critical for finding fungi since Gomori's methamine silver stain (GMS) yields fungi in merely 82% (n > 101) of patients with chronic eosinophilic airway inflammation (Braun et al., 2003; Ponikau et al., 1999). Using a chitin based immunofluorescence staining technique, fungal elements are found in 100% of airway mucous specimens (Taylor et al., 2002). Thus, this newer fungal detection technique demonstrates significant fungi in the upper airway mucus in almost every case. Another study confirms the existence of fungal DNA in the polypoid nasal tissue (n > 27) of all patients (100%) with chronic eosinophilic inflammatory mucosal changes (Gosepath et al., 2004). Interestingly, the patients compared to the healthy controls differed with respect to the presence of *Alternaria*-specific DNA which was detectable in all the patients but *Alternaria*-specific DNA was absent in all the healthy controls. This suggests that only patients (not healthy controls) process certain fungal DNA for antigen presentation (Gosepath et al., 2004). These findings regarding the ubiquitous presence of fungi is supported by a recent study from the National Institute of Environmental Health Science demonstrating that practically everyone (patients and normal healthy controls) is exposed to *Alternaria* fungal antigens in their homes (Salo et al., 2005). Certain fungal antigens, (from *Alternaria* species), are only secreted by the fungal cell wall when the fungal spore is germinating, or from the growing tip of the fungal hyphae. Despite improvements in the sensitivity of culture methods, *Alternaria* species were only growing in 40-50% of the specimens, which is an improvement over previous studies, but still lacks sensitivity compared to follow-up studies looking for *Alternaria*-specific DNA or the *Alternaria* antigen (Gosepath et al., 2004; Ponikau et al., 2005; Shin et al., 2004). Thus, with the newer detection techniques for fungi, *Alternaria* species can be found to be present in the nasal mucus secretions in almost every patient with nasal polyps and in normal healthy controls. Fungi are everywhere including the mucus of the upper airway of patients and healthy people. Hence, the mere presence of fungi in the nasal airway mucus is not diagnostic of any disease.

Fungi can induce an immune response in patients (not in healthy people):

There are a number of questions that require answering. What is the evidence that fungi can induce a chronic eosinophilic inflammatory response in patients but not in normal healthy control subjects? Why do the eosinophils exist and how do they fit into the pathology of CRS and nasal polyps? What mechanism triggers the parade of eosinophils from the bone marrow into the nasal tissue and finally into the mucus of the nasal airway? Each of these questions will be answered. It is known that some patients with nasal polyposis produce specific immunoglobulin E (IgE) against fungi (Feger et al., 1997; Mabry, Manning, 1995). Realize that an IgE-mediated reaction secondary to allergen exposure reliably results in allergic rhinitis and not in nasal polyposis. In fact, nasal polyposis occurs independent of the presence of an IgE-mediated allergy (Sánchez-Segura et al., 2000). Certain fungi, especially *Alternaria*, have the ability to induce symptoms of eosinophilic airway inflammation in the absence of an IgE-mediated systemic reaction (Krouse et al., 2004). Thus, patients with a chronic eosinophilic inflammation of the upper airway may have an IgE-mediated hypersensitivity to molds as a comorbid disease, but the underlying eosinophilic inflammation appears to be driven by a mechanism which is independent and separate from an IgE mediated mechanism. So if it is not an-IgE-mediated mechanism that drives the underlying chronic eosinophilic inflammation, what is the mechanism that drives or produces this eosinophilic inflammation of the human upper airway. Or, another way of asking the question is how can harmless ubiquitous inert fungi participate in triggering this upper airway inflammatory process? To answer these questions understanding that the human immune system is required. These inert fungi trapped in the mucus of the upper airway of patients (not healthy people) are determined as foreign by the immune system; therefore, the immune system reacts by recruiting eosinophils from the bone marrow secondary to cytokine production, which in turn regulates the eosinophilic inflammation. Cytokines activate eosinophils. It is the tissue-bound lymphocytes that are the primary source of cytokines in patients with chronic eosinophilic upper airway inflammation (CRS with or without nasal polyps) (Hamilos et al., 1995). Understand that the terms chronic eosinophilic upper airway inflammation and CRS are used interchangeably.

The eosinophilic inflammation induced by fungi:

Vascular cell adhesion molecule-1 (VCAM-1) has been identified in the vascular endothelium of patients with chronic eosinophilic inflammation of the nose and sinuses and this adhesion molecule is necessary to induce adhesion of eosinophils to the blood vessel wall, as well as the migration of the eosinophils through the blood vessel walls into the tissue (Hamilos et al., 1996). VCAM-1 specifically binds to VLA-4 (very late appearing antigen-4) on eosinophils, thereby causing selective adhesion and migration of eosinophils from the vasculature into the tissue (Hamilos et al., 1996). The cytokine that is directly associated with the expression of VCAM-1, independent from the allergy status of the patients, is IL-13 (Hamilos et al., 1995, Hamilos et al., 1996). Shin et al. recently demonstrated in patients with CRS that isolated peripheral blood mononuclear cells (PBMCs), which contain lymphocytes and antigen-presenting cells, are capable of producing large amounts of interleukin-13 (IL-13) when exposed to various mold extracts especially *Alternaria* species. This is in contrast to the lack of IL-13 production when PBMCs from healthy controls were stimulated with the same fungal extracts (Shin et al., 2004). IL-13 production in response to *Alternaria* may enhance the expression of VACM-1 by vascular endothelial cells, which in turn facilitates the "eosinophil exodus" from the vasculature into the tissues of the upper airway (Shin et al., 2004). IL-5 is the important cytokine responsible for eosinophil production, differentiation, activation, and survival. IL-5 is present in tissue specimens from patients with chronic upper airway eosinophilic inflammation and not in normal healthy controls (Durham et al., 1996; Hamilos et al., 1998; Lopez et al., 1988; Sanchez-Segura et al., 1998; 21). The majority of IL-5 staining cells are lymphocytes (68%) (Hamilos et al., 1998). A direct link to fungi is provided again in the study from Shin et al. where PBMCs from patients with CRS induced IL-5 production in 89% ($n > 18$) when exposed to *Alternaria* antigens (Shin et al., 2004). When *Alternaria* antigens were exposed to PBMCs from 15 healthy controls, none demonstrated any IL-5 production (Shin et al., 2004). Not surprisingly, elevated levels of specific IgE for *Alternaria* were detected in only 28% (5 of 18) of these patients. In other words, the increased IgE levels did not correlate with increased levels of IL-5 and did not explain the presence of eosinophils in nonallergic patients (Shin et al., 2004). The fact that this immune response was detected in peripheral blood lymphocytes is the evidence for CRS being a systemic disease. Mean serum IgG levels specific for *Alternaria* were increased fivefold in the CRS patients ($n > 18$) compared to the healthy controls and also correlated with the amounts of IL-5 produced in each individual patient ($n > 15$) (Shin et al., 2004). Since IgG levels usually indicate the amount of immunologic exposure, these results suggest a direct correlation between the exposure to *Alternaria* antigens and the degree of the immune response measured by the amount of IL-5 production. In summary, fungal organisms are present in the upper airway mucus of patients and healthy controls and that *Alternaria* induces the cytokine response in CRS patients (in contrast to healthy controls), this cytokine response is crucial for the recruiting of eosinophils from the bone marrow and eventually producing the eosinophilic inflammation seen in CRS patients and not in healthy controls.

Degranulation and the power of eosinophils to destroy fungi and tissue:

Once the eosinophils leave the blood vessels and enter the tissues, what happens to them? Newer studies and techniques focused on the preserving of the nasal airway mucus with its attachment to the tissue revealed the presence of eosinophils intact in the tissue and striking amounts of eosinophils clusters in the mucus (Braun et al., 2003; Ponikau et al., 1999). Therefore, this mucus was called eosinophilic mucus. This finding showed that the tissue eosinophils were merely in transit (migrating) from the tissue into the mucus of the nasal and sinus airway (lumen) (Braun et al., 2003; Ponikau et al., 1999). Once in the airway mucus (lumen) these eosinophils formed clusters. Furthermore, histochemical studies clearly demonstrated that once the eosinophils reached the airway mucus (lumen) they degranulated within these clusters (Ponikau et al., 2005). It was within the airway mucus (lumen) not within the tissue that these eosinophils degranulated releasing a toxic protein called major basic protein (MBP). It is the MBP from the granules contained within the eosinophil which is toxic to respiratory epithelium. And it is the MBP which is capable of producing the epithelial damage found in CRS (with or without polyps) (Wladislavosky-Waserman et al., 1984; Ponikau et al., 2005; Harlin et al., 1988; Hisamatsu et al., 1990). MBP levels exceeded the threshold for upper airway respiratory epithelial damage by at least 3,000-folds in each patient studied (Ponikau et al., 2005; Hisamatsu et al., 1990). These in vivo observations clearly explain the patterns of damage seen in patients with CRS, chronic upper airway eosinophilic inflammation. In addition, this epithelial damage to the lining of the nose and sinus mucosa produced by the released MBP from the eosinophil that allows for the secondary bacterial infections (acute exacerbations) seen in CRS patients. Eosinophilic inflammation has also been observed in tissues that contain large, nonphagocytosable parasites such as helminthes (Kita et al., 2003). Earlier reports documented the accumulation of eosinophils and their subsequent degranulation on the surfaces of these parasites suggest that it is the toxic MBP that kills these organisms. The toxic proteins released from the eosinophil granule (including MBP) produces damage to the outer layers of nasal airway tissue, indicating that the damage is inflicted to the airway side (airway lumen) of the airway mucosa and not from the inside tissue (Wladislavosky-Waserman et al., 1984; Ponikau et al., 2005; Harlin et al., 1988). Recent observations of eosinophil clusters in the mucus from CRS patients are reminiscent of the accumulations around parasites (Braun et al., 2003; Ponikau et al., 1999; Kita et al., 2003).

Can the inflammation in CRS be an immunologic defense against fungi?

Since fungal organisms are large (like helminthes) and cannot be engulfed and destroyed by phagocytosis the immune system is able to recognizes fungi as foreign (in sensitized people) and this immune defense system is capable of mounting an immunologic reaction against fungi by the following mechanism:

1. Eosinophil recruitment from the vasculature (leaving the blood vessels) by inducing IL-13 production.

2. Eosinophil activation and life prolongation by inducing IL-5 production.

3. Eosinophil migration (leaving the tissue) and directing the eosinophil to enter the mucus.

4. Eosinophils clustering around the fungi and degranulation with release of toxic proteins MBP)when eosinophil contacts the fungi, and by releasing toxic proteins such as MBP thereby destroying the fungi as well as causing collateral damage to the respiratory epithelium.

The above construction (the immunologic defense by eosinophils against fungi) is supported by direct immunohistochemistry evidence showing clusters of eosinophils and fungi within the airway mucus associated with intense eosinophil degranulation and the release of MBP. FIGURE 1 This evidence underscores the notion that the eosinophils move from tissue into the airway mucus (as the eosinophils do not move from the airway mucus into the tissue) specifically targeting fungi and these eosinophils by releasing MBP can induce CRS.

Can fungi (*Alternaria* in particular) induce eosinophil degranulation?

Are fungi capable of inducing eosinophil degranulation and are all fungi equally capable of activating eosinophils? When eosinophils were exposed to different fungal antigens, by far the most robust and consistent stimulation occurred when the eosinophils were exposed to an *Alternaria* extract. The *Alternaria* extract induced both eosinophil activation and degranulation (Lindsay et al., 2006). The antigen fraction from *Alternaria* that induced the eosinophil degranulation had a molecular weight of 61 kilo Daltons (kDa), was highly heat labile, and functioned via a G protein coupled receptor (Lindsay et al., 2006). Other fungal antigens, including *Aspergillus*, *Cladosporium*, and *Candida*, did not induce eosinophil degranulation. It must be pointed out that the fungus-induced degranulation of eosinophils is an innate response occurring automatically whenever fungi and eosinophils are brought together, regardless of whether they came from CRS patients or healthy controls. However, patients with upper and lower airway inflammation released about 60% more granular proteins per eosinophil compared to healthy controls (Inoue et al., 2005). Another study identified CD11b as the receptor through which the *Alternaria* antigen binds to the eosinophil and induces the degranulation (Yoon et al., 2008). Blocking this receptor completely inhibited the ability of eosinophils to degranulate in response to *Alternaria*. Interestingly, it was not the entire receptor but only the beta-2 integrin that was the binding site for the *Alternaria* antigen. The same study visualized that eosinophils degranulate on the *Alternaria* hyphae in vitro (Yoon et al., 2008). In contrast, when neutrophils were stimulated with *Alternaria* antigens, those neutrophils did not respond with degranulation or activation, which suggests that the presence of a specific fungal species (*Alternaria*) produces a specific innate immune cell type response to certain fungi in humans (Inoue et al., 2005). Thus, both innate (degranulation and activation of eosinophils) and acquired immune responses (cytokine production by lymphocytes, independent of IgE production) to environmental fungi, such as *Alternaria*, provide cellular activation signals necessary for the robust eosinophilic inflammation in CRS patients with or without nasal polyps.

In addition, murine models for CRS using fungal antigens have resulted in eosinophilic airway inflammation, in stark contrast to the neutrophilic inflammation seen when using bacteria as a stimulus (Lindsay et al., 2006). The displayed evidence supporting the role of fungi in CRS with or without nasal polyps brings up the role of antifungal therapy in CRS. While some trials demonstrated clinical benefit of topical Amphotericin B, including the reduction of both nasal polyposis and the eosinophilic inflammation, other trials did not show any patient improvement. The trials differed in the formulations used, the concentrations of the antifungal drug including the toxic dissolvent (desoxycholate), the delivery device, patient selection, and the outcome measures, making comparison between these studies virtually impossible (Ponikau et al., 2005; Corradini et al., 2006; Ebbens et al., 2006; Ponikau et al., 2002; Ricchetti et al., 2002; Ricchetti et al., 2002b; Weschta et al., 2004). It must be emphasized that the Amphotericin B for intranasal applications should not be formulated in glucose since the lack of an osmotic pressure gradient prevents diffusion of the drug into the airway mucin. Amphotericin B is incompatible with saline and this fact is clearly noted in the Physicians Desk Reference (PDR). Amphotericin B is toxic to respiratory mucosa (with desoxycholate as a desolvent) in concentration above 300 mg/mL. Amphotericin B must be correctly dissolved in sterile water (H_2O) and delivered through a devise capable of bringing an adequate amount of the drug into the obstructed nose and paranasal sinuses (Gosepath et al., 2002). While intranasal Amphotericin B continues to be in clinical development, approaches with the systemic antifungal itraconazole show promises (Rains, Mineck, 2003). A recently published randomized-controlled-trial showed that oral itraconazole of 400 mg/day for 32 weeks in patients with severe asthma and fungal sensitization (determined by positive skin prick or positive IgE RAST to fungi) demonstrated significant improvement in both the asthma symptoms, and significant improvement in the associated rhinologic symptoms (Denning et al., 2009). One wonders if the same treatment with itraconazole would be effective in patients who are not preselected for having an IgE-mediated allergy to fungi, but in all patients with eosinophilic airway inflammation regardless of their allergy status.

SUMMARY OF FINDINGS:

1. Fungi are present in the mucus of chronic rhinosinusitis (CRS) patients and normal healthy controls.
2. Fungi (especially *Alternaria*) induce the production of cytokines (IL-13 and IL-5) crucial for the eosinophilic inflammation. This immune response occurred only in CRS patients and not in healthy controls. (FIGURE 1)
3. Fungi induce an eosinophilic tissue airway inflammation in mammals (mice), in contrast to the neutrophilic response to bacteria.

4. Fungi can induce an eosinophilic airway inflammation with associated nasal congestion in patients.
5. Eosinophils, in vivo, target the fungi located in the nasal and sinus airway mucus in patients with CRS and associated nasal polyps.
6. Fungal antigens having a molecular weight of 61 kilo Daltons (kDa) cause activation and degranulation of human eosinophils (causing the release of major basic protein) via the beta-2 integrin on the CD11b receptor.
7. Clinically, antifungal drugs can reduce nasal polyps, improve findings (less inflammation seen) on computed tomography (CT) scans, and decrease levels of interleukin-5 (IL-5) and markers of eosinophilic inflammation. However, data obtained using different antifungal applications and different outcome measures have produced conflicting results.
8. CRS seems to be an immunologic disease triggered by inert fungi in the airway mucus and mediated by the activated and degranulating eosinophil.

FIGURE 1

(a) Mucus from a patient with CRS and nasal polyps that has been stained with hematoxylin and eosin (H&E). The image shows the typical cluster formation of eosinophils in the mucus of CRS patients (white arrows). Black arrow mark the eroded epithelium (H&E, original magnification ×400).

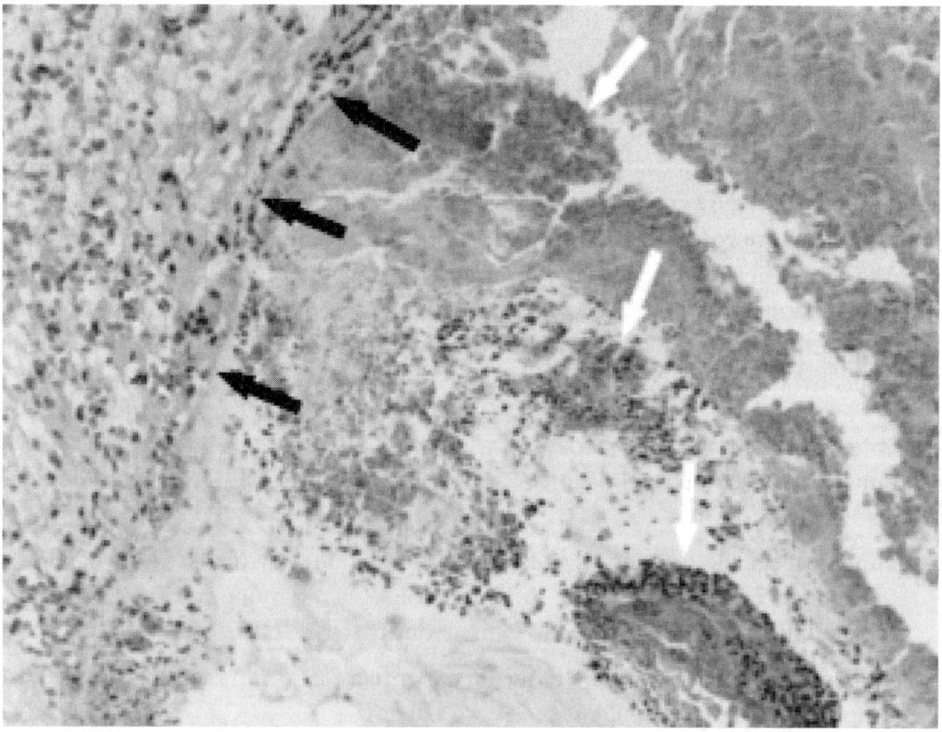

(b) Serial section of (a) is stained for eosinophilic major basic protein (eMBP). The diffuse release of eMBP demonstrates that eosinophils are degranulating. MBP release occurs only in the clusters in the mucus, not in the tissue (left side of the picture) where eosinophils are intact. This suggests that the eosinophils in the tissue are in transit toward their final target in the mucus (anti-MBP immunofluorescence staining; original magnification ×400).

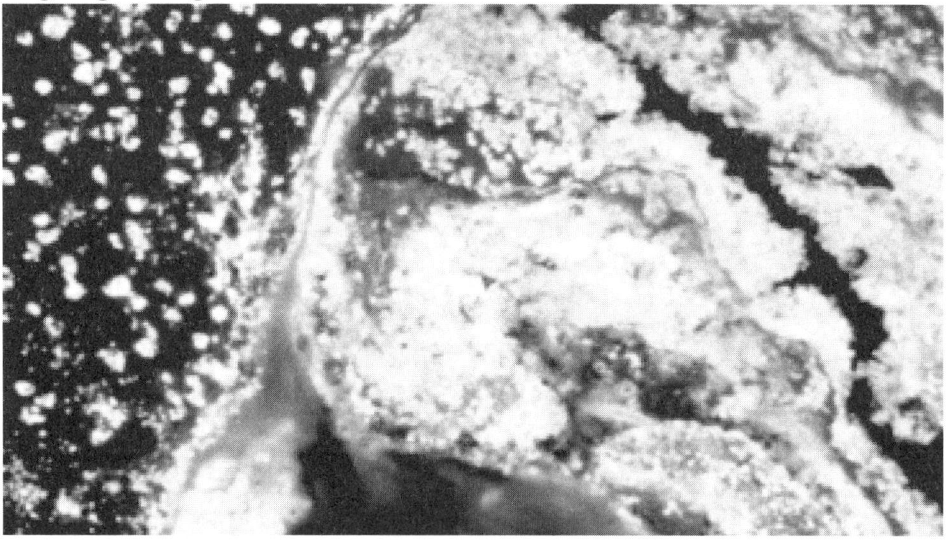

(c) Serial section of (a) and (b) is stained for *Alternaria* antigen. The clustering of eosinophils and the release of toxic eMBP occur at the exact location of the fungal antigens, suggesting that the eosinophils are targeting the fungus (anti-*Alternaria* immunofluorescence staining; original magnification×400).

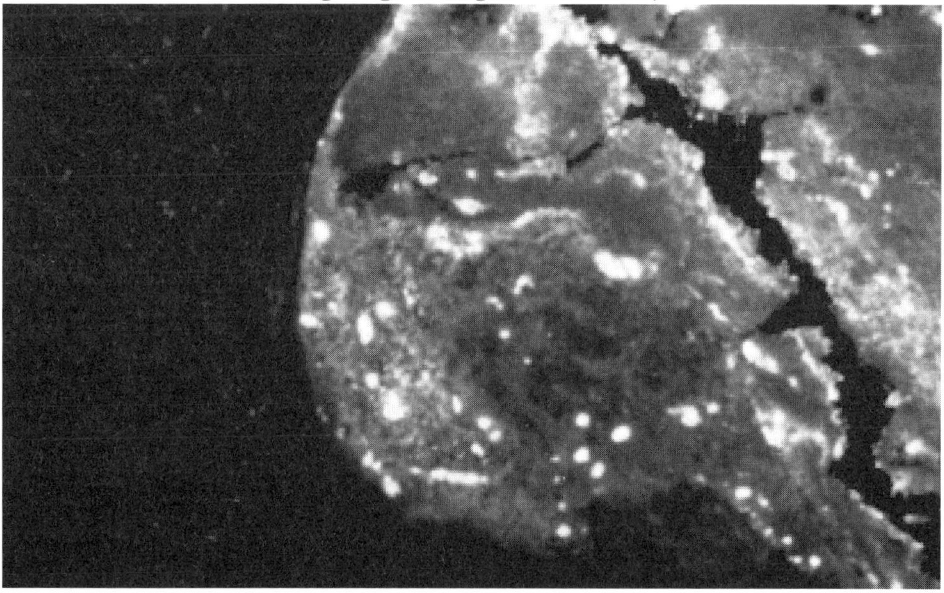

(d) Electron microscopy of mucus from a patient with CRS and nasal polyps showing a cluster of eosinophils. A fungal hyphae (black arrow) can be visualized in the cluster of eosinophils, suggesting targeting of the fungus by the eosinophils. Note the intimate relationship of the eosinophils on the the fungi (white arrow) in preparation to release their toxic MBP (transmission electron microscopy; original magnification ×5,275)

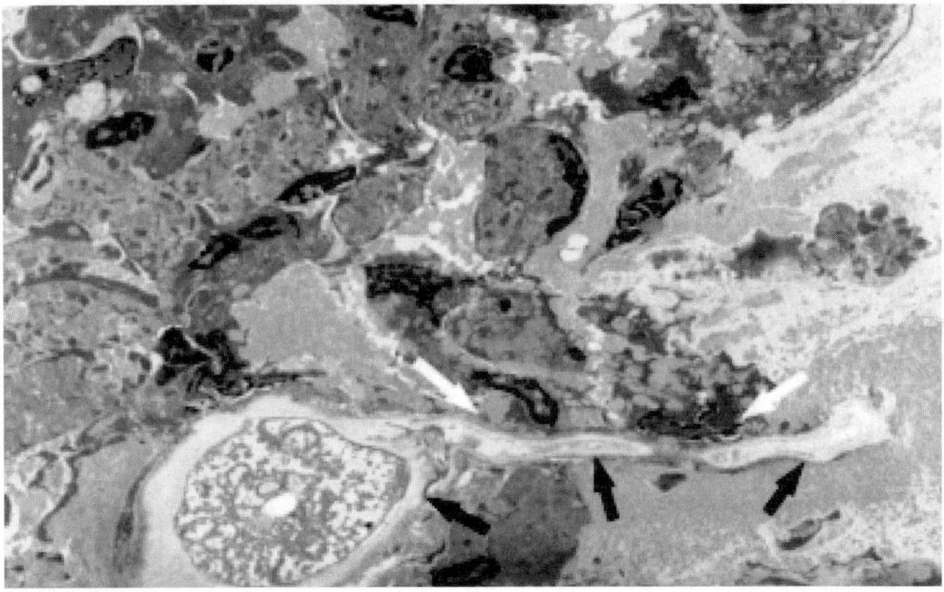

REFERENCES:

- Braun H, Buzina W, Freudenschuss K et al. "Eosinophilic fungal rhinosinusitis": a common disorder in Europe? *Laryngoscope* , 2003:113:264-269
- Corradini C, Del Ninno M, Buonomo A, Nucera E, Paludetti G, Alonzi C, Sabato V, Schiavino D, Patriarca G. Amphotericin B and lysine acetylsalicylate in the combined treatment of nasal polyposis associated with mycotic infection.*J Investig Allergol Clin Immunol*, 2006:16:188-193
- Denning DW, O'Driscoll BR, Powell G, Chew F, Atherton GT, Vyas A, Miles J, Morris J, Niven RM. Randomized controlled trial of oral antifungal treatment for severe asthma with fungal sensitization: The Fungal Asthma Sensitization Trial (FAST) study. *Am J Respir Crit Care Med*, 2009:179(1):11-18. Epub 2008 Oct 23
- Durham SR, Ying S, Varney VA et al. Cytokine messenger RNA expression for IL-3, IL-4, IL-5, and granulocyte/macrophage-colony-stimulating factor in the nasal mucosa after local allergen provocation: relationship to tissue eosinophilia. *J Immunol*, 1992:148:2390-2394
- Ebbens FA, Scadding GK, Badia L, Hellings PW, Jorissen M, Mullol J, Cardesin A, Bachert C, van Zele TP, Dijkgraaf MG, Lund V, Fokkens WJ. Amphotericin B nasal lavages: not a solution for patients with chronic rhinosinusitis. *J Allergy Clin Immunol* 2006:118(5):1149-1156

- Feger TA, Rupp NT, Kuhn FA et al. Local and systemic eosinophil activation in allergic fungal sinusitis. *Ann Allergy Asthma Immunol,* 1997:79:221-225
- Gosepath J, Brieger J, Vlachtsis K, Mann WJ. Fungal DNA is present in tissue specimens of patients with chronic rhinosinusitis. *Am J Rhinol,* 2004:18:9-13
- Gosepath J, Grebneva N, Mossikhin S, Mann WJ. Topical antibiotic, antifungal, and antiseptic solutions decrease ciliary activity in nasal respiratory cells. *Am J Rhinol,* 2002: 16(1):25-31
- Hamilos DL, Leung DY, Huston DP et al. GM-CSF, IL-5 and RANTES immunoreactivity and mRNA expression in chronic hyperplastic sinusitis with nasal polyposis (NP). *Clin Exp Allergy*, 1998:28:1145-1152
- Hamilos DL, Leung DY, Wood R et al. Evidence for distinct cytokine expression in allergic versus nonallergic chronic sinusitis. *J Allergy Clin Immunol*, 1995:96:537-544
- Hamilos DL, Leung DY, Wood R et al. Eosinophil infiltration in nonallergic chronic hyperplastic sinusitis with nasal polyposis (CHS/NP) is associated with endothelial VCAM-1 upregulation and expression of TNF-alpha. *Am J Respir Cell Mol Biol,* 1996:15:443-450
- Harlin SL, Ansel DG, Lane SR et al. A clinical and pathologic study of chronic sinusitis: the role of the eosinophil. *J Allergy Clin Immunol*, 1988:81(5 Pt 1):867-875. The first study linking the damage in CRS to the eosinophilic inflammation
- Hisamatsu K, Ganbo T, Nakazawa T et al. Cytotoxicity of human eosinophil granule major basic protein to human nasal sinus mucosa in vitro.*J Allergy Clin Immunol*, 1990:86:52-63
- Inoue Y, Matsuzaki Y, Shin S-H et al. Non pathogenic, environmental fungi induce activation and degranulation of human eosinophils. *J Immunol*, 2005:175:5439-5447
- Kita H, Adolphson CR, Gleich GJ. Biology of eosinophils. In: Adkinson NF Jr, Bochner BS, Yunginger JW (eds) Middleton's allergy principles and practice, 2003. 6th edn. Mosby, Philadelphia, pp 305-332
- Krouse JH, Shah AG, Kerswill K. Skin testing in predicting response to nasal provocation with alternaria. *Laryngoscope* , 2004:114:1389-1393
- Lackner A, Freudenschuss K, Buzina W et al. Fungi: a normal content of human nasal mucus. *Am J Rhinol*, 2005:19:125-129
- Lindsay R, Slaughter T, Britton-Webb J, Mog SR, Conran R, Tadros M, Earl N, Fox D, Roberts J, Bolger WE. Development of a murine model of chronic rhinosinusitis. *Otolaryngol Head Neck Surg* . 2006:134(5):724-730; discussion 731-732
- Lopez AF, Sanderson CJ, Gamble JR et al. Recombinant human interleukin 5 is a selective activator of human eosinophil function. *J Exp Med,* 1988:167:219-224
- Mabry RL, Manning S. Radioallergosorbent microscreen and total immunoglobulin E in allergic fungal sinusitis. *Otolaryngol Head Neck Surg,* 1995:113:721-723
- Ponikau JU, Sherris DA, Kephart GM et al. Features of airway remodeling and eosinophilic inflammation in chronic rhinosinusitis: is it the histopathology similar to asthma? *J Allergy Clin Immunol*, 2003:112:877-882

- Ponikau JU, Sherris DA, Kephart GM et al. Striking deposition of toxic eosinophil major basic protein in mucus: implications for chronic rhinosinusitis. *J Allergy Clin Immunol*, 2005:116:362-369
- Ponikau JU, Sherris DA, Kern EB et al. The diagnosis and incidence of allergic fungal sinusitis. *Mayo Clin Proc*, 1999:74:877-884
- Ponikau JU, Sherris DA, Kita H, Kern EB. Intranasal antifungal treatment in 51 patients with chronic rhinosinusitis. *J Allergy Clin Immunol*, 2002:110:862-866
- Ponikau JU, Sherris DA, Weaver A, Kita H. Treatment of chronic rhinosinusitis with intranasal amphotericin B: a randomized, placebo-controlled, double-blind pilot trial. *J Allergy Clin Immunol*, 2005:115:125-131 (AU2)
- Rains BM IIIrd, Mineck CW. Treatment of allergic fungal sinusitis with high-dose itraconazole. *Am J Rhinol*, 2003:17(1):1-8
- Ricchetti A, Landis BN, Giger R, Zeng C, Lacroix JS. Effect of local antifungal treatment on nasal polyposis. *Otorhinolaryngol Nova*, 2002-2003:12:48-51
- Ricchetti A, Landis BN, Maffioli A, Giger R, Zeng C, Lacroix JS. Effect of anti-fungal nasal lavage with amphotericin B on nasal polyposis. *J Laryngol Otol*. 2002b:116: 261-263
- Salo PM, Yin M, Arbes SJ Jr et al. Dustborne *Alternaria* alternate antigens in US homes: results from the National Survey of Lead and Allergens in Housing. *J Allergy Clin Immunol*, 2005:116(3):623-629
- Sánchez-Segura A, Brieva JA, Rodriguez C. T lymphocytes that infiltrate nasal polyps have a specialized phenotype and produce a mixed TH1/TH2 pattern of cytokines. *J Allergy Clin Immunol*, 1998:102:953-960
- Sánchez-Segura A, Brieva JA, Rodríguez C. Regulation of immunoglobulin secretion by plasma cells infiltrating nasal polyps. *Laryngoscope*, 2000:110(7):1183-1188
- Shin S-H, Ponikau JU, Sherris DA et al. Rhinosinusitis: an enhanced immune response to ubiquitous airborne fungi. *J Allergy Clin Immunol*, 2004:114:1369-1375
- Simon HU, Yousefi S, Schranz C et al. Direct demonstration of delayed eosinophil apoptosis as a mechanism causing tissue eosinophilia. *J Immunol*, 1997:158:3902-3908
- Taylor MJ, Ponikau JU, Sherris DA et al. Detection of fungal organisms in eosinophilic mucin using a fluoresceinlabeled chitin-specific binding protein. *Otolaryngol Head Neck Surg*, 2002:127:377-383
- Weschta M, Rimek D, Formanek M, Polzehl D, Podbielski A, Riechelmann H. Topical antifungal treatment of chronic rhinosinusitis with nasal polyps: a randomized, double-blind clinical trial. *J Allergy Clin Immunol*, 2004:113(6):1122-1128
- Wladislavosky-Waserman P, Kern EB, Holley KE, Eisenbrey AB, Gleich GJ. Epithelial damage in nasal polyps. *Clin Allergy*, 1984:14:241-247
- Yoon J, Ponikau JU, Lawrence CB, Kita H. Innate antifungal immunity of human eosinophils mediated by a beta 2 integrin, CD11b. *J Immunol*, 2008:181(4):2907-2915

RHINOSINUSITIS AND MOLD AS RISK FACTORS FOR ASTHMA SYMPTOMS IN OCCUPANTS OF A WATER-DAMAGED BUILDING

Ju-Hyeong Park

ABSTRACT

Mold exposure has been associated with nasal symptoms and asthma development in damp building occupants. However, progression of building-related (BR) rhinosinusitis to asthma due to mold exposure is poorly understood. We examined risk of asthma development in relation to prior BR-rhinosinusitis symptoms and microbial exposure in occupants of a damp building. We conducted four cross-sectional health and environmental surveys on occupants of a water-damaged office building. We defined BR-rhinosinusitis symptom and comparison groups from participants' first questionnaire responses. We compared the odds for asthma development and BR-asthma symptoms between the two groups over the last three surveys. We used logistic regression models adjusted for demographics, smoking status, year of building occupancy, and initial exposures to culturable fungi, endotoxin, and ergosterol in floor dust. The BR-rhinosinusitis symptom group had higher odds for developing BR-asthma symptoms [odds ratio (OR)=2.2; 95% confidence interval (CI)=1.3-3.6] and self-reported physician-diagnosed current asthma (OR=2.0; 95% CI=0.7-5.4) in any of the follow-up surveys. The highest tertile mold exposure group had an OR of 3.5 (95% CI=1.6-7.5) for developing BR-asthma symptoms as compared to the lowest tertile exposure group. The BR-rhinosinusitis symptom group with higher mold exposure within the building had an OR of 7.3 (95% CI=2.7-19.7) for developing BR-asthma symptoms, compared to the lower exposure group without symptoms. Our findings suggest that rhinosinusitis associated with occupancy of water-damaged buildings may be a sentinel for increased risk for asthma onset in such buildings.

INTRODUCTION

Exposure to indoor dampness and mold is a risk factor for chronic rhinosinusitis, allergic fungal sinusitis, and development of asthma (IOM, 2004; WHO, 2009). Comorbidity from nasal symptoms and asthma supports the "unified airways disease" concept which suggests interaction between upper and lower respiratory tracts (Derebery et al., 2008; Krouse et al., 2007). Epidemiologic studies of the general population show that rhinitis is an independent risk factor for asthma (Guerra et al., 2002; Leynaert et al., 1999; Togias 2003). However, there is little information about the degree of risk for progression of rhinosinusitis symptoms associated with being in damp buildings during the workday (building-related) to asthma. Using the baseline and three subsequent cross-sectional survey data collected from a water-damaged building, we examined: 1) whether occupants who reported building-related

(BR) rhinosinusitis symptoms in the baseline survey were more likely to develop BR-asthma symptoms or physician-diagnosed asthma in the subsequent surveys than those with no BR-rhinosinusitis symptoms; and 2) how microbial exposures interacted with the presence of BR-rhinosinusitis symptoms at the baseline survey in the development of asthma symptoms in the building occupants over the three later surveys.

METHODS

In 2001, the National Institute for Occupational Safety and Health (NIOSH) received a health hazard evaluation request concerning building-related diseases in a water-damaged 20-story office building built in 1985 and located in the northeastern United States. In response to the request, NIOSH staff conducted an initial cross-sectional health questionnaire survey in September 2001 and an initial environmental survey in April 2002. Current building occupants reported that new onset respiratory and dermatological conditions that they perceived to be building-related had occurred as early as initial occupancy in 1994. After major remediation efforts from 2002 through early 2004, we conducted three subsequent cross-sectional surveys in August 2004, 2005, and 2007 to examine the effect of remediation on occupants' health.

We defined building-related (BR) symptoms as those that improved when away from the building. We defined BR-rhinosinusitis symptoms as one or more BR nasal symptom (stuffy, itchy, or runny nose), sneezing, or sinusitis or sinus problems occurring one or more times per week in the last 4 weeks. We defined BR-asthma symptoms as two or more asthma-like symptoms (wheeze/whistling, chest tightness, attack of cough, attack of shortness of breath, awakened by attack of breathing difficulty) occurring one or more times per week in the last 4 weeks with at least one symptom being building-related. The non-BR-rhinosinusitis symptom and non-BR-asthma symptom groups' definitions were as above but with no symptoms which improved when away from the building. The same self-administered questionnaire was used in all four cross-sectional surveys, with the first page indicating that consent to participate was implied by completing the questionnaire, as approved by the NIOSH Human Subjects Review Board.

We selected all participants who did not have BR-asthma symptoms in the initial (baseline) cross-sectional survey and who participated in any subsequent cross-sectional survey. These selected participants were categorized into a 2001 BR-rhinosinusitis symptom group (a risk group) and a 2001 comparison group (those without BR-rhinosinusitis symptoms) for investigation of the development of BR-asthma symptoms or physician-diagnosed asthma in the subsequent surveys.

We collected floor dust samples from the selected occupants' workstations from all four environmental surveys. Each dust sample was collected by vacuuming a 2 square meter carpeted floor area in a standardized way as described previously (Park et al., 2006). Fungal colonies were cultured with malt extract (selective for a broad spectrum of fungi), cellulose (selective for cellulolytic fungi), and dichloran 18% glycerol (selective for xerophilic fungi) agars at room temperature for 7–10 days, identified to species level, and enumerated. If a species grew on more than one medium, a standardized laboratory protocol was used to select which medium would be the basis for the reported colony count results. Ergosterol was analyzed with gas chromatography-mass spectrometry (Sebastian and Larsson, 2003). Endotoxin was analyzed using the kinetic quantitative chromogenic Limulus amoebocyte lysate (KQCL) method (Chun et al., 2002). Using the rank order of floor-specific geometric means of microbial levels per square meter, we categorized all 15 occupied floors into tertile (low, medium, and high) exposure groups for each agent in each survey (Park et al., 2006). We also used binary (low versus medium/high tertiles) fungal exposure variable to examine interaction between the exposures and the presence of BR-rhinosinusitis symptoms in the initial survey on development of BR-asthma symptoms.

We used logistic regression models to examine associations between the onset of BR-asthma symptoms (in 2004, 2005, 2007, or any one of these subsequent surveys) or physician-diagnosed asthma in any of the subsequent surveys and the presence of BR-rhinosinusitis symptoms in the 2001 survey as an explanatory variable. The models were adjusted for age, gender, race, smoking status, building tenure, and baseline survey tertile exposures to culturable fungi, ergosterol, and endotoxin in floor dust. We also performed logistic regression modeling with an outcome variable of current physician-diagnosed asthma in any one of the subsequent surveys.

We used polytomous logistic regressions to examine associations among BR- and non-BR-rhinosinusitis symptoms (explanatory variables) and asthma symptoms (a three-level outcome variable: asymptomatic, non-BR, and BR). All data analyses were performed with SAS® 9.2 (SAS Institute Inc., Cary, NC). Statistical significance was set at the P-value ≤ 0.05.

RESULTS

Approximately 70% (623/888) of the 2001 survey participants participated in at least one of the three subsequent surveys. Excluding 84 participants with BR-asthma symptoms in the initial survey and 47 participants with missing information left 492 participants for inclusion in the logistic regression models. The prevalences of BR-rhinosinusitis symptoms alone ranged from 22% to 30%, with the highest prevalence in 2004. Prevalences of BR-rhinosinusitis symptoms alone were always higher than those of BR-asthma symptoms alone. Prevalences of BR-asthma symptoms ranged from 14% to 22%, with the highest prevalence in 2005.

In unadjusted analyses, the 2001 BR-rhinosinusitis symptom group was approximately twice as likely to develop BR-asthma symptoms in 2004, 2005, or any one of the three subsequent surveys, compared to the 2001 comparison group (Table 1).

The results were similar when the models were adjusted for demographics, or for both demographics and baseline survey exposures to culturable fungi, ergosterol, and endotoxin (main effect model). The 2001 BR-rhinosinusitis symptom group had also twice the odds of developing current physician-diagnosed asthma in any one of the subsequent surveys than the 2001 comparison group in both unadjusted and adjusted analyses. The size of effect was similar to those found for the development of BR-asthma symptoms, but they were not statistically significant at $\alpha=0.05$.

Table 1. Crude and adjusted odds ratios for developing BR-asthma symptoms by subsequent survey in the 2001 BR-rhinosinusitis symptom group compared to the 2001 comparison group

Year of follow-up survey	Crude Odds Ratio (95% Confidence Interval)	Adjusted OR (95% CI)	
		Demographics*	Demographics and Environmental Exposure**
2004	2.11 (1.11-4.00)†	2.00 (1.02-3.92)†	2.10 (1.03-4.31)†
2005	2.30 (1.30-4.08)†	2.26 (1.23-4.15)†	2.28 (1.19-4.36)†
2007	1.74 (0.87-3.51)	1.76 (0.85-3.63)	1.54 (0.73-3.25)
Any follow-up survey	2.00 (1.27-3.14)†	2.25 (1.39-3.66)†	2.24 (1.34-3.72)†

The 2001 building-related (BR) rhinosinusitis symptom group includes those who reported BR-rhinosinusitis symptoms but no BR-asthma symptoms in the initial 2001 survey, and the 2001 comparison group includes those who had neither BR-rhinosinusitis nor BR-asthma symptoms in the initial 2001 survey.

* Race, gender, age, smoking status, and building tenure.

** Tertile exposure (low/medium/high) based on rank order of floor-specific geometric means (per m^2 area) of culturable fungi, ergosterol, and endotoxin measured in the initial 2002 environmental survey.

† P-value < 0.05.

Polytomous logistic regression analyses showed that 2001 BR-rhinosinusitis symptoms were a significant risk factor [OR=2.4 (95% CI=1.4-4.1)] for development of BR-asthma symptoms, but not for non-BR-asthma symptoms in any one of the subsequent surveys; on the other hand, the group with non-BR-rhinosinusitis symptoms tended to have increased, but not significant, odds [1.8 (0.7-4.8); p=0.2] for developing non-BR-asthma symptoms, but not for BR-asthma symptoms.

In the main effects model, increasing fungal tertile exposure significantly increased the odds of developing BR-asthma symptoms in any one of the subsequent surveys in an exposure-dependent manner [medium: OR=3.1 (95% CI=1.4-6.7); high: 3.5 (1.6-7.5)]. The odds ratios for fungal exposure were larger than that for 2001 BR-rhinosinusitis symptoms. However, measurements of ergosterol or endotoxin in the initial environmental survey did not increase the odds of developing BR-asthma symptoms. The odds ratio for developing BR-asthma symptoms with the presence of both BR-rhinosinusitis symptoms and higher fungal exposure within the building at the initial survey was much higher [OR=7.4 (95% CI=2.8-19.9)] than that for the comparison group with higher exposure [3.4 (1.3-8.6)] or that for the BR-rhinosinusitis symptom group with lower exposure [2.4 (0.6-10.0)].

Average levels of each of the three fungal exposure groups in the initial survey were similar to those in 2004 and 2005, but the levels substantially increased in 2007. The levels of ergosterol were similar in 2002 and 2004 but increased in 2007. The endotoxin levels steadily increased until 2005 and stayed similar in 2007 (Figure 1).

DISCUSSION

Our study showed that the risk for development of BR-asthma symptoms was higher in the group of occupants with existing BR-rhinosinusitis symptoms and that the peak prevalence of BR-rhinosinusitis symptoms preceded the peak prevalence of BR-asthma symptoms. It is possible that BR-asthma symptoms are an indication of undiagnosed asthma or of future asthma development in this building population. Our findings suggest that occurrence of BR upper respiratory illness in water-damaged buildings may presage future asthma. Furthermore, occurrence of BR upper respiratory illnesses in such buildings should motivate building evaluation and remediation to eliminate dampness and related exposures. In addition, our finding of an interaction between the presence of BR-rhinosinusitis symptoms and fungal exposure for development of BR-asthma symptoms suggests that occupants of water-damaged buildings who are exposed to relatively higher levels of fungi and who have already developed rhinosinusitis have the highest risk for development of asthma.

Little information exists about the risk of development of asthma in association with damp building-related mold or other exposures in occupants who have already developed BR-rhinosinusitis symptoms.

Figure 1. Average levels of culturable fungi, ergosterol, and endotoxin by survey year and tertile exposure. Each tertile exposure category contains five floors by rank order of floor specific-geometric means of all 15 occupied floors in the building for each survey. Ergosterol was not measured in the 2005 survey. CFU=colony forming unit; ng=nanogram; EU=endotoxin unit.

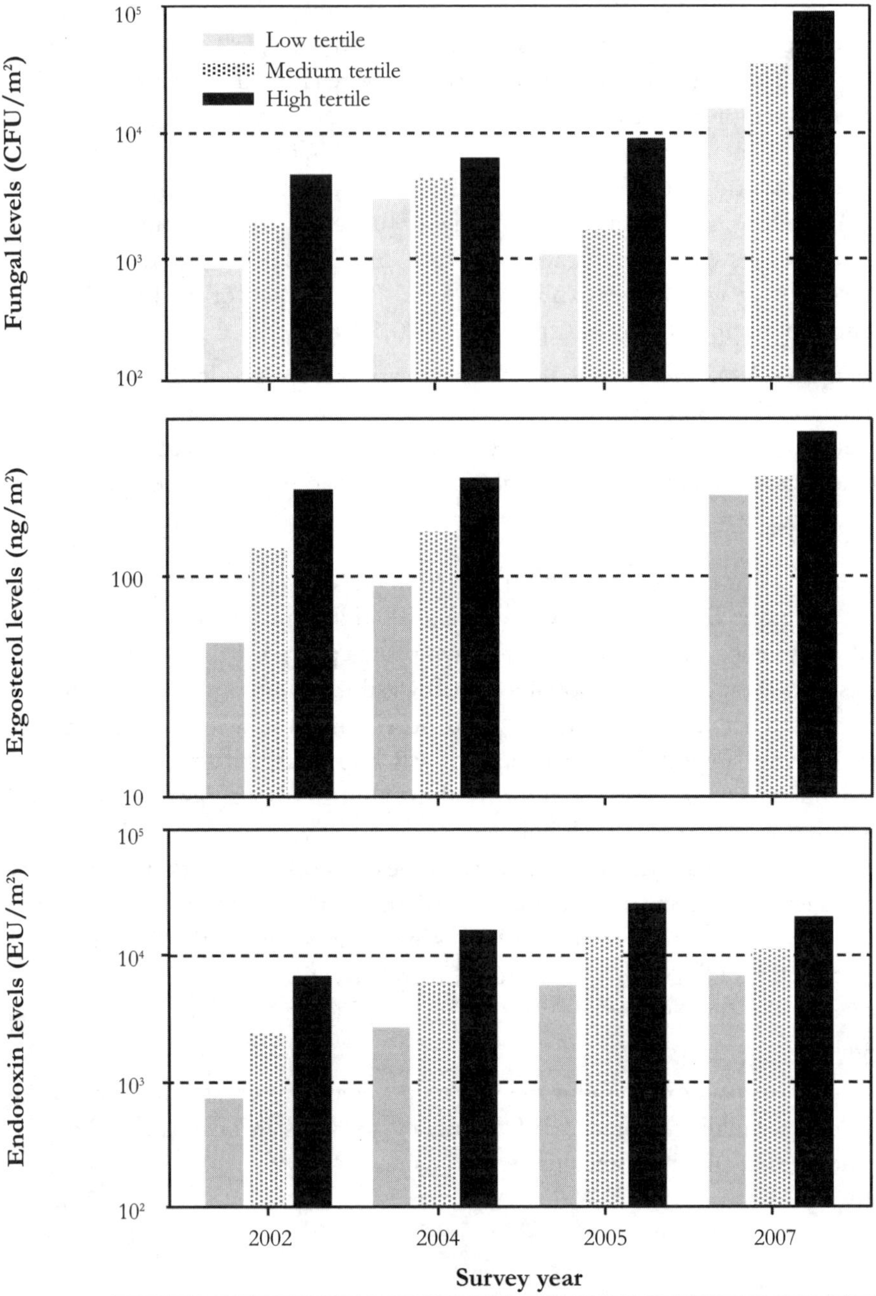

Our study indicates that the progression of BR-rhinosinusitis symptoms to BR-asthma symptoms and the progression of non-BR-rhinosinusitis symptoms to non-BR-asthma symptoms likely have independent associated exposures. Microbial or some other exposures related to the damp building environment may uniquely contribute to the progression of BR-rhinosinusitis symptoms to BR-asthma symptoms. Our findings imply that the building-related exposures were not involved in the progression of non-BR-rhinosinusitis to non-BR-asthma symptoms; these latter conditions are associated with exposures outside the building.

In conclusion, our study findings suggest that occurrence of BR upper respiratory illnesses such as rhinitis, sinusitis, or rhinosinusitis in damp building occupants may be a warning for increased risk of developing BR lower respiratory illnesses such as asthma in the future. Thus, early recognition of increased BR upper and lower respiratory symptoms, identification of water infiltration and damaged areas, and timely effective remediation may reduce development of asthma and minimize the burden of respiratory diseases in occupants of water-damaged buildings.

REFERENCES

- Chun DT, Chew V, Bartlett K, Gordon T, Jacobs RR, Larsson BM et al. Second interlaboratory study comparing endotoxin assay results from cotton dust. *Ann Agric Environ Med.*, 2002:9, 49–53.
- Derebery JD, Meltzer E, Nathan RA, Stang PE, Campbell UB, Corrao, M. and Standord R. Rhinitis symptoms and comorbidities in the United States: Burden of rhinitis in America survey. *Otolaryng. Head Neck*, 2008:139, 198-205.
- Guerra S, Sherrill DL, Martinez FD and Barbee RA. Rhinitis as an independent risk factor for adult-onset asthma. *J. Allergy Clin. Immun.*, 2002:109, 419-425.
- Institute of Medicine of the National Academies of Science. Human health effects associated with damp indoor environments: Damp indoor spaces and health. First edition (ed.) Washington D.C.: National Academies Press. 2004
- Krouse J, Brown R, Fineman S, Han J, Heller A, Joe S, Krouse H, Pillsbury H3, Ryan M and Veling M. Asthma and the unified airway. *Otolaryng. Head Neck*, 2007:136, S75-S106.
- Leynaert B, Bousquet J, Neukirch C, Liard R and Neukirch F. Perennial rhinitis: An independent risk factor for asthma in nonatopic subjects: results from the European Community Respiratory Health Survey. *J. Allergy Clin. Immun.*, 1999:104:(2 Pt 1), 301-304.
- Park JH, Cox-Ganser JM, Rao CY and Kreiss K. Fungal and endotoxin measurements in dust associated with respiratory symptoms in a water-damaged office building. *Indoor Air*, 2006:16, 192-203.
- Sebastian A, Larsson L. Characterization of the microbial community in indoor environments: a chemical-analytical approach. *Appl Environ Microbiol.*, 2003:69, 3103–3109.
- Togias A. Rhinitis and asthma: Evidence for respiratory system integration. *J. Allergy Clin. Immun.*, 2003:111, 1171-1183.
- World Health Organization. Health effects associated with dampness and mould. In: World Health Organization (ed.).WHO Guidelines for Indoor Air Quality: Dampness and Mould Copenhagen, Denmark: WHO Regional Office for Europe. 2009.

CONCLUSIONS ON HEALTH IMPLICATIONS OF AIRBORNE MOLDS: ANALYSIS OF AIRBORNE MOLDS IN 11 CONTAMINATED HOUSES USING A NEW METHOD OF EVALUATION

Urban Palmgren, Judith Müller

KEYWORDS: airborne CFU, airborne total cell count, indoor mold, total cell count, new method of detection, airborne

ABSTRACT

Airborne molds in 11 contaminated houses were investigated using a new method of evaluation.

There was a fluctuation of colony forming units/m³ from < 1% to > 90% when compared to the levels of total cell count method (tcc). The results of the outdoor samples are always higher as the indoor samples.

The investigation with a cfu/tcc evaluation showed the influence of time on the results in 5 ways:

In an older microflora, many molds have not survived the elapsed period of time and can because of this not be confirmed by the determination of colony forming units/m³ (cfu/m³). Older mold sources emit more spores and mycelia parts to the air as younger sources. Through an air flow close to the contaminated surface of the building material the mold spores and fragments will get airborne and these microorganisms will show the same age as those on the building material. The comparison of cfu/tcc can be used as an indication of the age of the source of the airborne contamination with molds (cfu/tcc-evaluation). Outdoor and indoor samples are seldom of the same age and this indicates that the indoor and outdoor sources of airborne molds have come from different places and should consequently not be used as reference measurements.

Out of 29 measurements of just cfu/m³ there were no information indicating an indoor mold source. When additional methods of confirmation were used (cfu/tcc-evaluation) there were 18 samples indicating indoor mold sources.

The toxic and allergenic potential of airborne molds is dependent on the amount of biomass/m³ or on the total number of microorganisms/m³ and independent of the number of colony forming units/m³ (cfu/m³). In conclusion with all results of this investigation an attempt to correlate the number of airborne colony forming units to any types of health problems is irrelevant.

Introduction/ Goal of study

Increased awareness of the danger to human health caused by airborne molds in housing has triggered an interest in reliable methods of determining a potential mold exposure. In many publications, the correlation between living in damp houses and negative health effects could be established (Bornehag et al., 2001; Bornehag et al., 2004; Fisk, Lei-Gomez and Mendell, 2007; Institute of Medicine, 2004; Mudarri and Fisk, 2007; WHO Guidelines for Indoor Air Quality, 2009; Tischer, Chen and Heinrich, 2011; Norbäck, 2001) but the correlation with the colony forming units (cfu/m³) of airborne mold could not (WHO Guidelines for Indoor Air Quality, 2009; Malmberg et al., 1986; Malmberg et al., Norbäck et al., 1993). By the risk assessment of airborne molds, with emphasis on the airborne colony forming units (cfu) as the cause of illness, the standard method of detection is the measurement of cfu/m³ air.

Molds grow only if the relative humidity of the surroundings is high enough to support it (> 70%). When for example building materials dry out after the cause of the water damage has been repaired, the relative humidity is lowered and the mold growth stops. With time more and more molds on the material are so inhibited in there metabolism that they can not even start to grow as cfu on the nutrient rich agar media in the labs. In this low activity status the mold can not be found with laboratory methods based on plate count techniques, they can just be detected with methods which count the mold although they do not grow (for examples microscopic counting).

The diminishing capacity of the mold to grow over time (and be detected as cfu) in combination with a constant remaining total cell count, give microbiologists the possibility to estimate how old a special surface contamination could be.

The aim of this investigation is to test the hypothesis that airborne microorganisms from contaminated surfaces will have a different age pattern between different rooms and houses as well as when compared to microorganisms in air samples from outside the building.

METHOD

Air samples, were initiated in damp houses where the presence of a high concentration of a mold growth on surfaces already had been proven. The evening before the sampling day all houses were ventilated through windows for > half an hour and then the rooms were locked and reopened at the time for sampling the next day.

The sampling was made with the filtration method (Camnea method). Polycarbonate filters with a pore size of 0.4µm and a diameter of 37mm were placed on support pads in sterilized filter cassettes (Millipore). With an air flow of 2 liters per minute over 4 hours, air was sucked through the filter holder with an air sampler (SKC).

After the sampling, the filter and the cartridge were washed with a sterile washing liquid. 100 mµ of the sample was dyed with acridine-orange (fluorescens dye) and filtrated through a black polycarbonate filter. Thereafter the microorganisms on the filter surface were counted under a fluorescence microscope. In the next step, colony formed units (cfu) were plated out on three standard media, incubated at 25 °C and counted as well as differentiated after seven days.

Total cell count (tcc) of microorganisms stained with fluorochromes such as acridine orange has been used for a long time in different environments, such as marine research (Zimmerman & Meier-Reil 1974; Jones & Simon 1975), rapid determination in food samples (Pettipher et al., 1980; Pettipher & Rodrigues 1982) and counting of airborne microorganisms in highly contaminated environments (Palmgren et al., 1986).

Additional staining with FDA (flourochrome) shows the metabolic activity of microorganisms and gives the researcher information that makes a determination of the age of the mold growth on building materials possible. Labor Urbanus GmbH for example, uses this method of investigation to confirm if more than just one water damage has caused particular mold damage to a building.

In the rooms where air sampling was carried out, mold growth on wall surfaces was investigated with an adhesive tape method (mold spores and mycelia/cm^2).

RESULTS

The analyzes of airborne molds show a fluctuation in the levels of cfu/m^3 from < 1% to > 90% compared to the levels of airborne molds estimated with the total cell count method (fig. 1).

Fig. 1. Two methods of analyzes are compared in 11 houses.

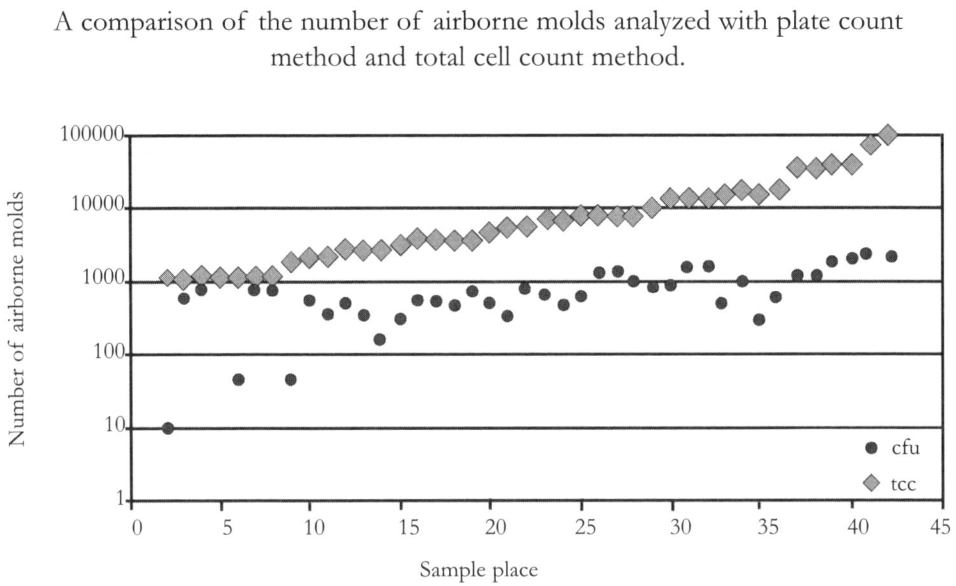

The variation is irregular. The results show that the cfu- and the tcc-levels can be low, compared with the outside reference air sample, all though there is a high concentration of microorganism generated inside the building (fig. 2, fig. 5). Both fig. 1 and fig. 2 show the expected results with lower numbers of cfu compared with tcc. These results would not confirm any indoor source of excessive mold contamination.

In fig. 2 the results of cfu and tcc at first sight looks as if they are not so far apart but that is caused by the logarithmic graph. Without this way to describe the results the column of the cfu values would be too small to make a comparison visible.

The results of the outdoor samples are always higher as the indoor samples. On the basis of this information from these samples, no estimation of expected health hazard can be determined.

It is important to point out that the microbial contamination of indoor air depends on a transport of molds from outside of the building to within or growth on surfaces inside of the building. The air itself is not a medium supporting microbial growth.

Fig. 2. Measurements of airborne molds outside compared to indoor air inside 11 houses.

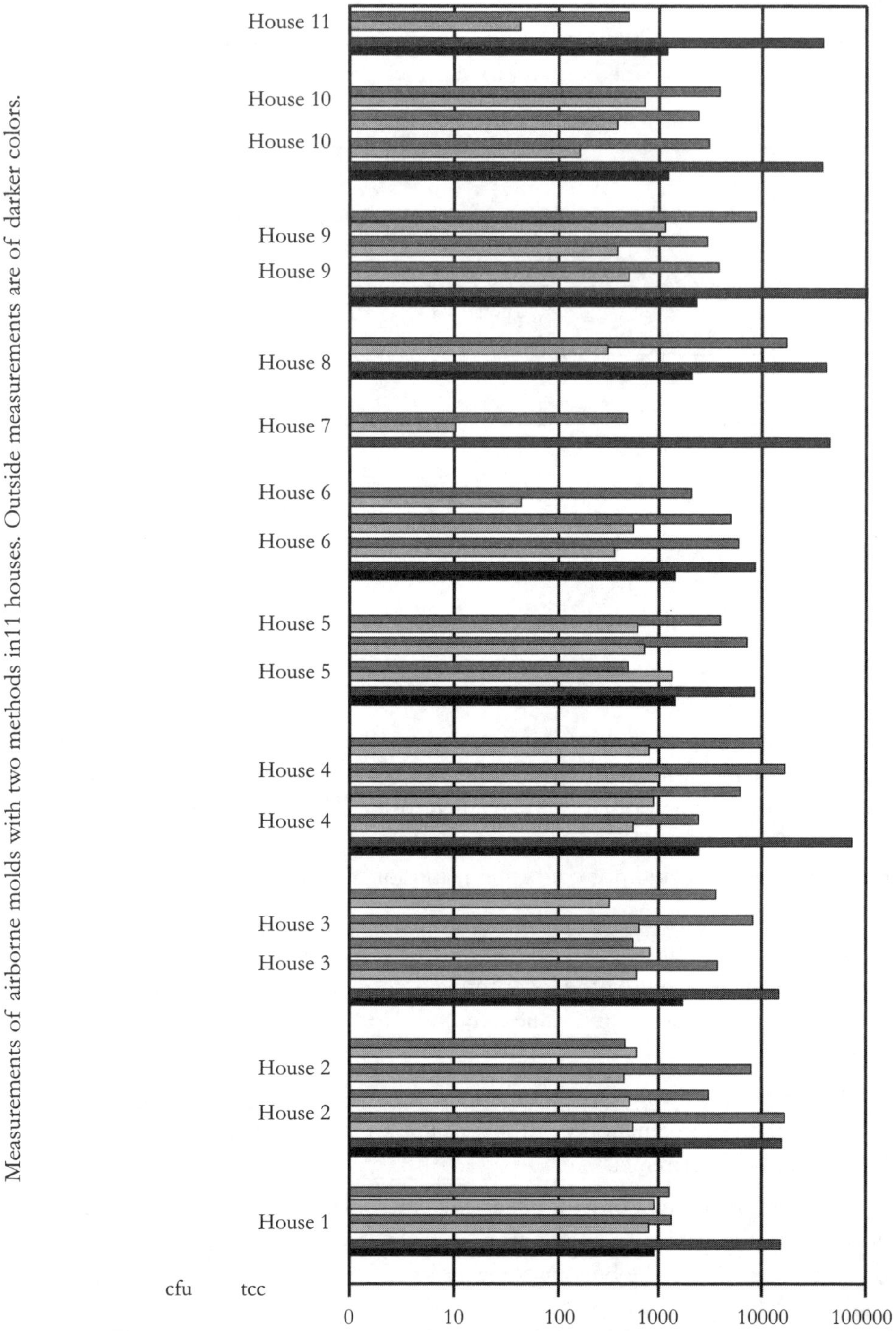

The graph in fig. 3 shows not only the differences between cfu and tcc, but also the comparison cfu/tcc, which can be used as an age indicator of the source of the airborne contamination with molds. The higher values of cfu/tcc on the left side of the graph confirm fresher sources of the airborne molds and lower values on the right hand side older sources. Additionally, fig. 3 shows that in an older micro flora, many of the older cfu´s of molds have not survived and can because of this not colonize the microbial plates in the lab.

These findings indicate that the results may depend on the time schedule of the release of mold spores and mycelia. The older the mold growth on a surface is, the more spores and mycelia parts are emitted to the air. This agrees with our own experiences, that new and visible mold growth may emit low numbers of airborne spores.

Fig. 3. The graph shows the differences between cfu and tcc but also the comparison cfu/tcc.

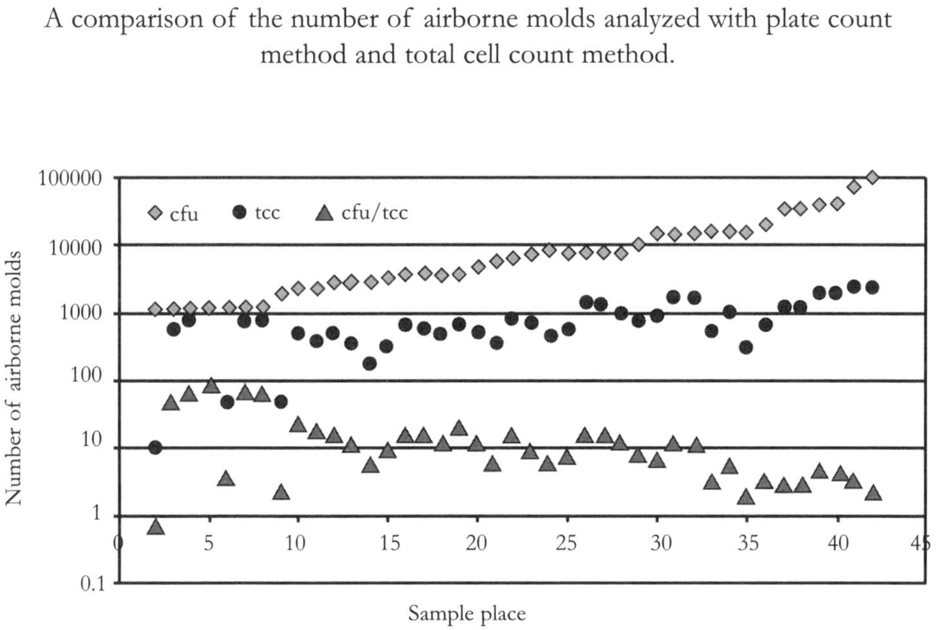

In fig. 4 the surface contamination (mold spores and mycelia/cm²) of the walls are compared to the resulting airborne molds. In some samples the airborne levels of molds are much higher then expected and in some samples much lower, but the visible resemblance between the patterns in the graph is intriguing.

Fig. 4. A comparison of molds from wall surfaces in the sampled room and the airborne molds in the air samples.

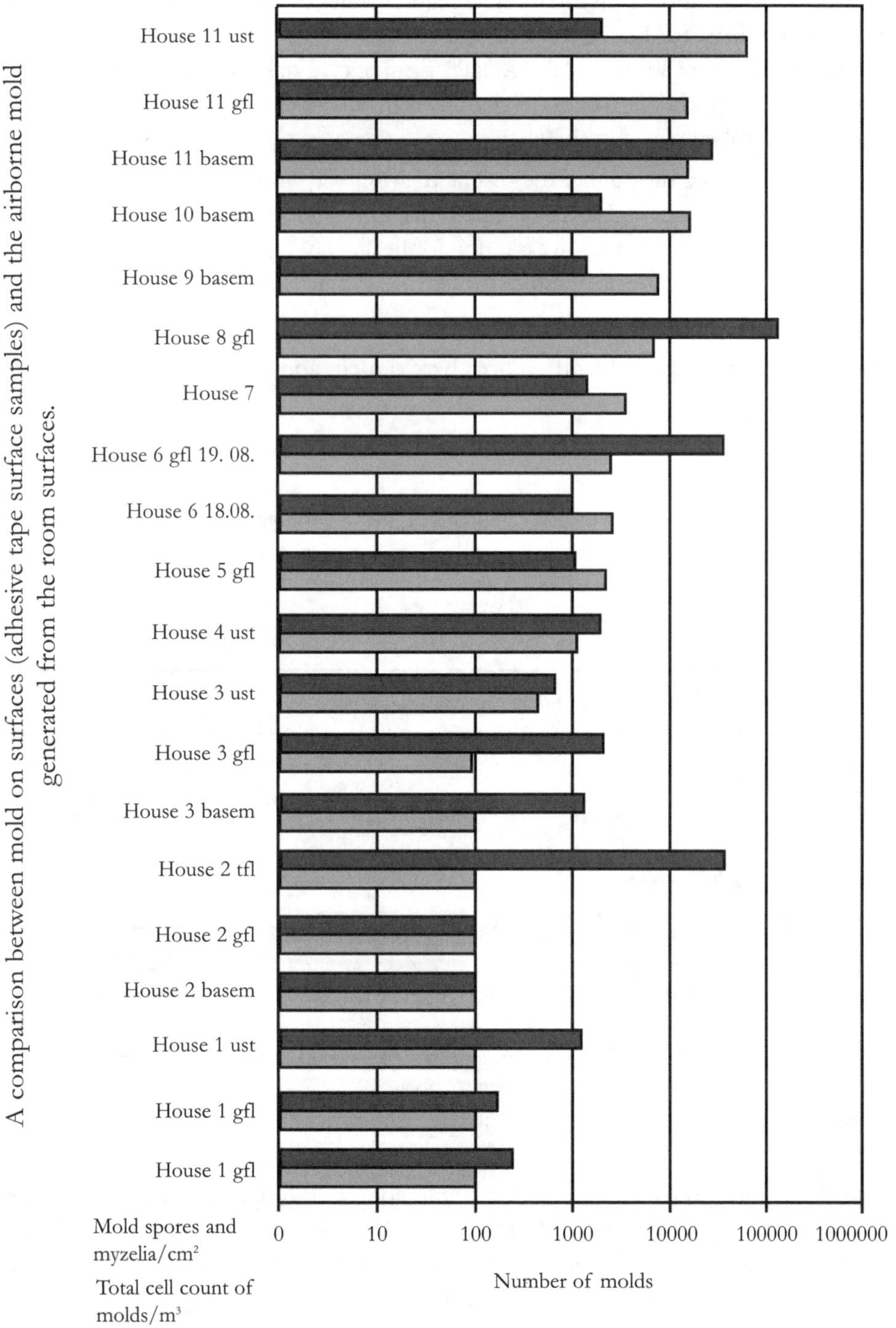

Fig. 5. In this graph the columns show the difference in age (cfu/tcc) between outdoor samples (dark grey) and indoor samples from different parts of the buildings.

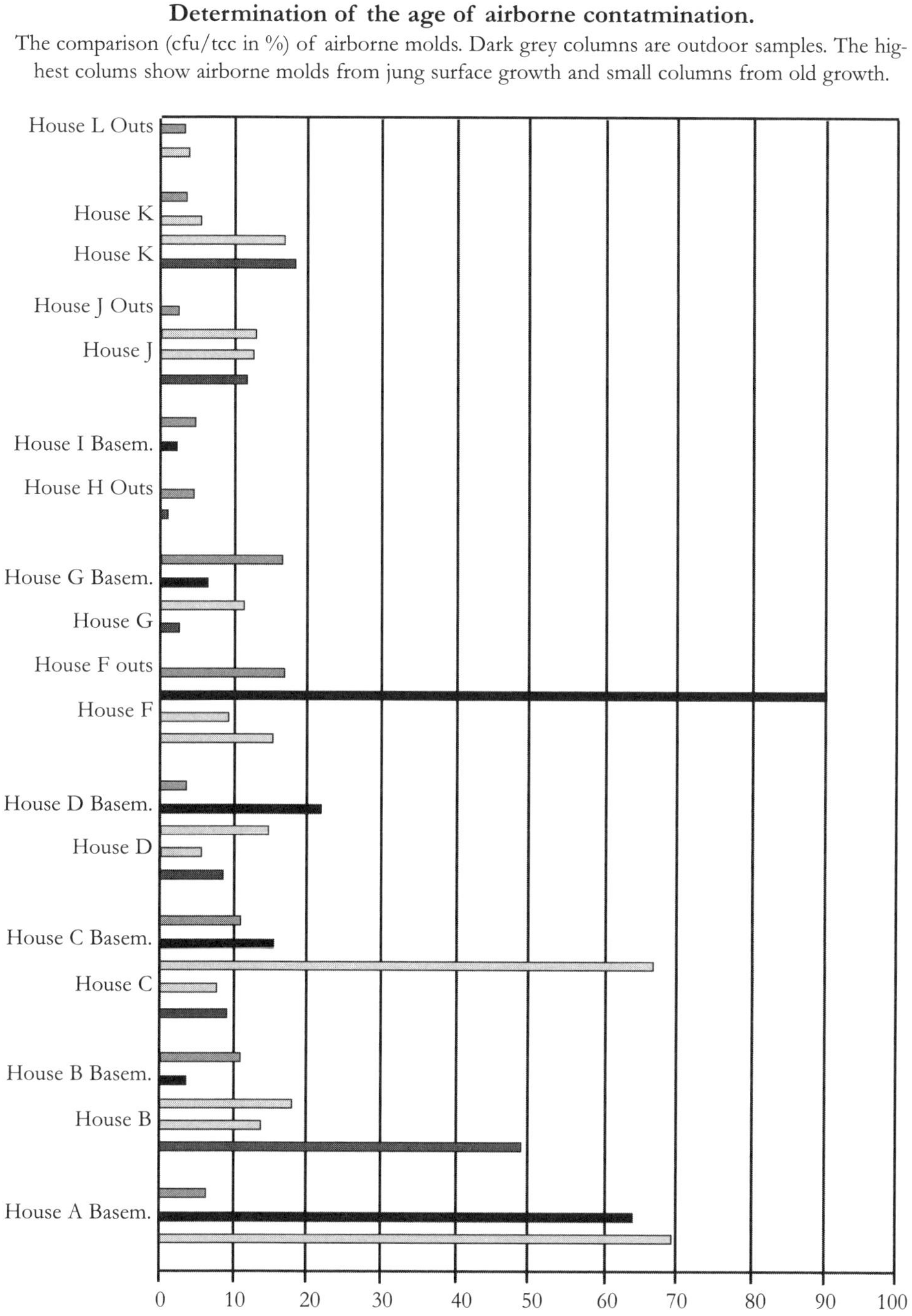

Through an air flow close to the contaminated surface of the building material the mold spores and fragments will get airborne and these microorganisms will show the same relation of cfu in percentage of the total cell count as those on the building material. The airborne molds can nevertheless naturally still have lower percentage due to them drying out in the dust before they get airborne a second time.

In fig. 5 the columns of outdoor samples (dark grey) and indoor samples are rarely of the same dignity and this indicate that the indoor and outdoor sources of airborne molds has come from different places. This information stresses the implications of the usefulness of relating indoor samples to not interfering outdoor samples.

In fig. 6. the importance of applying a reliable method of determining airborne mold becomes clear and the insufficiency of cfu analysis as sole information on airborne mold contamination in buildings is evident. Out of 29 samples with no indicative information, there were 18 indicative samples for indoor mold sources, when different methods of confirmation were used (cfu/tcc-evaluation).

Fig. 6. A comparison of the number of airborne molds analyzed with plate count method and total cell count method. Black dots are cfu values that fail to indicate an indoor mold source.This graph shows the unsuitability of using cfu levels as sole source of indication of airborne mold investigations.

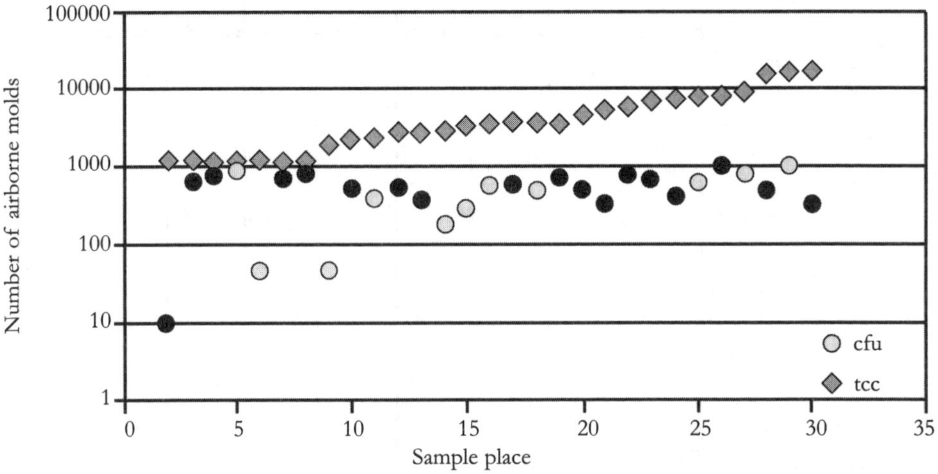

Possible health effects of airborne molds are infections, toxicity and allergies, of which only infections are dependent on the viability of organisms. The threat of molds invading a human body, like for example the more hazardous pathogenic microorganisms, are very faint. The harmfulness of molds depends on their toxic or allergenic potential and the bodily response to their presence. The toxic and allergenic potential of airborne molds is dependent on the amount of biomass/m^3 or on the total number of microorganisms/m^3. This toxic potential is however totally independent from the number of colony forming units/m^3 (cfu/m^3).

As a conclusion of all the results from this investigation, one can clearly distinguish that the attempt to correlate the number of airborne colony forming units to any types of health problems is unsuccessful. Even in the agricultural sector, no investigation has shown any correlation between the viability of microorganisms and the potential for inducing health problems. The correlation is just proven between total cell count/m^3 and allergic alveolitis (Palmgren et al., 1986).

REFERENCES:

- Bornehag CG, Blomquist G, Gyntelberg F, Jarvholm B, Malmberg P, Nordwall L, et al. . Dampness in buildings and health. Nordic interdisciplinary review of scientific evidence on associations between exposure to "dampness" in buildings and health effects (NORDDAMP). *IndoorAir* 2001 Jun,2001;11 (2):72-86.
- Bornehag CG, Sundell J, Bonini S, Custovic A, Malmberg P, Skerfving S, et al. Dampness in buildings as a risk factor for health effects, EUROEXPO: a multidiciplinary review of the literature (1998-2000) on dampness and mite exposure in buildings and health affects. *Indoor Air* 2004;14(4):243-57.
- Fisk WJ, Lei-Gomez Q and Mendell MJ. Meta-analyses of the associations of respiratory health effects with dampness and mold in homes. *Indoor Air* 2007;17(4):284-96.
- Institute of Medicine. Damp Indoor Spaces and Health. 2004.
- Malmberg P, Rask-Andersen A, Palmgren U, Höglund S, Kolmodin-Hedman B and Stalenheim G. Exposure to microorganisms, Febril and airway-obstructive symptoms, immune status and lung function of Swedish farmers. *Scand. J. Work. Environ. Health* 1985;11, 287-293.
- Malmberg P, Palmgren U, Rask-Andersen A. Relationship Between Symptoms and Exposure to Mold Dust in Swedish Farmers. *American Journal of Industrial Medicine* 1986;10:316-317.
- Mudarri D and Fisk WJ. Public health and economic impact of dampness and mold. *Indoor Air* 2007:17(3):226-235
- Norbäck D, Björnsson E, Widström J, Ström G, Edling C, Palmgren U, et al. Asthmatic and the home environment. Proceedings from The Conference on Building Design. Technology and Occupant Well Being in Temperate Climates, Brussels. 1993; 17-19
- Norbäck D, Björnsson E, Janson C, Palmgren U, Boman G. Current asthma and biochemical signs of inflammation in relation to building dampness in dwellings. *Int J Tuberc Lung Dis.* May; 1999;3(5):368-76.
- Palmgren U, Ström G, Blomquist G and Malmberg P. Collection of airborne microorganisms on Nuclepore filters, estimation and analysis, CAMNEA-Method. *J. Appl. Bact.* 1986;61, 401-406.
- Tischer, Chen and Heinrich. Association between Domestic Mould and Mould Components, and Asthma and Allergy in Children: a systematic Rewiew. 2011 in Press
- WHO. Dampness and Mould. WHO Guidelines for Indoor Air Quality, 2009.

OBSERVATIONAL EPIDEMIOLOGY AND WATER DAMAGED BUILDINGS

Joseph Q. Jarvis, and Philip R. Morey

INTRODUCTION

A NIOSH team reported their questionnaire based public health investigation of respiratory symptoms and asthma in employees in relation to damp indoor environments of two hospitals which was induced by the discovery of six sentinel cases of work-related asthma (Indoor Air, 2009). They achieved a participation rate of 64%, which they characterized as 'fairly good', but noted that employees who worked in departments with dampness problems were somewhat more likely to take part in the survey, raising some concern about a possible volunteer bias. They did not use medical records or medical testing to validate participant reports of physician-diagnosed asthma, except for the six sentinel cases. The NIOSH team concluded that follow-up of sentinel reports of building related illness is warranted, noting that new-onset building related asthma in relation to water damage can occur.

Outside of public health agency investigations of problem buildings, however, observational epidemiologic methods are not commonly employed in the United States in the setting of indoor air quality complaints. Our first experience with observational epidemiology in the setting of a water damaged building occurred 20 years ago. The subject building was a county-owned structure with nearly 600 occupants, located in a part of the US with a sub-tropical climate. The building had been opened for four years at the time of the questionnaire survey. Widespread moisture and fungal growth problems had been documented in the building practically since initial occupation after its construction. Our investigation was initiated after the local public health department had conducted its own evaluation of occupant complaints and concluded that building related symptoms were common and building related allergic respiratory disease could not be excluded. We began the investigation in standard public health fashion by conducting confidential interviews of approximately 50 volunteer subject building occupants. These interviews confirmed the findings of the local health department that building associated symptoms and building-related allergic respiratory disease may have occurred. We sought to establish whether rates of symptoms were higher than would be expected in the problem building by conducting a cross sectional survey of occupants of both the subject building and a neighboring county-owned building without known indoor air problems. 95% of subject building occupants and 93% of the occupants of a comparison county-owned building participated in the questionnaire survey. Using conservative epidemiologic definitions, subject building occupants had a fourfold increase in building related symptoms and a threefold increase in respiratory illness over rates found in the comparison building. A nested case/control study design confirmed that exposure to water damage and visible mold at the individual work area conferred higher risk for both building associated symptoms and building related respiratory disease. We conducted detailed clinical evaluations of 37 subject building occupants which confirmed 15 cases of building-related asthma. Based upon these data, the building owner chose to close the building for extensive remediation. Repeat questionnaire surveys were conducted four months after building evacuation, just before the building was re-occupied after remediation, and one month after the building was re-opened and significantly contributed to the protection of the comfort, safety, and peace of mind of the building occupants as well as the successful prosecution of construction defects litigation in their behalf. Details of both the environmental and epidemiologic aspects of this building investigation have been published (Applied Occ Env Hyg 16, 2001).

Since that experience we have been invited to participate in the evaluation of many other water damaged buildings large enough to consider the use of epidemiologic methods, including commercial and public office buildings, hotels, residential high rise buildings, and schools. More often than not, the search for a possible sentinel case found no health problem requiring follow-up investigation, and in accordance with standard public health practice, no observational epidemiologic study was attempted. When, however, building associated symptoms or building related illness may have occurred, as the NIOSH team stated, follow-up is indicated.

METHODS

The possibility of encountering bias, also as noted by the NIOSH team, should not deter the application of observational epidemiologic methods. Volunteer bias can be eliminated by assuring the participation of virtually all building occupants, as in the case of the county owned buildings just described. Selection bias occurs when the two populations being compared differ in a significant manner. Proper questionnaire design includes the questions necessary to identify selection bias, for instance by inquiring about smoking history, or other personal and workplace factors, which might affect the rate of symptom reporting. As was the case for the NIOSH team, the identification of possible selection bias can be managed through appropriate data analysis. Recall bias can be managed, at least in part, by conducting the environmental investigations in the comparison building necessary to verify its non-problem status in the open during hours when the building is occupied. Response and information biases can be avoided if the questionnaires are handled exclusively and confidentially by appropriately trained health personnel with no connection to the employers of the respondents. Questionnaire surveys of necessity contain private, personal health information and respondent confidentiality must be guaranteed. In addition, the group data which is derived from the questionnaire must be shared with the respondents from whom the data came. Analytic bias occurs when the investigator has a pre-determined outcome in mind, and is therefore totally avoidable.

Given a commitment to executing a properly designed observational epidemiologic investigation, the data derived therefrom can play a significant part in designing and carrying out a remediation project which protects the health of building occupants.

RESULTS

In this paper, we report on two separate building investigations wherein we employed observational epidemiologic methods. The first of these examples is a county owned office building located in a Pacific coast community, which opened in 1995 (Building A). Over the succeeding 5 years, occupants observed rainfall induced window leakage and water soaked carpeting in 59 separate locations around the perimeter (envelope) of the building. Some visible mold was observed on paper-faced gypsum board in the offices most affected by water intrusion. Destructive inspection of the building envelope on all elevations revealed the presence of extensive mold growth on biodegradable construction materials including paper faced gypsum board, asphaltic building paper, and cellulosic fireproofing. *Stachybotrys, Chaetomium, Penicillium* (esp. *Pen. chrysogenum*), and *Cladosporium* (esp. *Clad. sphaerospermum*) species were most commonly found growing on envelope construction materials (Table 1). Water indicator fungi such as *Penicillium chrysogenum* and *Rhizopus* were dominant in dusts from interior locations such as the upper surface of ceiling tiles, suggesting that building occupants were potentially being exposed to airborne fungi infiltrating from the envelope into offices.

TABLE 1: Dominant Culturable Fungi on Envelope Construction Materials and in Settled Dusts in Building A

Sample Description	Dominant Molds
Gypsum Board	*Stachybotrys, P. chrysogenum, Ulocladium, Chaetomium*
Building Paper	*Stachybotrys, P. chrysogenum, Clad.sphaerospermum*
Fireproofing	*P. chrysogenum, Stachybotrys*
Wall Cavity Dust	*P. chrysogenum, Stachybotrys*
Ceiling Tile Dust	*Penicillium species, Rhizopus, Ulocladium*

Culture on DG-18 and Cellulose Agars

In 2001 we interviewed 44 current or former occupants of the building concerning indoor environmental conditions and health. More than half of the interviewees described new onset building associated respiratory symptoms, such as cough, shortness of breath, wheezing, or chest tightness. Five of these building occupants had had no previous history of pulmonary problems but had developed respiratory symptoms since entering the subject building and had received a physician diagnosis of new-onset asthma. Four others had received a physician diagnosis of new onset acute or chronic sinusitis. As per NIOSH recommendation, a follow-up questionnaire survey was organized and conducted, consisting initially of a cross-sectional study of 286 occupants of the subject building and 108 occupants of a nearby county-owned office building thought to be free of indoor air quality problems. The questionnaire survey instrument consisted of a standardized respiratory symptom questionnaire supplemented with questions regarding other symptoms, discomfort complaints, previous physician diagnoses, medications, sick leave, physician consultation, job characteristics, job satisfaction, and work station factors. Participants filled out the questionnaire in the presence of a supervising public health physician. Questionnaire responses were analyzed using public domain software from the CDC. A probability of 0.05 or less was considered significant. Participation rates were nearly universal among the target populations. Occupants of the comparison building had a smoking rate 4 x higher than that of the subject building population. As expected, subject building occupants more often observed water damage and mold growth near their work station, though rates of reported building discomfort (temperature, humidity, noise, odor, and air movement) were high in both buildings. Coughing while at work was four times more common in the subject building, when comparing rates reported among non-smokers in both buildings. Non-smokers in the subject building also reported more bronchitis, sinus trouble, and recent chest illness which kept them home from work and were twice as likely as non-smokers in the comparison building to report having at least three of the following four chest symptoms: cough, phlegm production, wheezing, and shortness of breath. Using that same case definition and defining a control as a person who reported no respiratory symptoms, a nested case-control study of subject building occupants revealed that mold and water damage near the work station conferred an increased risk for respiratory disease of between 3 and 6 fold (see Table 2). Cases were far more likely than controls to have consulted a physician because of symptoms caused by their work environment, use physician prescribed asthma medications, and have experienced an episode of chest illness requiring time away from work. Clinical data in the form of 16 medical records were made available for my review, revealing six cases of building related allergic rhinitis, two cases of building-related exacerbation of pre-existing asthma, and one case of new onset building related asthma. Because building-related allergic respiratory disease had likely occurred among building occupants, the building owner changed the remediation plan for the building. In addition to the initial plan to rectify the water incursion problems and physical removal of moldy construction materials, following NYC (2008) and AIHA (2008) guidelines, an additional important aspect of building remediation included a thorough HEPA vacuum cleaning of interior surfaces to reduce residual dust and the spores that may be in the dust to the lowest feasible levels (AIHA, 2008, Chapter 18). Because of the risk of exposure during demolition and construction activities, the building owner opted to evacuate the building for the duration of the remediation project. Finally, to add a measure of assurance for building occupants, the building owner requested questionnaire surveillance just before and one month after the building was re-occupied.

TABLE 2: Estimating Risk for Respiratory Disease in Building A

	Odds Ratio	95% Confidence
Mold: 15 ft. of workstation	3.4	1.0 - 11.5
Water damage at workstation	3.7	1.5 - 9.3
Perceiving moldy odors at work	6.2	2.1 - 18

Our second example of the use of observational epidemiologic methods in the setting of a water-damaged building is a hotel located on a tropical island, which had been severely damaged by a typhoon with 200 mph sustained winds approximately six weeks before we arrived at the facility (Building B). Approximately 2 to 3 weeks after the typhoon most walls, ceilings, and carpeted floors were still water saturated. Visible mold growth was apparent on ceilings, walls, in wall cavities, and on furniture in many rooms. Air sampling by spore trap was carried out (see Table 3) and demonstrated that mold spores inside the hotel were predominantly *Penicillium* or *Aspergillus* species, with minor amounts of *Stachybotrys, Chaetomium,* and *Trichoderma*, all indicating indoor exposure to a damp and degraded indoor environment. Exposure to *Penicillium* and *Aspergillus* inside the hotel was about three orders of magnitude greater than that found outdoors. These exposure conditions are similar to spore exposures in composting and agricultural silos where organic dust toxicity syndrome or hypersensitivity pneumonitis may occur.

TABLE 3: Airborne *Penicillium-Aspergillus* in Building B 2-3 Weeks After Typhoon
Pen/Asp=Penicillium/Aspergillus; Burkard spore trap operating @ 10 L/min

Hotel Area	# Samples	Mean *Pen/Asp* per M3	% *Pen/Asp* of total count
Wing A	12	230,000	82%
Wing B (unoccupied)	8	1,680,000	96%
Wing B (occupied)	20	240,000	91%
Wing C	16	770,000	97%
Outdoors at Grade Level	16	250	14%

We interviewed 18 housekeepers who had been employed in mold clean-up efforts. Half of these persons reported building associated symptoms such as eye, nose, and throat irritation or headache, though some noted that the extra use of bleach during mold clean-up seemed to be an important cause of these symptoms. Three reported exacerbation of pre-existing asthma while in the building and 3 others reported new onset chest symptoms since the storm. Because of the possibility of clinically significant symptoms occurring among hotel housekeepers, we organized a questionnaire survey, using a comparison population of hotel housekeepers working for the same hotel chain on a different tropical island not affected by the typhoon. Participation rates were high (91% among subject building occupants, 97% among comparison building occupants). In this case, smoking rates were three times higher in the subject building population than for the comparison population. As expected given the extreme circumstances of the storm damage, water damage, mold growth, and moldy odors were reported far more commonly among subject building respondents. Among the chest symptoms, however, only cough and phlegm production were reported at a higher rate within the subject building population. Further analysis revealed that the excess rate of these symptoms was confined to the smokers and therefore not likely a building-related health problem. Eye, nose, and throat irritation symptoms were reported more commonly among subject building respondents, particularly among those with workplace exposure to mold and water damage. Because the cross-sectional study found no evidence supporting sentinel chest disease, a follow-up case/control study was not indicated. However, the finding of building associated symptoms associated with workplace exposure to mold and water damage indicated that clinically significant mold exposure was occurring among building occupants and that eventual acquisition of allergy and onset of allergic respiratory disease was possible. The building owner used the questionnaire data to target areas of the building associated with higher rates of BAS for early remediation. The building owner was advised that chronic ongoing exposure to atypical indoor mold antigen carried a risk for allergic chest disease which could be mitigated by surveillance.

CONCLUSION

In summary, observational epidemiologic investigations are a practical means for follow-up of sentinel health events among occupants of water-damaged buildings. High rates of participation can be achieved, eliminating volunteer bias. Other types of bias can be managed through proper questionnaire design and appropriate analysis. Our experience with water-damaged building investigations indicates that BAS and Building-related allergic respiratory disease are the principle health problems associated with atypical indoor mold exposure. Decisions about remediation projects, such as whether to evacuate the building, include mold antigen burden removal, or which parts of the building to prioritize, are better made when data from a public health investigation is acquired. In the relatively rare circumstance where building-related allergic chest disease has likely occurred and remediation has been undertaken after building evacuation, epidemiologic surveillance during building reentry can protect occupant health and peace of mind.

REFERENCES

- AIHA, Recognition, Evaluation, and Control of Indoor Mold. American Industrial Hygiene Association, Fairfax, Va. 2008.
- Applied Occ Env Hyg, 2001:16 (3): 380-388
- Indoor Air, Aug 2009:19(4):280-290
- NYC, Guidelines on Assessment and Remediation of Mold in Indoor Environments, New York City Dept. of Health, NY, NY. 2008.

MOLDS AND MYCOTOXINS: FACTORS THAT AFFECT EXPOSURE AND CONTRIBUTE TO ADVERSE HEALTH EFFECTS

Karin K. Foarde, Timothy Dean, Doris Betancourt, Jean Kim, Anthony Devine, Grace Byfield, and Marc Menetrez

INTRODUCTION

Fungal growth and the resulting contamination of building materials is a well-documented problem, especially after the reports from New Orleans and the US Gulf Coast following Hurricane Katrina. Inhalation is thought to be a major route of exposure; however, relating surface contamination to airborne levels of spores and subsequent mycotoxin exposure is challenging. Exposure to fungi may result in respiratory symptoms of both the upper and lower respiratory tract such as allergy and asthma. Everyone is potentially susceptible. However, of particular concern are children with their immature immune systems and individuals of all ages that are immunocompromised. While exposure too many of the fungi can be considered problematic and many fungi produce toxins, *Stachybotrys chartarum* is used as our representative mycotoxin producing organism. There are numerous reports demonstrating an association between exposure to *S. chartarum* and adverse respiratory health effects.

Our objectives were to:

1) quantify emissions of fungal spores from contaminated building materials

2) determine the mycotoxin content of the emissions

3) relate the level of emissions from contaminated building materials to exposure

4) use cell-free and cell based assays to begin relating the exposure results to potential adverse health effects.

METHODS

Chamber Studies

A room-sized test chamber was adapted to contain *S. chartarum* actively growing on gypsum board under controlled laboratory conditions allowing for the measurement of emissions from the surface of the active growth. Air samples were collected with Mattson-Garvin slit to agar samplers and Air-O-Cells and analyzed for culturable and total spores.

Levels of mycotoxin produced by *S. chartarum* spores grown on different substrates

Toxic activity was assessed using an in vitro luminescence protein translation inhibition assay. *S. chartarum* was grown at room temperature on MEA (malt extract agar) for 1 week or gypsum wallboard for 4 months. Spores were harvested from the surface of the material using sterile swabs and resuspended in sterile water. The spore preparations were centrifuged to remove any media or hyphal fragments and resuspended in water. Spore preparations were checked microscopically for purity (>99.99% spores with no visible hyphae).

Mouse macrophage cells exposed to *S. chartarum* spores

Toxic activity was also assessed by introducing the spores to cell culture using a mouse macrophage cell line (Raw 264.7). Spores were harvested from either wallboard or MEA plates at 1 x 10^5 spores/mL. A confluent monolayer of mouse macrophage cells (1 x 10^6 cells/well) were exposed to spores for 18 hours. Microscopic observations were made following the exposure (400 x magnification).

RESULTS

Chamber Studies

Spore emissions were measured over a 4 month time period. The levels of emissions were directly related to airflow and inversely related to relative humidity. When the airflow was abruptly increased, the spore emissions showed an immediate spike. The relationship between the culturable colony-forming units (CFUs) and total spores varied over time. The age of the growth had more of an influence on the duration of the emissions than the concentration of the surface growth. (Data not shown)

The humidity was decreased to a level where no further growth occurred. Both spore and CFU emissions slowly but steadily declined until around Day 50, when they appeared to remain at approximately 100 CFU through Day 120 when the experiment was terminated. The emission rates were calculated.

Using the emission rates shown in Figure 1, the airborne concentrations were calculated for both a large [100 ft^2 (9.3 m^2)] and small [4.3 ft^2 (0.4 m^2)] contaminated area of gypsum wallboard within a 1,500 ft^3 bedroom. The airborne concentration of spores in the bedroom was calculated to be 2 x 10^5 spores/L if there was a large contaminated area and 9 x 10^3 spores/L if the contaminated area was small.

Figure 1. Spore emissions quantified over 4 months ranged from over to 10,000 to 10 CFU/m^2/hr. Flow rate over surface of gypsum wallboard was 35 cm/s.

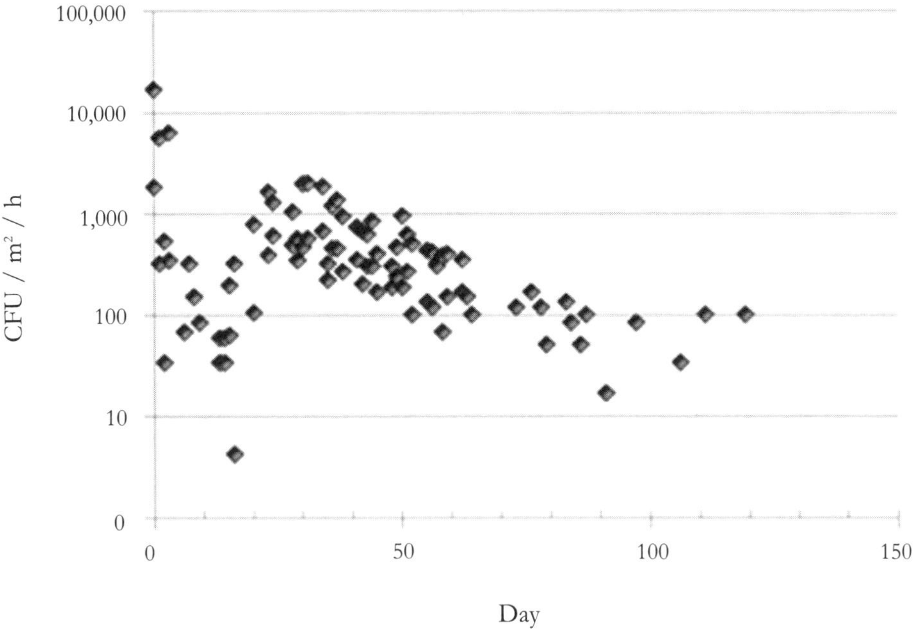

Using the minute ventilation for a child and his mother at rest from Table 1, the child would inhale ~ 3 x 10^4 spores and his mother would inhale ~ 7 x 10^4 spores in 1 minute in the bedroom with the small contaminated area.

Table 1. Sample respiratory measures for hypothetical child and adult pulled from multiple medical textbooks

Respiratory measures	Child	Adult	
	2 yrs old; 30 lbs (13.6 kg)	At rest	exercising
Tidal volume	0.05 - 0.11 L (4 - 8 mL/kg)	0.5 L	1.8 L
Respiratory rate	23-35 breaths/min	12- 20 breaths/min	30 breaths/min
Minute ventilation	1.2 - 3.9 L/min	7.5 L/min	50 L/min

Levels of mycotoxin produced by *S. chartarum* spores grown on different substrates

The results of the cell-free luminescence protein translation inhibition assay showed high levels of suppression when the fungi were grown on gypsum wallboard but minimal suppression when grown on traditional fungal media.

Figure 2: Mycotoxin measurement of 10^5 spores grown on either laboratory media or gypsum wallboard. Results are expressed as T2 equivalents.

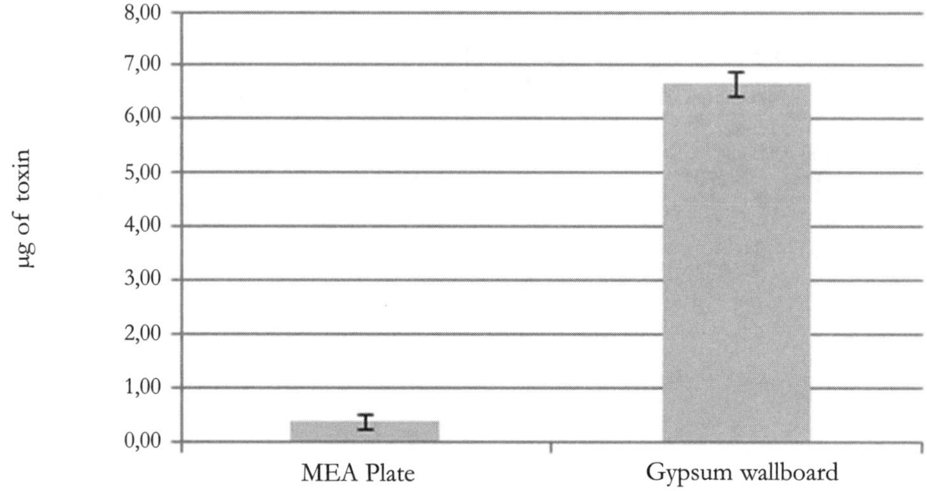

Mouse macrophage cells exposed to *S. chartarum* spores

The cell based assays to assess cell culture-defining biomarkers of adverse health effects showed elevated levels of inflammation as well as cell death by cytoxicity (Figures 3 -6).

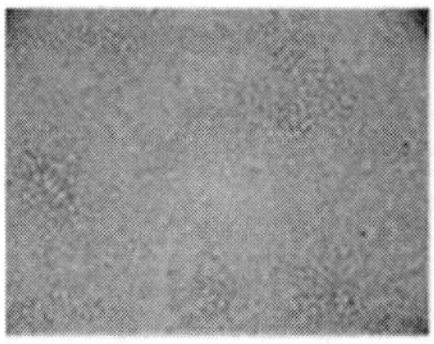

Figure 3. Raw 264.7 (mouse macrophage cells) only. No exposue to fungal spores.

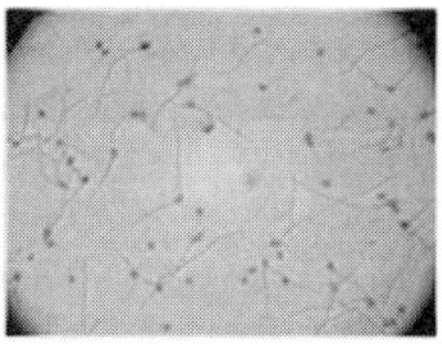

Figure 4. *Stachybotrys* spores only showing germination in tissue culture media.

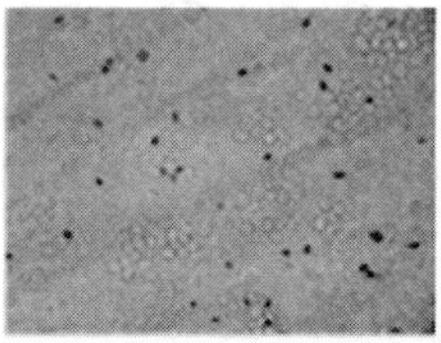

Figure 5. *Stachybotrys* harvested from MEA plates with mouse macrophage cells. No death of mouse macrophage and no germination of spores.

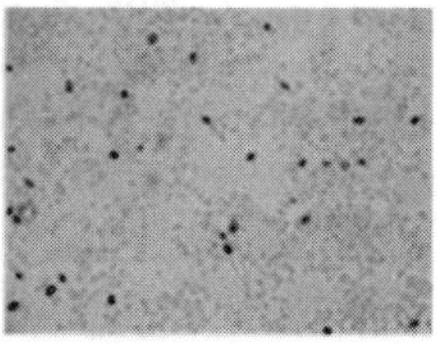

Figure 6. Mouse macrophage exposed to *Stachybotrys* harvested from gypsum wallboard. Note the cell death and some germination of fungal spores.

CONCLUSIONS

The estimation of potential human exposure to mold requires the collection of environmental samples. Understanding how a sample relates to exposure and subsequent dose is critical. Our results provide data on the relationship between surface contamination and airborne exposure. In addition to the fungal spore exposure, both mycotoxin levels associated with the spores as well as the biologic effect of the spores using cell culture have been quantified. These results begin to relate exposure to dose and help to more fully elucidate the health effects associated with mold exposure.

ACKNOWLEDGEMENTS

The authors would like to acknowledge the excellent technical assistance provided by Amy Evans, Jordan Espenshade, Lauren Harvey, Michael Herman, and Tricia Schwartz.

The materials toxicity and biologic effect research is preliminary data generated under an on-going program, "INDOOR BIOLOGICAL CONTAMINANT RESEARCH" collaboration between EPA and RTI. Parts of the emissions data and mycotoxin assay method are from earlier studies referenced below.

REFERENCES

- Black J, Foarde K and Menetrez M. Solvent comparison in the isolation, solubilization, and toxicity of *Stachybotrys chartarum* spore trichothecene mycotoxins in an established in vitro luminescence protein translation inhibition assay. *Journal of Microbiological Methods* 2006:6(2):354-361.
- Dean TR, Black JA, Foarde K and Menetrez M. Analysis of fungal spore mycotoxin and the relationship between spore surface area and mycotoxin content utilizing a protein translation inhibition assay. *The Open Mycology Journal,* 2008:2(6):55-60.
- Foarde KK and Menetrez MY. Factors relating to the release of *Stachybotrys chartarum* spores from contaminated sources. *Indoor Air* 2002:2:724-729.
- Menetrez MY and Foarde KK. Emission exposure model for the transport of toxic mold. *Indoor Built Environment,* 2003:13(1):75-82(8).

DOES REVERSIBILITY OF NEUROBEHAVIORAL DYSFUNCTION BY MONOSODIUM LUMINOL HAVE DIAGNOSTIC AND POSSIBLE THERAPEUTIC USE IN MOLD/MYCOTOXIN EXPOSED PATIENTS?

Kaye H. Kilburn

ABSTRACT

Five points led me to suggest to patients to try anti-inflammatory redox agents to intervene, to treat neurological inflammation from chemicals including those from molds and mycotoxins. First, functions of the human brain can be measured. Second, the number of total abnormalities can yield one number to quantify losses and recoveries. Third, neurologic inflammation-oxidation coincided with irritation of skin, nose, upper airway and lung. Fourth, the olfactory nerve delivers chemicals to the temporal lobes such as monosodium luminol and glutathione, just as it does pitressin and insulin. Fifth, the human brain responds in the same time frame as do animal and cell models.

Methods, as a trial of efficacy the first 10 patients received glutathione intranasally (IN) at 100 mg/ml, monosodium luminol (MSL) at 10 mg/ml or both at varied doses and frequencies and effects were measured at 24, 48, and 72 hours. There were no adverse effects. The second series of 35 patients received 10 times more IN MSL and improved function at 1.5 to 3.0 hours.

Results, Group 1: Seven mold/mycotoxin exposed patients and 3 chemically exposed received initial low dose luminol (0.1 to 0.2 mg twice a day) mean age 42.4, mean educational levels mean 15.5 years, had an abnormalities mean of 9.1, which decreased to 4.3 ($p < 0.01$) after 48 hours. Group 2: The 35 high dose luminol patients were mean age 44.2 years, educational level mean 14 years, and abnormalities averaged 8.4. After receiving 3.5 to 5 mg intranasally (IN), abnormalities decreased to 5.1 ($p < 0.001$) maintained for 24 hours.

Discussion, Balance improved to normal in most patients. In addition, verbal recall, problem solving, and long term memory usually improved. Subjectively, the 35 patients observed improved mental clarity, decreased headache, decreased tremors, decreased leg pain, improved steadiness, and improved recall.

Conclusions, There should be further exploration of MSL as a diagnostic test to determine the reversibility of neurobehavioral impairment from mold/mycotoxins and chemical exposures, and extended therapeutic trials should be considered.

INTRODUCTION

To augment advice to patients to avoid offending chemicals like chlorine, hydrogen sulfide and mold/mycotoxins, I considered evidence that argued to intervene. Five Points led me to suggest to patients to try anti-inflammatory redox agents to treat neurological inflammation from chemicals including those from molds and mycotoxins. (Holroyd et al., 1993; Ankarcrona et al., 1995; Doi and Uetsuka, 2011; Jiang et al., 2006) First, I had updated and developed physiological and adapted psychological measurements. (Kilburn, Thornton, Hanscom, 1998; Kilburn, 1998) Functions of the human brain can be measured: balance, reaction time, color discrimination, blink reflex latency, vibration, visual fields, grip strength and hearing that complement the psychological tests of cognitive functions, problem solving, vocabulary, verbal recall, perceptual motor speed, and long term memory. Second, the number of total abnormalities quantifies losses and recoveries. (Kilburn, 1999) Third, neurologic inflammation-oxidation coincided with irritation of skin, nose, upper airway and lung. (Meggs, 2009) Fourth, blood brain barrier does not exist for the olfactory nerve (Brenneman et al., 2000), so it delivers chemicals to the brain's temporal lobe, such as monosodium luminol and glutathione, just as it does pitressin and insulin. (Wang and Cynader, 2000; Scofield et al., 2009; Holroyd et al., 1993; Frey, 2002; Izzotti et al., 2009; Grady, 2003) Fifth, the human brain responds in the same time frame as do animal and cell models.

Glutathione is a tripeptide containing a sulfhydral group in cysteine is made in mammalian liver including that of man. It is a well known detoxifying agent for environment toxins in food in the gut. It was given by aerosol to protect lungs of patient receiving proteases to treat AIDS (Holroyd et al., 1993). In the brain, glutathione protects mitochondria against glutamate induced necrosis and apoptosis (Ankarcrona et al, 1995). Intranasal (IN) glutathione was suggested to me by Dr. Gordon Baker (personal communication) who found it improved fatigue and mental fogginess in patients with severe chemical intolerance. The rotten egg odor of glutathione nauseated some patients. I found IN monosodium luminol (MSL) improved neurobehavioral functions, particularly balance better than did glutathione. MSL alone equaled the effect to a mixture of both agents at doses of MSL at 10 mg/ml and glutathione at 100 mg/ml. So I simplified table 2 to show only the MS luminol effects.

What is Luminol chemically? $C_8H_6N_3O_2$ – 3 amino – 2,3 – dehydro – 1,4 – phthalazinedrone, molecular weight 177.16. Oxidation causes emission of light.

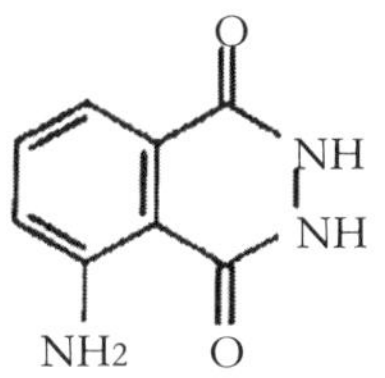

As sodium salt, monosodium luminol MW 199 $C_8H_5N_3D_2Na$.

What does luminol do in neurobiology?

1. It scavenges free radicals (ROS and RNS) and converts their energy into light.
2. The position of the amine group on the luminol phenolic ring enables luminol to enter the cell and act in the cytoplasm as well as the cell nucleus.
3. It is a redox buffering agent that modulates intracellular redox status.
4. Has antioxidant and anti-inflammatory effects at low doses, and has pro-oxidant effects at higher doses.
5. This process is reversible, and therefore the luminol molecule can be recycled.
6. Luminol is an iron, copper and cyanide chelator.
7. It is a proteolysis regulator.
8. It has proved non-toxic at established therapeutic levels.

Treatment with Redox Agents is Possible.

1. Luminol – Galavit
2. Glutathione
3. Vitamin C
4. Rebamipide
5. Phenylbutyric Acid
6. Heart failure, Isosorbital dinitrate and hydralazine (Taylor et al., NEJM 2004)

What is the rationale for the 2 hour luminol test? The demonstration of improvements in human neurobehavioral impairment in 48 hours seven years ago using nasal sprays of glutathione and of luminol prompted this project. The low to absent toxicity of luminol including tolerance of injected doses and short, 2 hour latent period made this therapeutic tests feasible. The exploration in patients with different responses to many different chemicals suggested the secondary diagnostic use to explore oxidation as the primary mechanism of poisoning by chemicals.

METHODS

Two groups of patients were studied one in 2004 and the second in 2009-2011. Performance testing is the key to profile the many brain functions that are not shown by imaging with CT scans, magnetic resonance images (MRI), position emission images (PET), or radioactive tracers for blood flow (SPECT) (Gaitonde et al., 1987; Simon and Rea, 2003). Similarly electrical impulses from brain surfaces, electroencephalography (EEG) shows convulsion, seizures and some mass lesions, but it does not reveal losses of performance. (Crapo et al., 2003) The methods which have evolved have been published previously, along with prediction equations (Kilburn, 1998; Kilburn, Thornton, Hanscom et al., 1998) will be reviewed briefly. .

Neurophysiological tests.

Simple reaction time (SRT) and a visual two choice reaction time (CRT) were measured from the appearance on the computer screen of a 10 cm block A to its cancellation by tapping a keypad A for simple and A or S for choice with a computerized instrument (Miller et al., 1989). The lowest median score of the last 7 in each of the two trials of 20 was accepted for SRT and for CRT. Body balance was measured with the subject standing erect with feet together. The position of the head was tracked by a sound-based position detector from a sound generating stylus on a head-band, processed in a computer and expressed as mean speed of sway in cm/sec (Kilburn and Warshaw, 1994). The minimal sway speed of three consecutive 20-second trials was the value used for sway with the eyes open and sway with the eyes closed.

The blink reflex was measured with surface electromyographic electrodes (EMG) from lateral orbicularis oculi muscles bilaterally (Shahani and Young, 1972, Kilburn et al., 1998a) after tapping the right and left supraorbital notches with a light hammer, which triggered a recording computer. Ten firings of R-1 were averaged to find the mean response for each side and failures to respond were recorded (Kilburn et al., 1998b). Color confusion index was measured with the desaturated Lanthony 15 hue test under constant 1000 Lux illumination (Lanthony, 1978) and errors scored by the method of Bowman (1982). Hearing was measured in left and right ears with standard audiometers (model ML-AM Microaudiometrics, So. Daytona, Florida, USA) at stepped frequencies of 500 to 8,000 Hertz. The sum of deficits for each ear was the hearing score.

Threshold testing of visual field performance used a computerized automated perimeter which mapped the central 30degrees of right and left eyes individually (Neuro-Test, Inc. Pasadena, California, USA). The visual field score for each eye was the sum of abnormal quadrants with two or more points detected at 2 or more steps of illuminance greater than others in that a circle (Anderson, 1987).

Neuropsychological tests.

Immediate memory or recall was measured with two stories from Wechsler's Memory Scale–Revised (1987). Culture Fair tested non-verbal non-arithmetical intelligence based on the selection of designs for similarity, for difference, for completion and for pattern recognition and transfer (Cattell, 1951; Cattell et al., 1941). Culture Fair resembles Raven's progressive matrices (Raven, 1988). The 46-word vocabulary test was from the multidimensional aptitude battery (Jackson, 1985). Digit symbol substitution from the Wechsler Adult Intelligence Scale–Revised (Wechsler, 1987) tested attention and integrative capacity. Information, picture completion and similarities were also from the WAIS-R, tested long-term, (embedded) memory. Time needed to place 25 pegs in the Lafayette slotted pegboard was measured as were times to complete Trail making A and B. These tests from the Halstead-Reitan battery (Reitan, 1958; Reitan, 1966) measured dexterity, coordination and decision making. Peripheral sensation perception was measured with fingertip number writing errors. Subjects' moods were appraised by responses to 65 terms describing emotional status for the week using the Profile of Mood States ((McNair, Lorr et al., POMS; 1981). Recall of the Rey 15 forms tested whether recall was appropriate or suggested malingering (Rey, 1964).

As a trial of efficacy the first 10 patients received glutathione intranasally (IN) at 100 mg/ml, monosodium luminol (MSL) at 10 mg/ml or both at varied doses and frequencies and effects were measured at 24, 48, and 72 hours. There were no adverse effects. Because cells in culture and rodent models showed improved function at 1.5 to 3 hours, the second series of 35 patients, received 10 times more IN MSL. They showed improvement in function at 1.5 to 3.0 hours. Regression analysis searched for patterns of improvement and time sequences from different exposure agents and duration and types of symptoms. Reversibility by mold/mycotoxin exposed patients was compared to that of chemically exposed patients.

All subjects gave informed consent and the protocol was approved by the Human Studies Research Committee of the University of Southern California, Keck School of Medicine.

RESULTS

The 10 patients who received IN MS luminol 10 mg/ml had exposures to mold/mycotoxin for 7, H_2S (2), and organophosphates 1, Table 1. They were 19 to 61 years old with education levels 11 to 20, mean 15.5, 5 were women and 5 men. Three had doctoral degrees. Abnormalities that were 7.5 to 19, mean 11.2, decreased by 6.6 to average 4.6, 48 hours after luminol. POMS scores were 29 to 117, averaged 81.1 and decreased to 46.4. Symptom frequency went from 3.2 to 10.2 with a mean of 6.0 down to 3.9, Table 2. Nine had abnormal sway eyes closed and 6 became normal. Of the 5 who did not improve, one had grand mal seizures, one chorioathetosis from mold/mycotoxins, the third was H_2S exposed from turkey manure in a tractor cab, and he could not stand without support, and two had long, 7 years exposure and one had neurosurgery twice to remove a focal epileptic lesion that contained mold DNA.

The 35 patients who received IN MS luminol 100 mg/ml and oral doses are summarized in Table 1. They were more than a decade older, a year less well educated, similar POMS scores and lower symptoms frequencies. Their average number of total abnormalities, 11.9 vs. 11.2. They presented in 4 groups in Table 3 by chemical exposure: first group, two to ethylene glycol, 3 to chorine and 3 to insecticides, malathione, diazanon, and pyrethroids; second group of 11 to mold/mycotoxins; third group of 11 to H_2S and reduced sulfur gases; and a fourth group of 5, each to a single major chemical. Vinyl chloride monomer, mercury, ingested arsenic, toluene with diisocyanates, and formaldehyde.

The man 1, and women 2, exposed to ethylene glycol had many abnormalities and no improvement from ML luminol. The first 3 chlorine exposed women, 3 to 5 all improved on MSL and the least abnormal one (5) dropped her abnormality score to 1. The malathione exposed man (14 abnormalities), and diazanon women (4 abnormalities) had 90 minute responses. She improved to normal. He had substantial improvement including loss of stutter and ability to draw pictures and walk a straight path. The pyrethroid exposed man was least improved and like others exposed to these chemicals indoors with much impairment , his many abnormalities seemed permanent.

The 11 patients exposed to mold/mycotoxins, 7 women, 1 child, and 3 men had 0 to 17 abnormalities, mean 9.2. Mean age was 49.9 EL 14.9, excluding the 7 year old. MSL decreased abnormalities to 2.5 or below in 5, so they became normal. One had no abnormalities, but was extremely intolerant to common chemicals. Five had other co-temporal exposures to formaldehyde for 1, hydrogen sulfide in 2 and 2 had years in chemical laboratories. Those who improved, but did not become normal used MSL as a crutch to help avoid effects of symptom triggering environments or agents. Five had POMS scores above 80 and 4 had symptom frequencies above 6 without correlations with POMS score or with numbers of abnormalities. Eleven patients exposed to mold/mycotoxins improved to be closer to predicted performance when compared to 11 patients exposed to hydrogen sulfide and reduced sulfur gases.

The 11 H_2S exposed, 3 women and 8 men, mean age 55.1, had 3 to 10.5 abnormalities, mean 11.8, and decreased to 5.5 with 5 at 3 or less. POMS scores were above 60 in 7, mean 74.4. Symptom frequencies ranges from 2 to 10.2 mean 4.7. That more men had been exposed to H_2S and more women to mold/mycotoxins was relatable to more workplace exposure to H_2S, similarly mold exposed people had to 2 more years of education, than the 13 years in the H_2S exposed.

A woman, age 35, and a man, age 57, exposed by inhalation to fumes or vapor of ethylene glycol, with 17.5 and 19 abnormalities showed least reversal by MSL. The 3 women exposed to chlorine, one time, one day, ran the gamut – the least impaired one who improved to normal was exposed from dental work. The youngest one who was exposed by a shower and improved almost to normal and maintained on oral MSL. The oldest one, a nurse chronically exposed to H_2S and episodically to chlorine that was freed from hypochlorite by phosphoric acid for bleaching at a paper mill was least improved.

The five single agent exposure patients numbers 31 to 35, averaged 51 years of age and 14.4 years of education had 13.4 abnormalities, lower POMS scores and only one, the formaldehyde exposed woman (patient 35) with a symptom frequency above 4. Abnormalities averaged 4 more than the mold/mycotoxin group. Impairment from exposures to mercury (patient 32) and arsenic (patient 34) did not improve with IN MS Luminol. However, the patient with abnormalities from chronic exposure to vinyl chlorine monomer, that appeared to dominate the new car smells list of 150 chemicals improved. Impairment from toluene and diisocyanates that dominated agents in DAP adhesives improved, but left patient 34 with constant headache and fatigue. Impairment in the patient who inhaled formaldehyde, the dominant neurotoxic ingredient in a hair de-frizzing agent called Brazilian Blowout, almost completely reversed.

DISCUSSION

The observations in the 35 consecutive patients illustrate the variable responses between reversibility and no effects on neurobehavioral function from redox intervention with MSL (Table 3). They begin to align with factors such as usage, total central nervous system failure, come at one extreme and convulsions on the other, nature of chemical interactions and reexposure, repeated episodes of damage. Adverse effects of MSL were minimal. Many people said MSL had a metallic taste that lasted a few minutes. One patient, whose impairment was not large, stopped taking MSL because she developed tension headaches.

TThis survey of experience with the trial of the monosodium luminol taken intranasally showed that it improves and sometimes reverses neurobehavioral impairment due to many chemicals. In contrast, it had no effect on some types of damage that are notoriously difficult to treat. Time from exposure to testing and intervention varied, but was rarely quicker than 12 months. Early intervention should be a goal. Methodical follow up of more patients is essential to sort patterns of response. A clinical trial is needed to study long term outcomes. The trial may need to have patients spend a few days in residence to establish a new "lifestyle" to change habits and beliefs and answer their questions and concerns. The present evidence shows that a redox agent MSL given intranasally improved neurobehavioral functions including balance, problem solving, grip strength and color discrimination in patients exposed to mold/mycotoxins and some patients exposed to other chemicals. After 30 days of treatment, stopping these agents returned several patients to pretreatment levels in 2 days. However, 3 patients maintained improvement after 3 months or more. Redox treatment with intranasal agents deserves further evaluation to determine the degree of improvement and whether after weeks or months of treatment, improved function persists after treatment is stopped.

It appeared that absence of balance impairment or large improvements of balance, defined by decreased speed of sway while standing with feet touching, within 3 hours of MSL may select those people who improve. One man exposed to malathione may times quit stuttering 105 minutes after starting MSL and felt mental fog clear. His ability to draw lines making corners for site maps and building plans returned.

Two patients with exposures to ethylene glycol, heated mist, had no improvement. An arsenic exposed man and one exposed to mercury did not improve or recover function with MSL. One 68 year old woman who was neighbor to a paper mill and exposed for 30 years of episodic chlorine added to H_2S generated in waste ponds showed modest improvement.

It is unclear when maximum benefit of MSL occurs, but my impression may take one week to four weeks. Several patients tried taking MSL before they had chemical exposures, thus prophylactically, which seemed to increase their tolerance to chemical triggering of symptoms and impairment.

Although the neurobehavioral test battery provides direction and profiles of the pattern of losses and the effects of MSL intervention, subtle changes in memory function reported by patients cannot be measured. More subtle tests are time consuming and need greater tester skill, but should be investigated to sharpen these appraisals. A clue about recall memory has been the spontaneous observation that reading a novel was possible again because characters and plot was retained in recent memory to "pick up the story" without starting over.

Cautions abound starting with the question: Do the long term benefits parallel the 90 to 180 minute effects in patients in each group? Secondly, do the neuronal switches get held closed by redox treatment (Kilburn et al., 2009) so the brain improves function by revitalizing energy electrons in high energy phosphates ADP to ATP, or in cytochrome enzymes that became oxidized and need reduced in neurons or in glia? (Roth et al., 1995; Ankarcrona et al., 1945; Izzotti et al., 2008) Third, are there replacements of cells, from stem cell induction or guidance of migration to substitute for dead or poorly functioning cells? (Jiang et al., 2006) How do these fit in special situations such a Purkinje cells in the cerebellum, retinal dendrictic cells (Izzotti et al., 2009) or nasal receptor cells in the olfactory organ or temporal lobe? (Kilburn, 1998; Frey, 2002; Brennerman et al., 2000)

Society's acceptance of chemical brain injury has not increased in the past 15 years in the United States. Neither have legal actions for disability been more successful. Medical, health insurance companies frequently deny claims for diagnostic function testing although they pay for brain scans and EEGs. When brain studies by CT scans, MRIs, PET and SPECT scans, and EEGs show no abnormalities, neurologists and other physicians conclude these patients have a somatization disorder or malingering, made psychiatric referrals. Triple SPECT using Te-99 show different blood flow/metabolic patterns that are side to side mismatches, temporal asymmetry and soft tissue shunting, but without changes during intervention so they were not specific or ineffectual (Simon and Rea).

Criminal actions against responsible parties, chemical polluters – individuals have been successful in California and Texas, but have not caused bankruptcy of major companies so have not changed companies' chemical handling behavior. Being forced into bankruptcy stopped manufacturing and use of asbestos in the United States before it was banned.

Chemical brain damaged patients with impairment are "cure shopping" and are entangled in fish nets of treatment supplements, detoxification, metal chelation, purging, dispuration, and psychotherapy. Some cannot decide to accept MSL even after the 3 hour post intervention improvements are shown. Frequently, feeling better and having less frequent symptoms after taking a month of MSL, but are not seeing enough change to motivate adherence to the daily regime.

This behavior raises the issue of whether they are competent to make decisions about their own care. This ethical problem is frequent in end-of-life resuscitation, and expensive treatment of chronic disorders in the face of rapidly terminal heart or lung disease or cancer.

This preliminary study of efficacy supports the idea that patterns of response to MS Luminol may be a tool, to help with diagnosis and assist consideration of the pathogenesis of brain damage. Which brain functions are susceptible and which are resistant to damage? How do they associate with a chemical or class of chemicals? So far, the many variable patterns have sorted into is balanced improved or not. Are problem solving of Culture Fair and digit symbol substitution improved or not and is verbal recall improved or not? Comparison long term memory tests, i.e. information, picture completion and classifying items or similarities has with vocabulary showed some patterns. However, more response patterns need examined to determine whether these predict prognosis or should this idea be abandoned.

Brain injury – damage to the brain arouses coping skills that recall the Elisabeth Kübler-Ross model : The Model of Coping with Dying (Kübler-Ross, 1969).

1. Denial
2. Anger
3. Bargaining
4. Depression
5. Acceptance – learning how to adapt to impairments, decision to live on despite desperate impairments

Adapt:

Activities appear important, doing things "use it or lose it," "wear out, don't rust out"

Therapeutic trial - release old habits, relearn, try new behaviors

Protect again chemical triggering – list to avoid

Interfere, block chemical brain injury

Prevent effects of chemicals – prophylactic use.

Facilitate brain plasticity – repair, rebuilding

Who should protect their brains with MS luminol?

It is hoped that these observations will tantalize and encourage critics to question mindsets and try teaching themselves, the old dogs, new tricks, and that this skepticism will carry over to these patients who need help desperately.

CONCLUSIONS

A single dose of intranasal monosodium luminol improved neurobehavioral functions including balance, problem solving, grip strength and color discrimination in chemically exposed patients with brain injury. After 30 days of treatment stopping these agents in 5 patients returned function to initial, pretreatment levels in 2 days, so treatment must be maintained. In several patients several months of MSL restored ability to hold a job. Further studies are needed to determine if improved function in 2 hours predicts long term results. It seems a useful tool.

Table 1 – Study Groups Summarized as Means

	2004 -2005	2010-2011
Number	10	35
Age	43	56.2
Education years	15.4	14
POMS score	81.1	75.5
Symptom Frequency	6	5.3
Abnormalities baseline	11.2	11.9
Abnormalities after MSL	6.6	3.6

Table 2 - Results on 10 Patients Treated with 10 mg/ml Luminol*

Patients			Abn		PO MS		Symp t		Ba- lance Cm / sec		
No./G ender	Expo sure	Age/ Educ	Be- fore	△	Be- fore	△	Be- fore	△	Be- fore	△	
1/M	m/m	61/20	7.5	5.0	103	77	7.7	4.3	3.7	2.2	1.5N
2/F	m/m	43/16	9.0	4.5	41	36	4.9	3.0	7.8	5.8	2.0
3 / F (1)	m/m	43/20	11.0	0.0	75	10	7.1	1.2	7.2	2.4	4.8
4/M	m/m	19/12	9.5	3.5	101	0	7.9	1.4	2.2	0.8	1.4N
5 / F (2)	m/m	48/16	13.0	4.0	69	5	5.2	0.0	1.3	0.5	0.8N
6 / F (3)	m/m	55/20	10.0	2.5	29	22	2.4	0.3	1.7	0.2	1.5N
7/M	m/m	28/16	8.0	4.0	117	50	7.5	2.7	4.0	2.0	2.0
8/F	OPS	40/12	14.0	11.5	109	25	8.3	4.2	1.5	0.4	1.1N
9/M	H_2S	43/11	11.5	10.5	56	60	4.6	3.4	5.4	4.5	0.9N
10 / M (4)	H_2S	50/12	19.0	1.0	111	62	4.3	1.5	21.0	9.0	12
Mean		43 / 15.4	11.2	4.6	81.1	46.4	6.0	3.9	3.9	-1.9	6N

*7 m/m exposed, 2 H_2S exposed, 1 OP

1. Seizures, some interactable to outpatient medications
2. Long exposure
3. Movement disorder – chorioathetosis
4. Movement disorder – could not walk

Table 3

No. / Gender	Age / Educ Ht/Wt	Exposure	Abn Base line	Abn post Rx	Δ	POMS	SF	Balance
Glycol								
1./M	51/16 ht/wt: 70/179	Glycol	19.0	18.0	-1.0	60	2.7	
2./F	37/18 ht/wt: 63/155	Glycol	17.5	16.0	1.5	36	3.9	
Chlorine								
3./F	41/20 ht/wt: 66/146	Chlorine	12.5	3.0	-9.5	35	36.0	
4./F	68/18 ht/wt: 63/118	Chlorine	13.5	9.5	-4.0	65/55	6.0	6.97/2.27
5./F	55/20 ht/wt: 61/120	Chlorine	5.5	1.0	-4.5	62	6.0	1.96/1.28
OP's/Pyrethroids/Chemicals								
6./M	47/16 ht/wt: 71/260	Malathione	14.0	7.0	7.0	60	8.1	
7./F	52/18 ht/wt: 66/294	OPS/ Diazanon	4.0	2 N	2.0	78	5.1	
8./M	37/13 ht/wt: 74/172	Pyrethroids	12.5	7.0	5.5	118	4.7	
Mean Group 1 (tiny groups, see results)	?	?	?	?	?	?	?	?

No. / Gender	Age / Educ Ht / Wt	Exposure	Abn Base line	Abn post Rx	Δ	POMS	SF	Balance
Hydrogen Sulfide (H_2S)								
20./M	55/12 ht/wt: 73/360	H_2S	19.5	14.0	5.5	133	5.6	
21./M	56/14 ht/wt: 69/211	H_2S / bunker fuel	16.5	5.0	16.0	99	4.0	
22./M	49/14 ht/wt: 70/200	H_2S	16.0	9.0	7.0	151	6.5	
23./F	58/12 ht/wt: 63/166	H_2S / dye	14 (20)	14.0	0	5	4.7	
24./M	71/14 ht/wt: 72/232	H_2S	14.0	3.0	11.0	34	3.9	
25./M	49/14 ht/wt: 70/236	H_2S	12.5	7.0	5.5	91	5.3	
26./M	42/10 ht/wt: 74/234	H_2S	11.5	7	4.5	80	4.7	
27./F	56/12 ht/wt: 63/280	H_2S	10.5	6.5	4.0	42	3.6	
28./F	60/14 ht/wt: 63/200	Chems / asphalt roof/H_2S	8.0	1 N	7.0	78	5.4	
29./M	56/14 ht/wt: 68/214	H_2S	4.5	0 N	4.5	94	3.6	
30./M	54/12 ht/wt: 68/214	H_2S, Cl	3 .0	0 N	3.0	11	4.8	
Mean Group 3	55.1 / 12.9		11.8	5.5	6.2	74.4	4.7	

No. / Gender	Age / Educ Ht/Wt	Exposure	Abn Baseline	Abn post Rx	Δ	POMS	SF	Balance
Vinyl Chloride								
31./F	60/16 ht/wt: 59/170	Vinyl chloride	7.0	3.0	-3.5	2	3.9	
Mercury								
32./M	45/14 ht/wt: 71/199	mercury	17	18	1	37	3.9	
Diisocyanates/Urethane, Toluene								
33./M	50/13 ht/wt: 66/166	Diisocyanates/urethane, toluene	11.5	4.5	7	52	3.2	
Arsenic								
34./M	51/15 ht/wt: 71/246	Arsenic	15.5	14.5	1	78	3.6	
Formaldehyde – Brazilian Blowout Hair Treatment								
35./F	47/16 ht/wt: 66/109	HCHO	16.0	3.0	13	64	7.1	
Mean Group 4	51.0 / 14.4		13.4	8.6	5.0	46.0	4.3	
ALL GROUPS MEANS:	52.2 / 14.0		11.9	5.1	5.4	75.5	5.3	

ACKNOWLEDGEMENTS

Funding was provided by participants. Monosodium Luminol (MSL) was provided by Bach Pharma Inc., 800 Turnpike Street, Suite 200, North Andover, MA 01845, email: info@bachpharma.com

REFERENCES

- Anderson DR. Perimetry: with and without automation. . 1987. 2nd ed. St Louis, MO: C.V. Mosby Co.
- Ankarcrona M, Dypbukt JM, Bonfoco E, et al. Glutamate-induced neuronal death: a succession of necrosis or apoptosis depending on mitochondrial function. *Neuron*, 1995:15:961-973.
- Bowman BJ. A method for quantitative scoring of the Farnsworth panel D-15. *Acta Ophthalmologica*, 1982:60:907-916.
- Brenneman KA, James RA, Gross EA, and Dorman DC. Olfactory neuron loss in adult male CD rats following subchronic inhalation exposure to hydrogen sulfide. *Toxicological Pathology*, 2000:28: 326-303.
- Cattell RB. Classical and standard score IQ standardization of the IPAT: culture free intelligence scale 2. *Journal of Consulting Psychology*, 1951:15: 154-159.
- Cattell RB, Feingold SN, and Sarason SB. A culture free intelligence test II evaluation of cultural influences on test performance. *Journal of Educational Psychology,* 1941:32: 81-100.
- Crago BR, Gray MR, Nelson LA, Davis M, Arnold L, Thrasher JD. Psychological, neuropsychological, and electrocortical effects of mixed mold exposure. *Arch Environ Health*; 2003:58:452-463.
- Doi K, Uetsuka K. Mechanism of mycotoxin-induced neurotoxicity through oxidative stress associated pathways. *Int J Mol Sci*, 2011:12:5213-5237.
- Frey WH, II. Bypassing the blood-brain barrier to deliver therapeutic agents to the brain and spinal cord. *Drug Del Tech*, 2002:5:46-49.
- Gaitonde UB, Sellar RJ, and O'Hare AE. Long term exposure to hydrogen sulfide producing subacute encephalopathy in a child. *British Medical Journal,* 1987:294: 614.
- Grady D. Study tests safety of a hormone nasal spray to curb hunger. *New York Times* Nov. 2003.
- Holroyd KJ, Buhl R, Borok Z, et al. Correction of glutathione deficiency in the lower respiratory tract of HIV seropositive individuals by glutathione aerosol treatment. *Thorax*, 1993:48:985-989.
- Izzotti A, Saccà SC, Longobardi M, Cartiglia C. Sensitivity of ocular anterior chamber tissues to oxidative damage and its relevance to the pathogenesis of glaucoma, *Invest Ophthalmol Vis Sci.* 2009:50:5251-5258. Epub 2009 Jun 10.
- Jackson DN. Multidimensional aptitude battery. Sigma Assessments Systems, Inc., 1985. Port Huron, MI.
- Jiang Y, Scofield VL, Yan M., et al. Retrovirus-induced oxidative stress with neuroimmunodegeneration is suppressed by antioxidant treatment with a refined monosodium alphaluminol (Galavit). *J Virol*, 2006:80:4557-4569.
- Kilburn KH. Exposure to reduced sulfur gases impairs neurobehavioral function. *Southern Medical Journal*, 1997:90:997-1006.
- Kilburn KH.. Chemical Brain Injury. New York: John Wiley. 1998.

- Kilburn KH. Evaluating health effects from exposure to hydrogen sulfide: central nervous system dysfunction. *Environmental Epidemiology and Toxicology*, 1999:1: 207-216.
- Kilburn KH, Warshaw RH. Balance measured by head (and trunk) tracking and a force platform in chemically (PCB and TCE) exposed and referent subjects.. 1994.
- Kilburn KH, Thornton JC, and Hanscom B. A field method for blink reflex latency R-1 (BRL R-1) and prediction equations for adults and children. *Electromyography and Clinical Neurophysiology,* 1998:38:25-31.
- Kilburn KH, Thornton JC, and Hanscom B. Population based prediction equations for neurobehavioral tests. *Archives of Environmental Health,* 1998:53: 257-263.
- Kilburn KH, Thrasher JD, Immers N. Do Terbutaline and mold associated impairments of the brain and lung relate to autism? *Toxicol Ind Health*, 2009:25:703-710.
- Kübler-Ross E. On Death and Dying. 1969. Simon and Schuster, New York.
- Lanthony P. The desaturated panel D-15. *Documents Ophthalmology*, 1978:46: 185-189.
- McNair DM, Lorr M, Doppleman LF, Profile of Mood States (1971/1981). San Diego, CA: Educational and Industrial Testing Service.
- Meggs WJ. Neurogenic Inflammation and Sensitivity to Environmental Chemicals. *Environ Health Perspect,* 1993:101: 234-238
- Meggs WJ. Epidemics of mold poisoning past and present. *Toxicol Ind Health*, 2009:25:577-582.
- Miller JA, Cohen GS, Warshaw RH, Thornton JC, and Kilburn KH. Choice (CRT) and simple reaction times (SRT) compared in laboratory technicians: factors influencing reaction times and a predictive model. *American Journal of Industrial Medicine*, 1989:15:687-697.
- Raven JC, Court JH, and Raven J. Standard Progressive Matrices. 1988. London, Great Britain: Oxford Psychologists Press.
- Reitan RM. Validity of the trail-making test as an indicator of organic brain damage. *Perceptual Motor Skills*, 1958:8: 271-276.
- Reitan RM. A research program on the psychological effects of brain lesions in human beings. In: NR Ellis (eds.) *International Review of Research* in Mental Retardation. 1966. New York: Academic Press, pp.153-216.
- Rey A. L'examen clinique en psychologle. 1964. Paris: Presses Universitaires de France.
- Roth SH, Skrajny B, and Reiffenstein RJ. Alteration of the morphology and neurochemistry of the developing mammalian nervous system by hydrogen sulfide. *Clinical Experimental Pharmacology and Physiology,* 1995:22: 379-80.
- Scofield VL, Yan M, Kuang X, Kim SJ, and Wong PK. The drug monosodium luminol (GVT) preserves crypt-villus epithelial organization and allows survival of intestinal T cells in mice infected with the ts1 retrovirus. *Immunol Lett*, 2009: 150-158.
- Shahani BT, Young RR. Human orbicularis oculi reflexes. *Neurology,* 1972:22: 149-154.
- Simon TR, Rea WJ. Use of brain imaging in the evaluation of exposure to mycotoxins and toxins encountered in Desert Storm/Desert Shield. *Arch Environ Health*, 2003:58:406-409.

- Taylor AL, Zeische S, Yancy C, et al. Combination of isosorbide dinitrate and hydralazine in blacks with heart failure. *N Engl J Med*; 2004:351:2049-2057.
- Wang XF, Cynader MS. Astrocytes provide cysteine to neurons by releasing glutathione. *J Neurochem,* 2000:74:1434-1442.
- Wechsler D. Adult Intelligence Scale Manual–Revised. The Psychological Corporation, 1981. New York.
- Wechsler D. A standardized memory scale for clinical use. *Journal of Psychology*, 1987:9: 87-95 (WMS-revised 1987).

Editors note:

The publication and the contents of this work of Dr. Kilburn are intended to advance general scientific research, understanding, and for discussion purpose only and are not intended and should not be relied upon as recommending or promoting a specific method, diagnosis, or treatment by physicians for any particular patient or professional expert recommendation. Any reader or patient should consult with their physician(s) regarding any information contained in this report.

PART II:

ASSESSMENT & REMEDIATION

ASSOCIATIONS BETWEEN VENTILATION AND MYCOLOGICAL PARAMETERS IN HOMES OF CHILDREN WITH RESPIRATORY PROBLEMS

Hans Schleibinger, Daniel Aubin, Doyun Won, Wenping Yang, Denis Gauvin, Pierre Lajoie

KEYWORDS mold, ventilation rate, field study, asthma, intervention

SUMMARY

The background of this contribution is the research into a healthy ventilation rate, especially for vulnerable parts of the population. This paper describes the results obtained during a randomized intervention study investigating the impact of ventilation rates on a wide range of indoor air quality (IAQ) parameters and the respiratory health of asthmatic children in Québec City, Qc, Canada. Following the ventilation intervention there was a marked decrease in the concentrations of a number of IAQ relevant parameters including mold spores demonstrating that ventilation interventions are effective at improving IAQ.

OBJECTIVES OF THE STUDY

The overall objective of this study was to determine whether increased ventilation will lead to a corresponding decrease in the number of respiratory symptoms in children, to correlate ventilation rates with IAQ parameters, and to address questions related to health based ventilation rates. Secondly, we wanted to determine if the recruited homes are under-ventilated and if contaminants, especially those triggering asthma-type symptoms, are prevalent in excessive concentrations. Furthermore we aimed at demonstrating, that mitigation by increased ventilation, specifically through heat or energy recovery ventilators (HRV or ERV), is possible and reliable.

MATERIALS AND METHODS

The first phase of the field study involved three residential home visits (two during the heating season (October to April) and one in summer), when a number of IAQ relevant chemical, biological and physical parameters were measured over a 6 to 8 day period. A series of questionnaires capturing information related to housing characteristics and occupant behaviour were also administered during the residential home visits. In the second phase, any home with one air exchange rate measurement below 0.25 h^{-1} or two in between 0.25 h^{-1} and 0.30 h^{-1} was considered eligible for the intervention. Of the 115 participants involved in the study, 83 were eligible for the intervention and these were randomly divided into two groups. One group of participants (n=43), had their ventilation rates increased by a combination of: installation of an HRV or ERV; modification of the existing ventilation system; and/or modification of occupant behaviour relating to the use and maintenance of the mechanical ventilation system in their home. The other group (n=40), did not have any modifications done to the ventilation rate and served as a control group. The monitoring in the second phase after the intervention was identical to the first phase and was used to assess the effectiveness of the intervention at improving the indoor air quality and respiratory health.

The average air exchange rates (AER) measured in the child's bedroom and living room over a 6 to 8 day period was determined using the perfluorocarbon tracer (PFT) technique developed by the Tracer Technology Group at Brookhaven National Laboratory (Dietz and Cote, 1982). The AER was also measured over a 4-5 hour period in the child's bedroom using SF_6 tracer gas decay according to the ASTM test method E 741-00 (ASTM, 2006) using an Innova 1312 or 1412 photo-accoustic field gas monitor. The building air tightness was also measured and the experimental details are available in Aubin et al. (2011). Flow rates for any pre-existing and subsequently added mechanical ventilation systems were measured at the inlet, outlet and registers. Particulate matter, formaldehyde, O_3, NO_2 and VOC's were measured over a 6-8 day period in the child's bedroom and the details for these experimental procedures are contained in Aubin et al. (2011). The relative humidity (RH) and temperature were measured with an Onset HOBO U12-013 data logger and the CO_2 concentration was measured with a Vaisala GMW21 CO_2 sensor.

Airborne mold spores were actively sampled using a Staplex model MAS-2 two-stage microbial air sampler for 5 minutes during each of the visits in the heating season in the child's bedroom, living room and/or playroom. The spores were collected onto both dichloran 18% glycerol (DG-18) and malt extract agar (MEA) petri dishes. In addition settling mold spores were collected for 4 to 5 hours in the child's bedroom onto both MEA and DG-18. Both the airborne and settling mold spore samples were incubated at 25 °C for 7 days then analyzed with a microscope. Dust samples were also taken for the analysis of SVOC's, endotoxins, β-1,3-D-glucan and ergosterol. The experimental methods and results for the analysis of parameters extracted from the house dust will be discussed in future publications.

Prior to conducting the individual ventilation intervention the precise description was modeled for several homes computationally and physically in NRC's Indoor Air Research Laboratory (IARL). Computational fluid dynamics (CFD) calculations were performed using Fluent and were conducted to optimize the geometry and initial placement of any supply and return air vents. The results of the computational models were then validated physically in a 1:1 scale model of the child's bedroom in the IARL. Measurements of SF_6 tracer gases, register flow rate, surface temperature, air temperature and velocity profiles throughout the model bedroom were used to validate the CFD modelling.

RESULTS

The vast majority of the homes examined during the pre-intervention phase were under-ventilated and 83% failed to reach our nominal ventilation goal of 0.30 h^{-1} during the heating-season. For this period, the geometric mean and median air exchange rate using the PFT technique were respectively 0.24 h^{-1} and 0.21 h^{-1} for the initial 114 homes involved in this study. The pre-intervention summer air exchange rates were roughly 30% higher for both the control and intervention groups, when compared to the winter values (see Table 1). Following the intervention, the PFT measured AER increased by over a factor of two in the intervention group in both the heating-season and summer.

Table 1: Ventilation measurements (air changes per hour, h^{-1}) from the child's bedroom for the intervention and control groups. Paired sample Wilcoxon signed rank test was conducted between the pre-intervention and post-intervention phase for each study group.

Parameter	Group	Pre-Intervention		Post-Intervention	
		Geometric Mean	Median	Geometric Mean	Median
Winter (PFT)	Control	0.20	0.19	0.21	0.20
	Intervention	0.17	0.17	0.36	0.36*
Winter (SF6)	Control	0.24	0.22	0.29	0.29*
	Intervention	0.26	0.27	0.42	0.42*
Summer (PFT)	Control	0.28	0.28	0.39	0.38*
	Intervention	0.23	0.25	0.56	0.54*

Note: Outliers above 1.5 h^{-1} were removed for analysis and * signifies that the difference between pre and post intervention phase is statistically significant to p=0.05

From Table 2 it can be seen that during the pre-intervention phase the highest median concentrations of airborne mold spores was measured in the child's playroom (often located in the basement), which were on average factor of 2-3 times higher than those measured in the child's bedroom and/or the living room. Following the intervention, decreases in the median concentration of airborne mold spores were observed in all three locations. The largest decrease was observed in the child's playroom.

Table 2: Median values of airborne mold spores measured for the pre- and post-intervention period at different locations of the home during the heating-season.
n: number of available measurements.

Location	Pre-Intervention (cfu m^{-3})		Post-Intervention (cfu m^{-3})	
	MEA	DG-18	MEA	DG-18
Child's Bedroom	67 (n=43)	64 (n=43)	60 (n=43)	27 (n=43)
Living Room	60 (n=28)	45 (n=28)	49 (n=42)	33 (n=42)
Playroom	110 (n=36)	174 (n=36)	49 (n=23)	51 (n=23)

Table 3 shows the spectra of the different mold genera: in the child's bedroom airborne *Aspergillus* was the most prominent genus, closely followed by *Penicillium*. In the intervention group airborne *Aspergillus* was reduced by nearly 75%, airborne *Penicillium* by 30 %. Concentrations of airborne *Cladosporium* spores were close to the detection limit, both before and after the intervention. Settling mold spores showed a similar pattern, both regarding the *Asp/Pen* ratio, as well as to the pre/post intervention behaviour. Total airborne and settling mold cfu's showed a moderate correlation (R = 0.37 to 0.76). The total airborne and settling mold spores concentrations both showed a statistically significant (p=0.05) decrease after the intervention.

Table 3: Median values for the genera of the airborne and settling mold spores in the child's bedroom during the heating season for the intervention group (DG-18).

Parameter	Genera	Pre-Intervention (n=43)	Post-Intervention (n=43)
Airborne (cfu m^{-3})	*Aspergillus*	19	5
	Penicillium	16	11
	Cladosporium	2	2
	Total mold spore	64	27
Settling (cfu h^{-1})	*Aspergillus*	0.33	0.14
	Penicillium	0.24	0.17
	Cladosporium	0.10	0.08
	Total mold spores	0.83	0.50

During the pre-intervention phase low relative humidities were observed in a number of homes during the heating-season. In many cases RH's were below the range of 30-50% recommended by Health Canada (1989) for winter. ERV's, which can transfer moisture from the exhaust air into the incoming fresh air supply stream, were installed in the homes with RH's lower than 35-40% and with other factors present, such as low heating-season indoor temperature, low number of occupants and presence of mechanical ventilation, all factors potentially contributing to low indoor humidity levels during the heating-season. Overall, the intervention group had 19 ERV's and 24 HRV's installed.

The HRV and ERV sub-groups experienced similar increases in their ventilation rates following the intervention. The median air exchange rate after the intervention was increased by 0.20 h^{-1} and 0.21h^{-1} for the HRV and ERV sub-groups respectively. By introducing ERV's to parts of the intervention group, the relative RH could be maintained; in fact, there was a (negligible) increase by 0.4 % on average during the heating-season. In contrast, the sub-group receiving an HRV experienced a further RH decrease by 6.2 % after the intervention. This further decrease was not unexpected due to the increased supply of dry outdoor air during the heating-season.

The concentration of the mold spores were found to be correlated with a number of the measured environmental parameters. From Table 4 it can be seen that the concentration of the airborne mold spores and the air exchange rate were negatively correlated, although the correlation is weak.

Table 4: Spearman rank order correlations (R values) between mold spores concentrations and selected environmental parameters during the heating season (combined for the control and intervention groups).

Mold Spores	Total airborne		Total settling	
Medium	MEA	DG-18	MEA	DG-18
AER (SF6)	-0.04	-0.07	0.01	-0.04
AER (PFT)	-0.15*	-0.14*	-0.10	-0.16*
T_{surf} of the coldest point in the child's bedroom	0.20*	0.11*	0.18*	0.09
RH_{surf} of the coldest point in the child's bedroom	0.43*	0.22*	0.39*	0.30*
T_{surf} of the coldest point in the house	0.31*	0.15*	0.28*	0.14*
RH_{surf} of the coldest point in the house	0.42*	0.23*	0.41*	0.35*

Note: $*p < 0.05$, T_{surf} = surface temperature, and RH_{surf} = surface relative humidity

The negative correlation with the ventilation rates suggests that the collected mold spores are from indoor sources. The airborne and settling mold spores were significantly, but moderately positively correlated with the RH measured close to the coldest point in the child's bedroom ($p<0.05$, $R = 0.22 – 0.43$). The correlation between the airborne and the settling mold spores was high ($R = 0.77$, $p < 0.001$, $n = 615$) for the combined data sets for the control and intervention groups.

DISCUSSION

1) Characterizing the Initial Ventilation Rates

The individual short-term (SF6; 4 to 5 hours) and long-term (PFT; 6 to 8 days) air exchange rate measurements cannot substitute each other for a given home visit, even though the aggregate geometric mean and median values across the different groups were close to each other (Won et al., 2011). Results of both ventilation rate measurements along with the infiltration rate data obtained by the Orifice Blower Door test showed, that the use of all methods including a careful characterization of any existing mechanical ventilation system - sometimes including air flows at supply air grills - is required to accurately characterize the current ventilation scenario, and in order to reliably increase air exchange rates in under-ventilated homes during the intervention phase. Overall, the homes in this study were under-ventilated, and the intervention group benefited from the introduction of HRV or ERV systems.

2) Ventilation Interventions and Mold Spore Concentrations

The amount of total airborne mold spores, and even more so the concentrations of spores of the genera Aspergillus and Penicillium could be significantly reduced by improved ventilation. It was observed that there was no increase in mold concentrations due to higher ventilation rates, which might have resulted in higher linear air speeds, and consequently the re-suspension of already settled mold spores was not observed. The reduction in airborne mold spores is obviously mainly due to dilution of spores by outdoor air being low in terms of airborne mould spores. It can be assumed that outdoor concentrations of mold spores during Québec's heating season are low. This is also substantiated by the fact, that *Cladosporium* concentrations, typically originating from outdoor sources, are low. The reduction of the airborne mold genera *Aspergillus* and *Penicillium* went in parallel with the concentrations of settling mold spores. Due to the high volume of samples the species under the genera *Aspergillus* and *Penicillium* could not be speciated, which limits partly the value of this aspect of the study; however, this was not a mold study, and mold growth was hardly observed in the homes under investigation, likely because the parents were sensitized to the importance of good IAQ (see bullet 4).

3) Association Between (low) Ventilation Rates, Temperature and IAQ Parameters

The analysis of the pre-intervention phase showed clear links between low ventilation rates and elevated concentrations of VOC's (Aubin et al., 2011). Pronounced seasonal variations in the concentrations of a number of VOC's was observed as well: in winter, where the ventilation rates were typically lower, we found higher concentrations of volatile alcohols, aromatics and alkanes. In summer, when indoor temperatures were typically higher, we found higher concentrations of aldehydes, especially formaldehyde, most likely due to the higher emission rates caused by higher indoor temperatures and relative humidities.

4) Bias Due to Knowledgeable Parents

Because the children all had physician diagnosed asthma the parents were likely sensitized to the importance of good ventilation and IAQ. This may explain why we encountered in the field:

- a larger percentage of homes having mechanical ventilation than that found in previous studies (Gilbert et al., 2006);
- less carpeting than expected, and instead of this mostly hard-wood flooring, thus limiting the amount of collectible dust, which reduced our capability of finding many microbial indicators in house dust; and
- many children having dust mite covers for their mattresses.

On a positive note, parents, and also children's interest in participating in the study was very high, resulting in a comparatively low (~4%) drop-out rate across a nearly three-year study for each individual.

CONCLUSION

1) It was clearly demonstrated that under-ventilation can be corrected through a careful retrofit with HRV or ERV systems, and that the concentrations of potentially health relevant microbial and chemical IAQ parameters can be decreased by increased ventilation, without revealing any negative side effects like re-suspending particles or mold spores due to higher air flows.

2) Effects of further reducing low RH's in cold climates could be mitigated by choosing ERV's over HRV's.

3) The study sample may neither be representative for Québec City nor for homes in cold/continental climates, as we did not randomly recruit homes. Physicians from the partnering Centre hospitalier universitaire du Québec were recruiting asthmatic children, with parents being owners of detached family homes (to avoid problems with landlords) and open to scientific studies.

4) The final answer as to whether the improved ventilation or air distribution in the homes led to a corresponding improvement in the child's health, i.e. reduced

self-reported asthma symptoms as well as objective respiratory health improvements, can be given when all health related observations have been evaluated. The analysis of the health results, including the epidemiological data, are expected by late summer 2012.

ACKNOWLEDGEMENTS

The authors would like to thank Karine Barriault (NRC), Kristina Boros (NRC), Jean-Francois Chouinard (INSPQ), Don Fugler (CMCH), Simon Godbout (INSPQ), Véronique Gingras (INSPQ), Pierre-Luc Lamarre (NRC), and Stephanie So (NRC) for their support in field and laboratory work. This work was funded by the Government of Canada's Clean Air Agenda, Québec's Ministère de la santé et des services sociaux and the Canada Mortgage and Housing Corporation and Health Canada. We would also like to thank the research ethics boards from both the Centre hospitalier universitaire du Québec and the National Research Council of Canada for their professional support.

REFERENCES

- Aubin DG, Won D, Schleibinger H., Gauvin D, Lajoie P, Fugler D. Seasonal variation in indoor air quality from a field study investigating the impact of ventilation rates on the health of asthmatic children in Québec City. 12th International Conference on Indoor Air Quality and Climate, Austin, Paper #93; 2011.
- ASTM, 2006. ASTM Test Method E 741-00: Standard test method for determining air change in a single zone by means of a tracer gas dilution. West Conshohocken, American Society for Testing and Materials. 2011.
- Dietz RN and Cote EA. Air infiltration measurements in a home using a convenient perfluorocarbon tracer technique, *Environment International*, 1982:8(1-6), 419-433
- Gilbert NL, Gauvin D, Guaya M, Heroux ME, Dupuis G, Legris M, Chand CC, Dietz RN, Levesque B. Housing characteristics and indoor concentrations of nitrogen dioxide and formaldehyde in Québec City, Canada, *Environmental Research*, 2006:102, 1-8
- Health Canada. Exposure Guidelines for Residential Indoor Air Quality, Publication H46-2/90-156E, Ottawa Canada. 1989.
- Won D, Aubin DG, Schleibinger H., Gauvin D, Lajoie P, Fugler D. Comparison of Air Exchange Rates Measured with Different Methods and Influencing Factors: Preliminary Results of a Field Study Involving Asthmatic Children. 12th International Conference on Indoor Air Quality and Climate, Austin, Paper #125. 2011.

THE PENETRATION OF MOLD INTO FIBROUS HVAC INSULATION MAKES CLEANING IMPOSSIBLE

Thomas G. Rand and Phil Morey

ABSTRACT

Fibrous insulation is often installed along the airstream surfaces of air handling units and air supply ducts for thermal and acoustical purposes. Glass fibers in new fiberboard and new fibrous liners are characterized by an apparent absence of dirt and mold growth. Dirt and dusts that accumulate on or within fibrous insulation are hydrophilic. During the air-conditioning process the relative humidity downstream of cooling coils can consistently approach 100% and mold spores present on dirt accretions can germinate and grow. In some cases, the airstream surface of the insulation becomes covered with microfungi. In this study, we examined samples of new and mold contaminated airstream insulation, including moldy samples archived in dry storage for 18 yrs., using differential interference (DIF) and low vacuum scanning electron microscopy (SEM). None of the new insulation samples showed evidence of mold contamination. However, abundant hyphae, conidiophores, and spores were present on the mold colonized insulation surfaces, as well as in porosities several millimeters below the liner surface. Archived samples also contained hyphae, conidiophores, and spores, and although mold structures were sometimes melanized, they were still easily recognizable on and in the glass fiber matrix, despite their time in storage. As mold structures were on and around subsurface glass fibers, we then tested the hypothesis that attempts at physical cleaning would not be successful in removing embedded mold. To test this hypothesis, moldy, non-archived fiberboard was thoroughly vacuum-cleaned so that mold encrustation on the liner surface was visually removed. DIF and SEM examination of fiberboard after cleaning showed that subsurface hyphae, conidiophores, and spores were still present on and around glass fibers. It was concluded that vacuum cleaning of moldy fiberboard does not remove mold embedded in the fibrous matrix. Collectively, this study shows that moldy HVAC insulation cannot be physically cleaned and should be discarded.

INTRODUCTION

Fibrous insulation is often installed along the airstream surfaces of air handling units and air supply ducts (HVAC system) because of its low cost, ease of installation and for thermal and acoustical purposes. Fibrous insulation or liner present on HVAC airstream surfaces readily accumulates dirt and dust entrained from the outside, which can support a relatively high viable fungal spore load constituting a diversity of species (Rand, 2005). While new (unsoiled) insulation and duct liner made from glass fibers is hydrophobic (West & Hansen, 1989), the dirt and dust that accumulates on HVAC airstream surfaces is hydrophilic (West & Hansen, 1989) and its inorganic and organic constituents provide nutrients for mold growth when moisture is non-limiting (Morey & Williams, 1991). During the air conditioning process, the relative humidity in AHUs and supply air ducts can approach 100% and the soiled insulation and duct liner can become a mold amplification site. A period likely to promote mold growth is defined as any 24 hr. period during which there was 100% relative humidity (RH) or a RH greater than 70% following a period of 100% RH at the surface being tested (Rand, 2005).

Mold growth can be readily recognised as isolated to confluent cotton-white to brownish-black mycelial patches on or completely covering HVAC surfaces. Because mold growth is visually manifest on surfaces, some investigators have suggested that mold contaminated HVAC can be vacuum cleaned. Using microscopy, this study was undertaken to determine if mold growth on the airstream surface of HVAC system fibrous insulation is accompanied by penetration of hyphae into the subsurface fiberglass matrix. Furthermore, it was to determine whether mold-contaminated fibrous insulation vigorously vacuumed can be cleaned and whether contaminated insulation archived under dry storage for over nearly 18 years still supported mold structures.

METHODS

Unused HVAC fiberboard was purchased commercially. Moldy fibrous insulation was obtained from air handling units and air supply ducts from buildings located in Florida and California. Samples collected by one of us (PM) between 1989 and 1995 were thereafter archived in dry storage in an air conditioned office environment until examined in this study. For light microscopy, samples were collected from airstream surfaces of unused (n = 10) and moldy fibrous insulation (n = 20) using cellotape, and from 1-mm3 glass fiber samples excised 1-5mm subsurface by gently prying away surface lining material using a sterile scalpel. Cellotape and glass fiber samples were mounted in lactic acid on glass microscope slides and examined using a Leica DMRE microscope equipped with a Hamamatsu Photonic digital camera system and using differential interference microscopy (DIF). Images were captured using PCImage analysis software (PCI Geomatics). For scanning electron microscopy (SEM), cylinders (5-mm W x 5-mm L x 15-mm H) of unused, un-vacuumed and vacuumed moldy fibrous insulation samples were cut using a sterile scalpel, oriented on SEM stubs so that airstream surface to sub-surface regions could be examined. Cylinders were examined uncoated using a Hitachi S3500 Variable Pressure SEM operated at 80 Pa and 15 KV. To evaluate vacuuming efficacy, approximately half of a 10 cm^2 mold contaminated non-archived fibrous insulation sample was thoroughly vacuum-cleaned so that mold encrustation on the liner surface was visually removed. To facilitate microscopic examination (especially SEM), the sample was then cut into approximately 3 cm^2 squares so that both vacuumed and un-vacuumed areas were equally represented and could be examined using both DIF and SEM as described above.

RESULTS AND DISCUSSION

Examination of surface and sub-surface samples of new, unused fibrous insulation using both DIF and SEM revealed that none of the samples supported any apparent mold contamination (Figures 1 a-d) and that only 2/10 samples supported modest surface dirt and/or dust accumulation (Figure 1 b). However, for non-archived mold-contaminated fibrous insulation DIF (Figures 2 a-b) revealed that fine dirt (PM 2.5), dust, hyphae, condiophores and high spore concentrations were accreted to surface and subsurface glass fibers. SEM (Figures 2 c-d) also revealed that contaminated fibrous insulation supported abundant hyphae, conidiophores and spores on fibers and in porosities several millimeters below the liner surface. Interestingly, both DIF and SEM also showed that some of the accreted hyphae were entwined around the glass fibers.

Compared to un-vacuumed insulation SEM revealed that thoroughly vacuum-cleaned non-archived fibrous insulation supported less mold encrustation on the liner surface (Figure 3 a-b). However, not all surface mold was removed by vacuuming. DIF and SEM examination of fibrous insulation after cleaning showed that subsurface hyphae, conidiophores, and spores were still present on and around glass fibers (Figure 3 c-d).

These results revealed that all contaminated insulation samples examined supported mold growth that was associated with both the airstream surface and subsurface glass fiber matrix. Importantly, the continued presence of mold structures on and around surface and subsurface glass fibers of vigorously vacuum cleaned airstream surfaces points to the futility of physical cleaning as a means of mitigating mold from fibrous HVAC insulation. The inability to remove all mold structures by physical cleaning supports the recommendations in the New York City Guidelines (2008) to discard insulation or porous liner materials that are visually moldy. While recognising that fibrous insulation has desirable acoustical and thermal properties, alternatives to its use in HVAC could be used (Morey et al., 2009). These might include: coating the fibrous insulation with mastic thus making its airstream surface semi smooth; use of a closed cell foam liner with a smooth airstream surface and/or eliminate the use of fiber-based duct liner completely. The smoother surfaces of the un-insulated airstream wall would minimize the potential for collection of dirt and dust and not provide porosities into which moisture and mold could penetrate. The elimination of internal liner from the airstream would require that closer attention be paid to air system acoustical control. The design of acoustic control in an air distribution system that is not internally lined should be completed with a more holistic approach that includes equipment selection, sizing, and location and proper air distribution design (Morey et al., 2009).

It may be assumed that mold structures diminish or disintegrate on colonized surfaces as the material ages in a dry state. In order to test this hypothesis moldy insulation samples that had been held in archived storage for up to 18 years were examined using DIF and SEM for the presence of surviving mold structures. Inspection of archived samples showed that they still supported abundant hyphae, conidiophores, and spores (Figure 4 a-b). Interestingly, while hyphae and other mold structures were physically intact they were sometimes melanized and exhibited "copper penny morphology". Nevertheless, they were still easily recognizable on and in the glass fiber matrix, despite their time in storage.

Figure 1a & b. New insulation. Differential interference microscopy (DIF) (a) and scanning electron microscopy SEM (b) showing the apparent absence of mold structures on fibrous insulation surfaces. The flaky particles between glass fibers in figure 1b are likely from a surface mastic coating on the insulation; **Figure c & d.** DIF (c) and SEM (d) showing the apparent lack of mold structures on sub-surface fibrous insulation.

Figure 2a & b. Insulation recently colonized by mold. DIF revealing that fine dirt, dust, hyphae, condiophores and high spore concentrations were accreted to surface (a) and subsurface glass fibers (b) of non-archived mold contaminated fibrous HVAC insulation; **Figure c & d.** SEM showing abundant hyphae, conidiophores and spores on surface fibers (c) and in sub-surface (d) porosities of contaminated fibrous insulation. Note the presence of hyphae accreted to and entwined around insulation fibers (a, b, d).

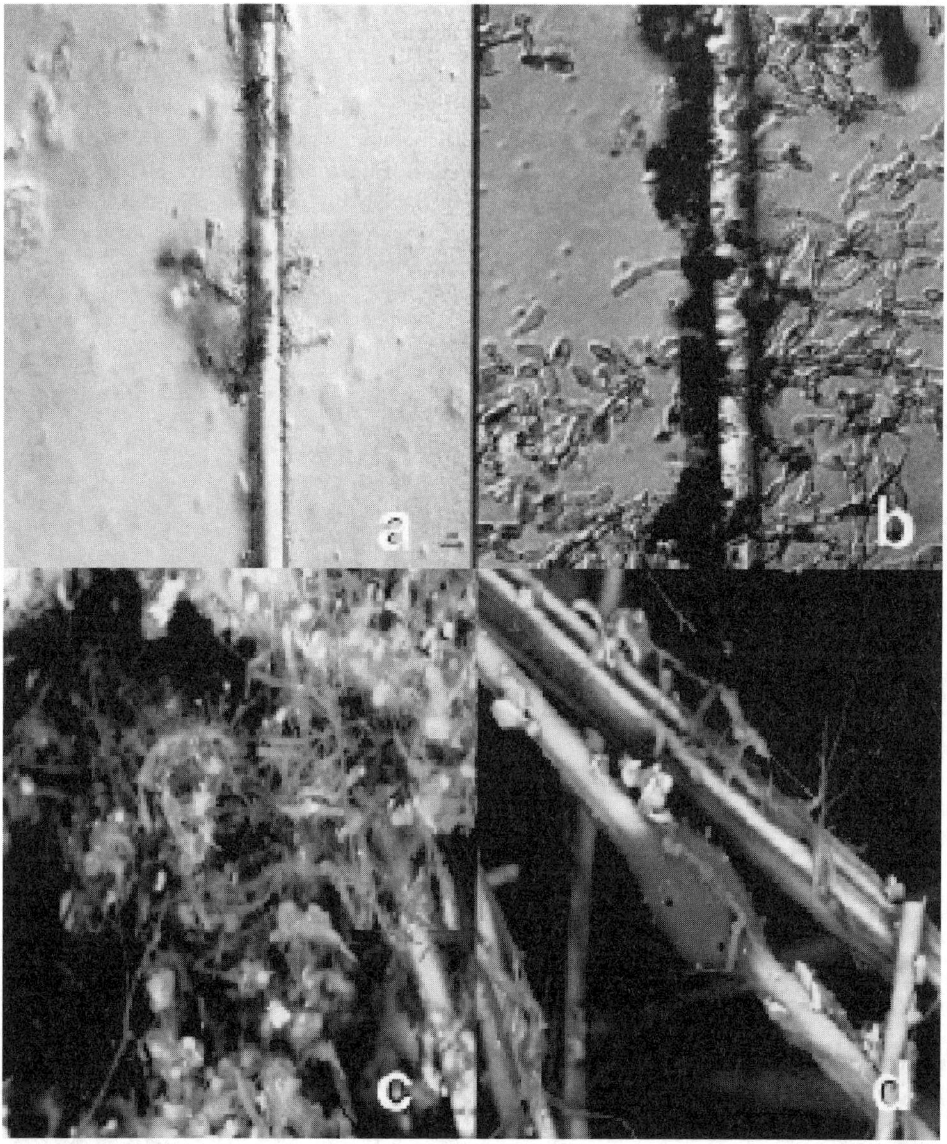

Figure 3a & b. Cleaned and un-cleaned moldy insulation. SEM revealing un-vacuumed (a) and vacuumed (b) areas of the surface of a non-archived mold contaminated insulation sample. Note the remarkably higher concentration of mold structures on the un-vacuumed surface area compared to that of the vacuumed area. **Figure c & d.** SEM revealing mold structures in sub-surface areas of un-vacuumed (c) and vacuumed (d) non-archived, mold contaminated insulation sample.

Figure 4a & b. Archived moldy insulation. DIF revealing the presence of copper penny hyphae, *Cladosporium*-like (a, b) and *Penicillium*-like (b) conidiophores and spores associated with fibrous insulation archived under dry conditions since 1995.

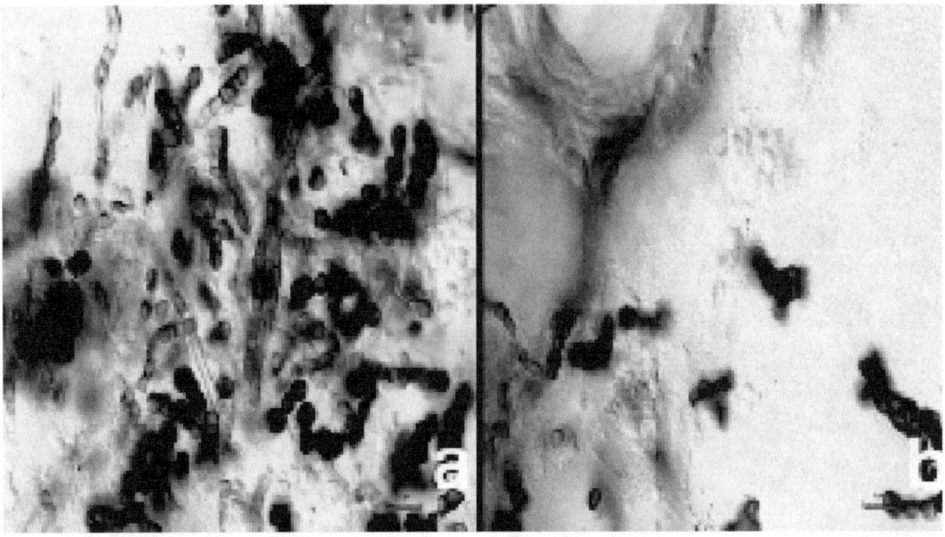

Acknowledgements: We thank C. Leggiadro and D. O'Neil, NRC Institute of Marine Biosciences for their assistance and excellent microscopy technical support.

REFERENCES

- Morey P and Williams C. Is Porous Insulation Inside a HVAC System Compatible With a Healthy Building? In: *IAQ 91 Healthy Buildings*, ASHRAE,1991. pp. 128-135.
- Morey P, Rand T and Phoenix T. On The Penetration of Mold Into Fiberboard Used in HVAC Ductwork. In: Proceedings of the 9th International Healthy Buildings Conference, 2009. Syracuse, paper #110, Pl-100, 4 p. 2009.
- NYC, *Guidelines on Assessment and Remediation of Mold in Indoor Environments*. New York Dept. of Health. 2008
- Rand TG. Ecology of molds in building environments. In. American Industrial Hygiene Association Field Guide for the Determination of Biological Contamination in Environmental Samples (eds. Hung L-L, Miller JD and Dillon K). AIHA Press. Washington, D.C. 2005. pp. 29-38.
- West M. and Hansen E. Determination of Material Hygroscopic Properties That Effect Indoor Air Quality, In: *IAQ 89 The Human Equation: Health and Comfort,* ASHRAE, 1989. pp. 60-63.

PAST, PRESENCE AND FUTURE OF IMMUNOASSAYS FOR MYCOTOXIN TESTING

Erwin Maertlbauer

I like to give you some information about immunoassay for mycotoxins. I will present a short history of mycotoxins so you better understand the history of immunoassays for mycotoxins. We do not know very much about mycotoxins problems in the past centuries but I guess that during that time, in the medieval ages, insulation of houses was not considered to be a major issue. We are certain, that ergotism was an important cause for thousands of victims in these ages and later on. You may have seen the famous painting by Pieter Bruegel illustrating the possible effects of intoxications shown in the picture with the title "the beggars".

PICTURE 1: Pieter Bruegel: The Beggars (source: http://www.wikipaintings.org/en/pieter-bruegel-the-elder/the-beggars-1568)

Ergotism (characterized by convulsive and gangrenous symptoms) occurs after rye becomes contaminated with Claviceps. This disease was also called 'St. Anthony's Fire' or 'Holy Fire' at that time. Later on, the 20th century became the century of mycotoxins, particularly the second half of the century. Certain observations about toxicosis in humans were made for example in Russia, which, retrospectively, were thought to be caused by trichothecenes causing a condition called 'alimentary toxic aleukia'. Furthermore, during the search for antibiotics a number of mycotoxins were discovered, such as Glutinosin as it was named at that time. It was later determined that it was a mixture of Verrucarin A &B. The main research about mycotoxins really started at the beginning of the '60s when an enormous veterinary crisis occurred close to London, U.K.. There probably up to 100,000 turkeys died after consumption of contaminated feed. The agent that caused the intoxication was identified as Aflatoxin B and G. In 1965 the mycotoxin 'ochratoxin was discovered. One of the most recent described new mycotoxins are the Fumonisins, described at the end of the '80s. Some people call these years, starting in the '60s and '70s, the 'mycotoxin gold rush' because a significant research funding become available during that time and a many scientists started to do research on mycotoxins.

Considering the time history of immunoassays, we can see here similar developments. You may remember the first radio immunoassay (RIA) for Insulin was described in 1959 by Solomon Berson and Rosalyn Yalow. The major breakthrough came with the development of enzyme immunoassays at the beginning of the '70s by Engvall and Perlmann. There are other assays which played an important role in mycotoxin testing such as luminescence immunoassays or fluorescence polarization immunoassays. But the major impact on development of analytical methods in that field was the invention of the enzyme linked immunosorbant assays (ELISA) in the beginning of the '70s. It remains the major detection method for mycotoxins today. Later, many other detection methods were developed and introduced, but even today the major assays used in the industry and research are enzyme immunoassays and fluorescence immunoassays.

The first immunoassays for mycotoxins were developed for Ochratoxins and Aflatoxins. At the end of the '70s, the research shifted to trichothecene mycotoxins because of reports about the alleged use of these agents as biological warfare agents in Southeast Asia. You might have read about the 'yellow rain' incident. It was likely a fake. Thereafter, in the beginning of the '80s, high grain contamination with Deoxynivalenol (DON or Vomitoxin) were reported, a mycotoxin produced by certain species of *Fusarium*, the most important of which is *F. graminearum*. Favorable weather conditions for the growth of *Fusarium* molds in the winter wheat occurred in Canada during a very wet summer. At that time research concentrated on ways to easily detect *Fusarium* mycotoxins such Deoxynivalenol and Zearalenone. Such contaminations still represent a worldwide problem in food safety. Typically grains, such as wheat and barley, are contaminated with these mycotoxins.

At the end of the '80s, our study group described for the first time an assay for the detection of macrocyclic trichothecene, Roridin A. It is a similar assay which is used today for detection of Satratoxins in the indoor environment. In 1992 the first Fumonisin EIA was described.

Figure 1: Early Years of Immunoassays for Mycotoxin Testing

Early Years of Immunoassays for Mycotoxin Testing

1975	RIA	Ochratoxin A
1976	RIA	Aflatoxin B1
1977	EIA	Aflatoxin B1

1979	RIA	T-2 toxin
1983	RIA	Zearalenone
1984	RIA	Diacetoxyscirpenol
1988	EIA	Deoxynivalenol
1988	EIA	Roridin A
1992	EIA	Fumonisin

Saratoga Springs 2011

If we look at the trend at that time, these tests were mainly based on rabbit antibodies. The first monoclonal antibodies came in the '80s. By the end of the '80s we saw a remarkable rise in the use of monoclonal antibodies, which are the major antibody source today. We have used rabbit antibodies because it's very easy and inexpensive to produce antibodies in rabbits. Even today there are many assays successfully using rabbit antibodies. The only drawback with rabbit antibodies is the antiserum from one rabbit is not the same as from another rabbit. As a result you have slight variations from batch to batch, if you produce assays based on rabbit antiserum. This was overcome by monoclonals. In principle, the properties of antibodies for mycotoxin testing, either rabbit antibodies or mouse monoclonal antibodies are very similar, i.e. they have similar sensitivities and specificity. The advantage of the monoclonals is that once you have a clone, theoretically, you should have it forever if you'll store it carefully.

There have been a number of publications about recombinant antibodies in the literature. To my knowledge, no commercially available assays for mycotoxins use recombinant antibodies today because there is not a real advantage at the moment. Furthermore, there are still problems with the stability of recombinant antibodies because if you for example try to express the variant fragments in *E. coli* or in yeast, you still have to chemically link it.

One of the most important aspects in Mycotoxin analysis is to consider the size of the molecule. A mycotoxin such as Aflatoxin B_1 has a diameter of approximately one nanometer. The antibody is much bigger, about 15 or more nanometers. That means if we put it on a comparison scale, Aflatoxin is roughly the same size as the binding side of the antibody. This is a very important point to consider because it may explain problems which may occur with new assays. The first important fact is that you cannot use marker-free methods. Usually we have to label the toxin or the antibody to detect the binding of the toxin to the antibody. That's very important. The label of course was in the beginning "radioactivity" on enzymes or fluorescent markers and many others. Most of the assays used these days are so-called heterogeneous assays, which means they use a solid phase to separate bound from unbound reagents. The only exception from that is fluorescence polarization assays which is a homogeneous assay. You need no separation of the bound from the free in a homogeneous assay because the fluorescence polarization changes if the labeled toxin is bound by the antibody. Ninety-nine percent of the assays are a heterogeneous assay and use a solid phase. Even new developments like *Surface Plasmon Resonance* for *Immunoassays* use a solid phase and usually use labeled reagents.

The labels used today may be fluorescent or luminescent, but in most cases, enzymes or gold particles. You may recall that pregnancy tests are a common example for rapid on-site tests. There are only a few such rapid tests available for mycotoxin testing. Usually gold labeled antibodies or mycotoxins are utilized. One should note the large size of the label and the small size of the mycotoxin. As a consequence you can detect mycotoxins only with assays called competitive assays. You have a free mycotoxin, labeled mycotoxin, and you have your antibody. What actually happens in the assay follows a chemical equation. It is a biochemical reaction which goes back and forth, and the forward rate and backward rate is dependent on the affinity constant of the antibody. And this is the most important parameter.

The other consequence will be, if you draw an standard curve, you can see that the result you get is indirectly related to the assay response. That means that the higher the assay response is, the lower is the concentration of the free mycotoxin because the labeled mycotoxin is measured.

Nowadays, there are now many of these assays commercially available which are primarily designed for food and feed testing. The major applications of immunoassays are for mycotoxin testing such as Aflatoxin, Ochratoxins, and others. You can notice that the detection limits are quite low, in the nanogram/g range or below. For Aflatoxin M_1 for example, there are assays which detect one or two picograms/ml. These are in the femtomoles per assay range. So that's very low concentrations. The laboratory procedures used are either microtiter plate assays or rapid assays such as capillary flow. In the later assays all the steps which you have to do by hand or by machine on the microtiter plate are included on a membrane. With this simple assay, you just have to put a drop your test sample on it and wait five minutes and then get your result.

The second important key point is specificity of your antibody. One example for this is the assay for Roridin A. The antibodies were generated using Roridin A chemically linked to a protein carrier for the immunization of rabbits or mice. Usually the antibodies recognize the parent compound very well and that reaction is set to 100% for calculation of cross-reactivity. However, the antibodies may also recognize other substances that are chemically similar but differ only slightly in their structure. One can notice that there is no big difference in the specificity (% cross-reactivity) between rabbit antibodies and monoclonals. This is quite common with assays for mycotoxin testing. As a result, if your sample may contain several of these toxins, you never can say for sure what specific toxin you are measuring at the moment. In addition, the test result is not quantitative in that case. For example, if you compare the ELISA results with the fungal contamination of samples or results from cytotoxicity tests, there may be a relatively good correlation. However, you should note that you are measuring Roridin A equivalents. If you talk about equivalents that means that you do not exactly know what you're measuring. If your field sample has Roridin A only, then the result would be correct. If your sample contains Satratoxins, the concentrations will be seven or eight times higher. If it contains a mixture of toxins, you do not know. That's the problem.

TABLE 1: Specificity of Antibodies to Roridin A (see also Märtlbauer et al., 1988).

	% Crossreactivity	
Mycotoxin	**Mouse mAb**	**Rabbit pAb**
Roridin A	100	100
Roridin J	44	41
Verrucarin A	17	15
Satratoxin H	19	15
Satratoxin G	4	7
Diacetylverrucarol	0.05	0.15
Verrucarol	0.09	0.05

TABLE 2: Results of environmental samples (see also Dietrich et al., 1999).

Sample	*S.atra*	MTT-assay	EIA ng/g RoA*
Wall paper	+	+	7
Wooden ceiling	+	+	10
Wall paper	+	+++	28
Insulation	++	++	31
Gysum board	+++	+++	70
Wall paer	++++	++++	2500
Insulation	+++(+)	++++(+)	300-1100

*Roridin A equivalents

It becomes much, much more complicated if you want to test specimens or samples from humans. You need to know a lot of facts and details before hand, most of those we still do not know. First of all, how long is the biological half-life? It may be weeks. For example, we know for Ochratoxin A that this toxin has a very high affinity for serum albumin. But usually the biological half-life of many mycotoxins are just a few minutes to several hours. In a recent publication by the group of Pestka et al. the kinetics and excretion of Satratoxin G following intranasal exposure in mice was described.

FIGURE 2: Kinetics of Satratoxin G tissue distribution and excretion following intranasal exposure in the mouse. (Amuzie et al, 2010)

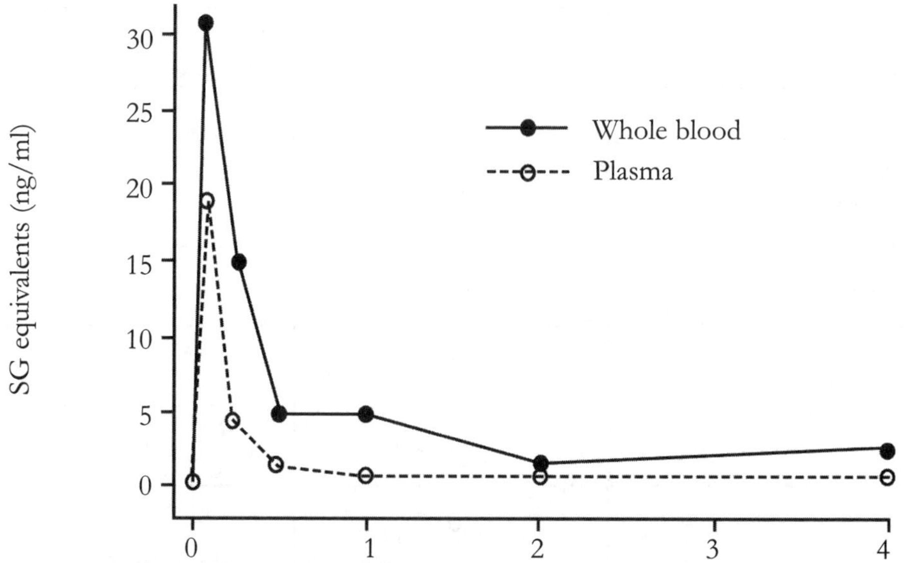

The next question is, is there a main metabolite that is known. We know for example in the case for Ochratoxin A that the metabolites are (4R)-4-hydroxyochratoxin A and Ochratoxin alpha. We know that probably every human will have some low levels Ochratoxin A in the blood. No toxicologist can probably tell you at the present time what health consequences this may have. You'll probably have also low levels of Deoxynivalenol (DON) in your blood. In several studies from the United Kingdom it was reported that DON levels in urine were correlated to the cereal intake of the individual. This is not a new finding. That means one has to be very carefully in interpreting such results in respect to environmental exposures. Such findings may be more likely related to food intake than to environmental exposures.

Some comments about test performance issues. As mentioned before, these assays were designed and validated for food and feed testing. And typically, as shown on the standard curve of this slide, it is a non-linear response, which means that results close to the end of the curve have a greater variability than in the middle of the curve. The assay response (Y-axis) usually keep the same variability but going to the nonlinear curve (good or very low variation at the middle), but it results in higher variation at the end of the curve. In the middle of the curve you may get a variation coefficient of 5%, but at the end it may be up to 20%. If you're measuring the right end of the standard curve, there may be no significant difference between 100 or thousand, for example.

FIGURE 3: Quantitative Immunoassay: Performance Characteristics depending on toxin concentration

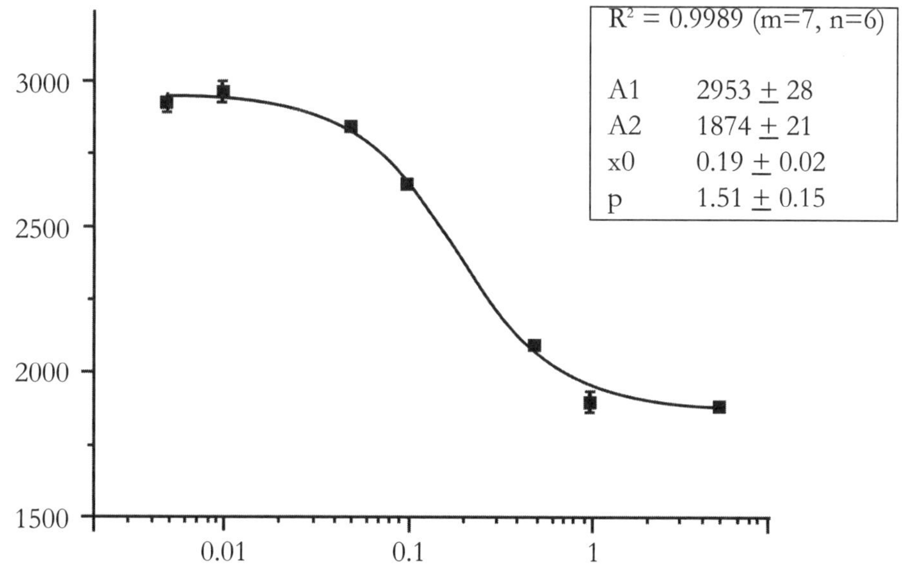

Some comments about specificity. If your assay cross reacts with some other substances and you do not know what is really in your test sample, your result will not be quantitative. That is the reason why some organizations provide validation of immunoassays, such as the AOAC, the International Union of Pure and Applied Chemistry (IUPAC) and others – and define a variety of parameters for these assays. In my opinion, the most important parameter is the test sample matrix you want to check. This has a major effect on the test assay. If everything is OK in your test assay, you´ll get a good result. If anything in your sample disturbs this reaction, you get a very high false positive result, because of the competitive assay format. You should be aware of that. So what can you do in these situations? You should confirm your results with other methods. So if you use a validated assay, it can be quantitative or semi-quantitative. If you use a non-validated immunoassay for new sample matrices, you must confirm the results by any other reference methods.

A comments about biosensors. Everybody's talking about biosensors these days and there are an enormous number of publications about biosensors in any field, particularly in the analytical field. IUPAC gave a definition what a biosensor is, such as: "A biosensor – in contrast to biotic sensors or biotests – is a self-contained, integrated receptor-transducer device, which is capable of providing selective quantitative or semi-quantitative analytical information and which uses a biological recognition element (bio-receptor) and a transducer in direct special contact (IUPAC, 1996)". So it should be a biological receptor, usually an antibody, connected with a physical transducer which gives a reversible signal, in a kind of on-off reaction, allows multiple measurements, and is a small and cheap device. Strictly speaking, no such immune-assay or immune-sensor for mycotoxins does exist. There are some assays, which come close to that but no one is fulfilling all these requirements. In the majority of cases (99%) the biological receptor are antibodies. Occasionally there are alternatives. The signal transduction is usually optical. There are some assays using electrochemical transduction but these are usually less sensitive than optical methods. The major advantage of electrochemical is that they are very cheap, very easy to handle, no need for complicated equipment, you just need a volt meter costing about $60. The major problem of this method is its reproducibility, a problem seldom mentioned in the publications. The principle behind it is the same as any immunoassay. It's a competitive assay usually using labels. It is a very fast method, it will take only a few minutes, it is very inexpensive, but the reproducibility is still a problem.

What is the outlook for the future? There are already some publications about toxin arrays. That, in principle, sounds fascinating because you could produce an array that may be an antibody-array or you can coat the surface with the toxin, which is also possible and detect, let's say, 20, 30 or more mycotoxins as long as you have antibodies available. Initially that may sound very fascinating, however, if you do not specifically know what you are analyzing, what is in your test sample, than you may get a lot of unexpected results with these arrays. It's quite unlikely that you may have 20 different mycotoxins present in one sample, and if you choose a non-suitable sample with these arrays, you will get a lot of "positive" results. So you have to confirm the results. Essentially such an array could contain 50 antibodies for 50 mycotoxins. If you do not validate or confirm your results, it may falsely suggest that your sample contains Aflatoxin, Ochratoxin, DON or many others. Such result would probably be nonsense. In other words, not everything which can be done makes scientific sense.

Today, if one considers all commercially available assays, we can find, that the majority are (about ninety percent) micro-titer plate assays, followed by a few lateral flow assays, one or two commercial biosensor applications for mycotoxin. There is not a great variety available. In the next 10 years we may see an increase in the rapid assays and particularly in the mycotoxin arrays. Additionally, we may see that all of these test devices will become smaller. The nanotechnology will get into this analytical field. And this development will make these methods or these devices considerably smaller and hopefully also cheaper. In about 10 years or so, we maybe be also seeing a shift to rapid assays and immuno-sensors. Although, my guess is that we still have a majority with microtiter plate assays because the big advantage of microtiter plate assays is that they are so cheap and on one plate you can process 30 to 40 samples parallel. If you have a single sample to test, other methods may be better suited. But if you do two, three or more samples the microtiter plates have a great advantage.

Well, everything beyond this is in my view fiction, maybe even science fiction. But I don't know. Thank you for your attention.

REFERENCES:

- Amuzie CJ, Islam Z, Kim JK, Seo JH, Pestka JJ. Kinetics of satratoxin g tissue distribution and excretion following intranasal exposure in the mouse. *Toxicol Sci.* 2010;116(2):433-40.
- Berzofsky JA, Schechter AN. The concepts of cross-reactivity and specificity in immunology. *Mol. Immunol.* 1981:18, 751-763.
- Bunch DS, Rocke DM, Harrison RO. Statistical design of ELISA protocols. *J. Immunol. Meth.* 1990:132, 247-254.
- Dietrich R, Schneider E, Usleber E, Märtlbauer E. Use of monoclonal antibodies for the analysis of mycotoxins. *Nat Toxins.* 1995:3, 288-293.
- Edwards R. Essential data series: Immunoassays, New York, Wiley-Liss. 1996.
- Halfman CJ, Schneider AS. Optimization of reactant concentrations for maximizing sensitivities of competitive immunoassays. *Anal. Chem.*1981:53, 654 658.
- Jackson, TM, Ekins RP. Theoretical limitations on immunoassay sensitivity. Current practice and potential advantages of fluorescent Eu3+ chelates as non-radioisotopic tracers. *J. Immunol. Meth.*1986:87, 13-20.

- Köhler G, Milstein C. Continuous cultures of fused cells secreting antibody of predefined specificity. *Nature.*1975:256, 495-497.
- Märtlbauer E. Enzymimmuntests zum Nachweis antimikrobiell wirksamer Stoffe. Ferdinand Enke Verlag, Stuttgart. 1993.
- Märtlbauer E. Antibody techniques, in: Ashurst PR, Dennis MJ. (Hrsg.): Analytical Methods of Food Authentica tion, Chapman & Hall, London. 1998: 241 pp.
- Peters JH, Baumgarten H. (Hrsg.): Monoclonal antibodies. Springer Verlag, New York. 1992.
- Roitt I, Brostoff J, Male D. Immunology. Gower Medical Publishing, London, New York. 1985.

PICTURES:

- Pieter Bruegel: The Beggars (source: http://www.wikipaintings.org/en/pieter-bruegel-the-elder/the-beggars-1568)

Question from the audience:

Question: I have patients who come now with "mycotoxin test results" that they get done somewhere in Texas in a lab that claims to be able to test mycotoxins in human – such as in urine, hair, or other body parts as an indicator for mycotoxin exposure from the indoor environment. Could you comment on this and what are your thoughts about this?

MAERTLBAUER: As far as I know, these laboratory offers testing for Aflatoxins, for Ochratoxin, and for trichothecenes. And in regard to the trichothecenes, these are not specified. It remains unclear what a "positive" results means, because there are hundreds of trichothecenes. You cannot know what that means. And from the one slide I showed you, I would guess testing for trichothecenes, except maybe for Deoxynivalenol, does not make much sense to me. For several reasons: For example these chemicals have a very, very short half-time. We do not know exactly how these are metabolized in humans, we do not know what metabolites we should look for. And if you have a continuous exposition to Deoxynivalenol DON in the food, you will find Deoxynivalenol in the urine or other samples. Same applies to Ochratoxin A mycotoxin. If you have a continuous exposure to Aflatoxin, you will find Aflatoxin in the serum or in tissue samples. We know that. But I doubt that this is related to indoor environment. In the majority of cases I would say it's directly related to food intake.

Question : A few years back - this question is not necessarily directed to the current speaker but the group at large - Dr. Richard Lemon, formerly of NIOSH, published a very compelling paper addressing the question of whether chrysotile asbestos can cause mesothelioma. And the way he addressed it was by utilizing the

Bradford-Hill criteria for medical causation and going one by one through all nine of them. Has anybody given any thought to or actually done such an analysis as it relates to mycotoxins? Particularly in the lower concentrations that you would typically find in the indoor environment and the effects that they have or are postulated to have on human beings. Or is the state of the science not just there to even undertake such an analysis?

MAERTLBAUER: I am not aware of this.

Question: We talked about the biological half-life of some of the mycotoxins. I believe you referred to orally and intra-nasally. Has there been research done on environmental? Just sitting in the environment of biotoxins, as far as how long they last?

MAERTLBAUER: I'm not aware of that.

Thank you.

HEALTH RISKS OF SURFACE DISINFECTION IN HOUSEHOLDS WITH SPECIAL CONSIDERATION ON QUATERNARY AMMONIUM COMPOUNDS (QACS)

Axel Kramer, Harald Below and Ojan Assadian

ABSTRACT

Due to their antimicrobial mode of action the use of disinfectants requires a careful risk benefit assessment in order to minimize potential side effects on humans and in the environment as far as possible. Particularly the regular use of household disinfectants needs careful assessment, as the risk of contracting infection in household is less frequent than in a health care environment, because pathogen organisms are not prevalent in such high frequencies, and the immune status of most household residents is not decreased when compared to patients. Additionally, using disinfectants in household may be associated with unwanted health risks, such as allergy, sensitization, or intoxication. Therefore, risk assessment in presence of reduced risk for infection and an increased risk for other health affecting side effects may result in different conclusions for health care settings or households.

If, after careful consideration of a feasible indication or necessity for e.g. cleansing and surface disinfection of a household area with fungal growth, is given, selection and choice of a suitable disinfectant is imperative. Because of toxicological reasons, aldehydes are to be rejected. Quaternary ammonium compounds (QACs) are considered to show a low-level antimicrobial efficacy and so far, are considered harmless. Therefore, QACs are most probably the most widely used antimicrobial compounds in households in Europe and the US. Based on a critical risk benefit assessment and first reports of adverse effects after chronic use in household cases, it may be assumed that a long-term use of QACs on surfaces or even a single nebulization in households is associated with occupational and household health risks. To clarify this association, further research needs to be conducted.

There are, however, a number of alternatives to QACs. For repeated use on smaller surface areas, alcohols may be used. On larger surfaces, surface disinfectants on basis of organic carbonic acids or oxidants are suitable. Instead of using products on basis of QACs, various H_2O_2 formulations without environmental persistence (if the dispensable critical catalysator silver is not used) are to be preferred.

KEY WORDS Quaternary ammonium compounds, benzalkonium chloride, healthy risks, non-critical alternatives

INTRODUCTION

Because of their antimicrobial action, the use of disinfectants requires a careful risk benefit assessment to minimize potential side effects in humans as well as in the environment. This is particularly important for their application within the household setting.

Independently from the setting where they are used, when disinfectants are applied, the following premises need to be considered:

- Is there a justifiable indication?
- Is the selection of the antimicrobial spectrum of the active compound based on the expected or present microbial contamination?
- Is the active compound selected characterized by a favorable bio-compatibility index and an acceptable environmental compatibility depending of the indication?

Following indications for surface disinfection in household including offices may be meaningful and justified:

- Gastrointestinal infections if in the household seriously immune suppressed or susceptible persons such as immunocompromised newborns lives, and if the focus is hand disinfection (Kampf, Kramer, 2004]) usually supplemented by disinfection of sanitary and food preparation area (Kampf, Dettenkofer, 2011)
- a family member's (or co-worker in an office) diseased caused by a highly contagious pathogen, e.g. noroviruses, tuberculosis or EHEC; especially if the infectious doses is very low such as for noroviruses a virucidal surface disinfection is indicated after contamination with faeces or vomit (Kramer et al., 2006)
- if a patient dismissed from the hospital returns home still with invasive devices or wound dressing present (Kampf, Dettenkofer, 2011)
- prevention of re-colonization of MRSA carriers during the period of antiseptic eradication is performed, which includes hand disinfection in combination of disinfection of relevant contact surfaces (Neely, Maley, 2000; Goodman et al., 2008)
- although frequently immobilized QACs are added to carpets, wallboards, HVAC filters, etc. in the US and Europe, fungicidal disinfection for the prevention of health effects due to molds after moisture damage and during critical mold exposure (Kramer 1998; Baudisch, 2010) should be regarded as a limited justified indication. It must be pointed out that dead or alive mold presents the same allergic risk to people in a building. Only superficial circumscribed fungal decayed areas are eradicable by direct application of fungicidal disinfectants. For decontamination of indoor air only nebulization of the agent is effective (Koburger, Below, Kramer, 2011). Following fungicidal disinfection, however, a thorough cleaning of the affected area is essential to remove the dead, but still allergenic molds.

The necessity for disinfection in such situation results of the persistence of viruses, bacteria and fungi which differ between 1 week and several months (table 1). (Kramer, Schwebke, Kampf, 2006, Hübner et al., 2011).

Table 1: Persistence of clinical relevant pathogens on dry inanimate surfaces (modified after Kramer, Schwebke, Kampf, 2006)

Species	Persistence	
Bacteria		
Klebsiella spp.	2 hours -	> 30 month
P. aeruginosa	6 hours -	16 month
E. coli	1,5 hours -	16 month
S. aureus incl. MRSA	7 days -	7 month
Enterococcus spp. incl. VRE a. VSE	5 days -	4 month
M. tuberculosis	1 days -	4 month
S. typhimurium		50 month
C. jejuni		6 days
S. pneumoniae	1 day -	20 days
Fungi		
T. glabrata	102 days -	150 days
C. albicans	1 day -	120 days
C. parapsilosis		14 days
Viruses		
Adenovirus	7 days -	3 month
Vaccinia-Virus	3 weeks -	20 weeks
Herpes-simplex-Virus, Typ 1 a. 2	4,5 hours -	8 weeks
HAV	2 hours -	60 days
Papovavirus		8 days
HBV, HIV		> 1 week
Norovirus a. Felines Calici-Virus	8 hours -	7 days
SARS-associated Corona -Virus	3 days-	4 days
Influenzavirus	1 day -	2 days

Contrary to aldehydes and phenolic compounds, quaternary ammonium compounds (QACs) are to be considered as safe. The consequence are a non-critical misuse of QACs for surface disinfection in health care institutions as well as increasingly in households including for decontamination of homes in situations of fungal growth , if health complaints exist. Because QACs are stabile in indoors and are characterized by high surface activity, it is to be assumed that QACs are generating a visible film on surfaces with layer-wise enrichment. This film cannot be removed with usual cleaning methods. From theoretical point of view, it is to be assumed that dried particles can be detached from this film by running on the floor and can inhaled directly by air flow or indirectly with in dust accumulated QAC. Due to the high surface activity of QACs, it also may be assumed that inhaled dust particles attack the surfactant layer of the lung whereby the development of chronically obstructive lung illness (COPD) can be caused and/or promoted. QAC are compounds with a strong cationic charge that can act with negative charged compounds or surfaces.

Therefore, the aim of the present work was to evaluate the literature for evidence of possibly toxic risks of QACs with special focus of inhalation risks with the aim of a benefit risk analysis. Because benzalkonium chloride (BAC) is the prototype of QACs, mostly this agent was evaluated.

METHOD

On a basis of a case report and an analytical study to proof residues of benzalkonium chloride after disinfection on surfaces (Thimm, Feld, 2005) the literature was systematically reviewed on the internet site of the National Library of Medicine without language restrictions. The search was done on 20 July 2011 and covered all years available in PubMed. The following search terms were applied: quaternary ammonium compound, quat, benzalkonium chloride, benzethonium chloride, Didecyldimethylammonium chloride, cetrimide and for each compound connected with one of the terms toxicology, toxicity, side effects, mutagenicity, carcinogenicity, teratogenicity, cytotoxicity, sensibilisation, allergy. In addition, the citations in each study found during the main search were reviewed for potential relevance. Finally, standard textbooks on antimicrobial agents were examined for information. All reports with experimental evidence were included. Information from textbooks was also included, even if the chapter itself did not contain experimental evidence. At least two of the investigators decided on the relevance of each report. Reports were not blinded to the investigators so that they knew the names of the authors of all studies.

RESULTS

Case reports

Own observation: A female employee in a central sterilization service department (CSSD) reacted during full protective clothing repetitive with contact eczema. The patch test with the used disinfectant on base of benzalconium chloride reacts positively with skin reddens and slightly swells which implies a type IV allergy. Also after contact with a surgical mask or gloves the sensitized person reacts with eczema. Only after transfer of the employee to a not-contaminated ward the health care worker became symptom free. This puts the suspicion on airborne sensitization on the dermal positively tested disinfectant on QAV basis.

Case report from literature: In December 2006, after dampness damage in a private house in a large area, QAC was applied. Thereafter, the family was evacuated in a hotel. In April 2007 the family returned in their home. Again, in May 2007 evacuation of the family in the hotel was necessary due to strong health complaints.

In September 2007 and a second time in 2010, chemical analysis in the home demonstrate killing of fungi, but not sufficient decontamination of destroyed fungi elements and a significant increase of QAC in the house dust (exceed the 95% percentiles up to 75-fold), which was probably the main cause of health complaints (Kroczek, Thumulla 2011).

Residues of benzalkonium chloride on surface after disinfection

The accumulation of benzalconium chloride on polypropylene surfaces was analyzed using Time-of-Flight Secondary Ion Mass Spectrometry (ToF-SIMS) and Attenuated Total Reflection (ATR) - Fourier-Transformation-Infrared (FTIR) analytic. For this, per application of a surface disinfectant on the basis of benzalconium chloride (0.1% active ingredient in the disinfectant suspension) 400 μL were applied repeatedly 32-fold on the test surface (polypropylene plates, 150 x 150 mm). After every application and drying on air over 1 hour, the prepared test plates were washed off with 1 ml of water. It could be demonstrated that under this test condition, an accumulation of benzalconium chloride by the factor of $2.3\text{-}4.3 \times 10^{17}$ molecule/cm^2 does occur (Thimm, Feld 2005).

Cytotoxicity and biocompatibility index (BI)

The BI is defined as quotient from IC_{50} and log reduction factor (RF) $\geq$lg 3 both within 30 min and tested in FBS (Müller, Kramer, 2008). Is the quotient >1, the microbicidal efficacy is higher than the corresponding cytotoxicity and in opposite. Because the quotient of benzalkonium chloride is distinct <1 (table 2), the agent is more toxic than microbicidal effective.

A concentration of 0.01 (Klöcker, Rudolph, Verse, 2004) resp. 0.005% BAC (Mickenhagen et al., 2008) resulted in an irreversible ciliostatic activity for nasociliary epithel, which reduced the conservation of nasalia by BAC. As alternative for chemical conservation of nasalia the 3 K Systems was introduced which is microbiologically safe without conservation (Klöcker et al., 2004).

Table 2: Biocompatibility index (BI) of selected antimicrobial agents, tested with L929 cells of mouse fibroblasts

Antimicrobial Agent	BI (30 min)	
	L929/E. coli	L929/ S. aureus
Octenidine	1.73	2.11
Polihexanide	1.51	1.36
Chlorhexidine digluconate	0.83	0.98
Benzalkonium chloride (BC)	0.63	0.79

Inhalative toxicity: With aerogenic exposure of rats with 30 mg BAC/m3 for 6 h over 3 days (LC_{50} ~ 53 mg/m^3 for 4 h) strong inflammatory irritation was induced. BAC may be classified to class I of acute inhalation toxicity (Swiercz et al., 2008). In humans 600 µg BAC can induce bronchial obstruction in patients with bronchial asthma (Lee, Kim, 2007).

Results of a postal questionnaire survey among 239 pig farmers and 311 rural controls among pig farmers showed that the use of disinfectants (prevalence odds ratio 9.4, 95% confidence interval 1.6-57.2 for QAC) were associated with asthma symptoms. However, pig farmers reported fewer allergies to common allergens ($p<0.001$ for pollen) and fewer symptoms of atopy in childhood ($p<0.05$ for one or more of four symptoms) (Vogelzang et al., 1999).

Sensitization potency: In animals BAC is characterized by low sensitization potency (Fuchs et al., 1993). In humans BAC is a weak allergen. Anaphylactic reactions and eczema by contact as well as by inhalative exposure are described (Uter et al., 2008; Hann, Hughes, Stone, 2007; Anderson et al., 2009; Fräki et al., 1993; Schnuch, 1997).

Dermal absorption: For mecetronium etilsulfate, another QAC, dermal absorption at rat within 24 hours under occlusive conditions was about 2 % (Bode Chemie 2002). However, it has to be pointed out that at present no data on systemic risk exist. Analogously for BAC dermal absorption occur because dermal LD50 was determined with 3700 mg/kg at mouse (Kramer et al., 1985).

Chronic toxicity: The data for chronic toxicity show no hazardous potential; NOEL resp. LOEL for BAC oral 95days/rat 1000 ppm, applied by gastric tube Beagle NOEL 400 ppm, LOEL 1200 ppm. 0.25 % in alimentary diet 2 years/rat no side effects (Widulle et al., 2008). Toxicological data for didecyldimethyl ammonium chloride are in the same range (Widulle et al., 2008).

For all QACs no hint for mutagenicity, teratogenicity and carcinogenicity (Widulle et al., 2008).

Risk benefit assessment: Formulations on base of QAC have the advantage of good cleaning efficiency. Otherwise, BAC is characterized by the following disadvantages:

- Limited efficacy against Gram-negative bacteria and fungi, lack of activity against mycobacteria, bacteria spores and non-enveloped viruses (Widulle et al., 2008); the insufficient antimicrobial spectrum in vitro corresponds with inadequate disinfecting efficacy of QACs in bathrooms and toilets while an active oxygen based compound was effective (Dharan et al., 1999).
- Development of cross-resistance against gentamicin and chlorhexidine (Sidhu, Langsrud, Holck, 2001; Lambert, 2000; Russell, 2002).
- Incompatibility with natural rubber flooring.
- Non-sufficient toxicological characteristic (Widulle et al., 2008).

Additionally, sporulation of *Clostridium difficile* was enhanced when the epidemic strain was cultured in faeces exposed to non-chlorine-based cleaning agents such as QACs, concluding that the choice of disinfecting agent may have an effect on the persistence of bacterial spores in the environment (Wilcox, Fawley, 2000).

DISCUSSION

Undoubtedly, the prototype of QACs, the widely used agent BAC, is not harmless, particularly after repeated disinfection of large surfaces in indoors because of the enrichment of QACs on surfaces and their stability in the environment.

Considering the limited spectrum of antimicrobial activity and the risk of development of antibacterial resistance, QACs are dispensable for surface disinfection in hospitals. Suitable substitutes are H_2O_2, other peroxides and stabilized chlorine dioxide solutions (Callahan et al., 2010). Their spectrum includes vegetative bacteria, viruses, fungi as well as bacterial spores and stabilized chlorine dioxide solutions (Callahan et al., 2010). No inhalative or absorptive side effects are known and there is no risk of toxic residues on surfaces (Kramer et al., 2008). After sewage backflow, organic peroxides were highly effective for surface disinfection and are recommended as biocides of first choice in Germany (Below et al., in rev.).

Another alternative are disinfectants based on carbonic acids, because carbonic acids such as oxidants are without accumulation risk of residues on surfaces (Kramer, Widulle et al., 2008). For small surfaces alcohols are disinfectants of first choice. Their spectrum includes bacteria, fungi, enveloped viruses and for special formulations also non-enveloped viruses (Kramer, Below et al., 2008). Even at repeated dermal and inhalative exposition no critical blood levels are reached (<0.02 ‰) (Kramer et al., 2007; Below et al., 2011).

The same risk assessment is relevant for nebulization of QACs in indoors after dampness damage or fungal attack. A suitable non-toxic alternative for QACs is H_2O_2 (without silver as catalyst). By indoor nebulization of 5-6% H_2O_2 A. brasiliensis is reduced > 4 log on vertical and horizontal surfaces. As a load of 10^4 to 10^5 fungal spores is unlikely to occur on precleaned surfaces, this efficacy seems to be sufficient (Koburger, Below, Kramer, 2011).

Another option would be the use of chlorine dioxide, since it is effective for fogging without long term risk of residues. For most materials, deposition velocity decreased significantly over a 16-h disinfection period (Trinetta et al., 22011). Ozone seems to be unsuitable, because high concentrations (Li, Wang, 2003), high humidity > 90% (Sharma, Hudson, 2008) and long exposure is necessary (Aydogan, Gurol, 2006). Because of the toxic and mutagenic risks (Jorge et al., 2002, Tovalin et al., 2006) and suspicion of carcinogenicity (Victorin, 1996) of ozone a long term exposure is to reject.

Another health risk arises is given by the use of antimicrobial impregnation of carpets with QACs, wallboards, HVAC filters, textiles, etc., if the biocide cannot be completely be immobilized (direct release and by abrasion of material). The impregnation i.e. of wallboards can markedly delay regrowth of fungi, particularly of S. chartarum (Price, Ahearn, 1999), but it cannot exclude the risk of development of resistance as well as of long term exposure to released traces of the biocide in indoor air or by skin contact (Kramer, Guggenbichler et al., 2006).

Aldehydes shall not be used in hospitals as well as in household settings. Formaldehyde is sensitizing, carcinogenic, mutagenic and neurotoxic. Glutaral is classified as asthma inducing agent and is mutagenic. Glyoxal is sensitizing, mutagenic, possibly teratogenic (category 3) and classified under category 3B "possibly carcinogenic for humans" (Kramer et al., 2008).

Phenolic compounds are not as critical as aldehydes, but they lost their importance for surface disinfection because of high persistence in the environment, long term health risks by inhalation including reproduction toxicity, possible sensitization and limited antimicrobial spectrum (Kramer et al., 2008).

CONCLUSION

In health care settings, QACs should not be regarded as agents of first choice for surface disinfection.

In households, QACs are to decline as active substance or combination partner for continues surface disinfection. Even for single extensive sanitation of dampness damages in households QACs are to decline.

REFERENCES

- Anderson D, et al. Anaphylaxis with use of eye-drops containing benzalkonium chloride preservative. *Clin Exp Optom*, 2009:92(5) 444-6.
- Aydogan A, Gurol MD. Application of gaseous ozone for inactivation of Bacillus subtilis spores. *J Air Waste Manag Assoc* , 2006:56(2): 179-85.
- Baudisch C. Schimmelpilzbefall in Räumen und Exposition - Messkonzepte für kultivierbare Schimmelpilze im Hausstaub, im Sedimentationsstaub und in der Raumluft. Diss. rer. med. Med. Fak. Univ. Greifswald, 2010
- Below H, Partecke I, Kramer A et al. Dermal and pulmonary absorption of propan-1-ol and propan-2-ol from hand rubs. *Am J Infect Control*, 2012:40(3) 250-7.
- Bode Chemie. Internal information. Hamburg; 2002, cited by (Lee, Kim 2007).
- Callahan KL, Beck NK, Duffield EA, et al. Inactivation of methicillin-resistant Staphylococcus aureus (MRSA) and vancomycin-resistant Enterococcus faecium (VRE) on various environmental surfaces by mist application of a stabilized chlorine dioxide and quaternary ammonium compound-based disinfectant. *J Occup Environ Hyg* , 2010:7(9): 529-34.

- Dharan S, Mourouga P, Copin P, et al. Routine disinfection of patients' environmental surfaces. Myth or reality? *J Hosp Inf.* 1999:42 (2):113-7.
- Fräki JE, Kalimo K, Tuohimaa P, Aantaa E. Contact allergy to various components of topical preparations for treatment of external otitis. *Acta Otolaryngol,* 1985:100:414-18.
- Fuchs T, Meinert A, Aberer W, Bahmer FA, Peters KP, Lischka GG, Schulze-Dirks A, Enders F, Frosch PJ. Benzalkoniumchlorid -- relevantes Kontaktallergen oder Irritans? Ergebnisse einer Multicenter-Studie der Deutschen Kontaktallergiegruppe. *Hautarzt.* 1993:44(11):699-702.
- Goodman ER, Platt R, Bass R, Onderdonk AB, Yokoe DS, Huang SS. Impact of an environmental cleaning intervention on the presence of methicillin-resistant Staphylococcus aureus and vancomycin-resistant enterococci on surfaces in intensive care unit rooms. I*nfect Control Hosp Epidemiol.* 2008:29(7):593-9.
- Hann S, Hughes TM, Stone NM. Flexural allergic contact dermatitis to benzalkonium chloride in antiseptic bath oil. *Br J Dermatol* 2007:157(4)795-8.
- Hübner NO, Hübner C, Kramer A, Assadian O. Survival of Bacterial Pathogens on Paper and Bacterial Retrieval from Paper to Hands: Preliminary Results. *AJN* 2011:111(12) 2-6.
- Jorge SA, Menck CF, Sies H, et al. Mutagenic fingerprint of ozone in human cells. *DNA Repair,* 2002:1(5) 369-78.
- Kampf G, Dettenkofer M. Desinfektionsmaßnahmen im häuslichen Umfeld - was macht wirklich Sinn? *Hyg Med,* 2011:36 (1/2): 8-11.
- Kampf G, Kramer A. Epidemiologic background of hand hygiene and evaluation of the most important agents for scrubs and rubs. *Clin Microbiol Rev,* 2004:17 (4): 863-93.
- Klöcker N, Kramer A, Verse T, Sikora C; Rudolph P; Daeschlein G. Antimicrobial safety of a preservative-free nasal multiple-dose drug administration system. *Eur J Pharm Biopharm,* 2004:57(3): 489-93.
- Klöcker N, Rudolph P, Verse T. Evaluation of protective and therapeutic effects of dexpanthenol on nasal decongestants and preservatives: results of cytotoxic studies in vitro. *Am J Rhinol,* 2004:18(5) 315-20
- Koburger T, Below H, Dornquast T, Kramer A. Decontamination of indoor air and adjacent wall area by nebulization of hydrogen peroxide. GMS Krankenhaushygiene interdisz, 2011:6(1):Doc08 (20111215).
- Kramer A, Below H, Bieber N et al. Quantity of ethanol absorption after excessive hand disinfection using three commercially available hand rubs is minimal and below toxic levels for humans. BMC Infect Dis 2007:7 p117.
- Kramer A, Below H, Ryll S, et al. Immediate infection control measures and preventive monitoring after excessive water damage in an aseptic working area of a blood donation service center. *Int J Hyg Environm Health,* in rev.
- Kramer A, Galabov AS, Sattar SA, Dohner L, Pivert A, Payan C, Wolff MH, Yilmaz A, Steinmann J. Virucidal activity of a new hand disinfectant with reduced ethanol content: comparison with other alcohol-based formulations. *J Hosp Infect,* 2006:62(1): 98-106.
- Kramer A, Guggenbichler P, Heldt P, et al. Hygienic relevance and risk assessment of antimicrobial-impregnated textiles. *Curr Probl Dermatol* , 2006:33:78-109.

- Kramer A, Reichwagen S, Below H, et al. Alkohole. In: Kramer A, Assadian O (Hrsg) Wallhäusers Praxis der Sterilisation Desinfektion, Antiseptik und Konservierung. Stuttgart, Thieme, 2008:643-69.
- Kramer A, Reichwagen S, Heldt P, et al. Oxidanzien. In: Kramer A, Assadian O (Hrsg) Wallhäusers Praxis der Sterilisation Desinfektion, Antiseptik und Konservierung. Stuttgart, Thieme, 2008; 713-45.
- Kramer A, Reichwagen S, Widulle H, et al. Aldehyde. In: Kramer A, Assadian O (Hrsg) Wallhäusers Praxis der Sterilisation Desinfektion, Antiseptik und Konservierung. Stuttgart, Thieme, 2008; 670-86.
- Kramer A, Reichwagen S, Widulle H, et al. Organische Carbonsäuren. In: Kramer A, Assadian O (Hrsg) Wallhäusers Praxis der Sterilisation Desinfektion, Antiseptik und Konservierung. Stuttgart, Thieme, 2008, 690-710
- Kramer A, Reichwagen S, Widulle H, et al. Phenolderivate. In: Kramer A, Assadian O (Hrsg) Wallhäusers Praxis der Sterilisation Desinfektion, Antiseptik und Konservierung. Stuttgart, Thieme, 2008, 746-69.
- Kramer A, Schwebke I, Kampf G. How long do nosocomial pathogens persist on inanimate surfaces? A systematic review. BMC Infect Dis 2006:6: 130.
- Kramer A, Zbinden G, Stephan U, Koch S, Koch D, Junghans A. Aktue Toxizität von Antiseptika. In: Kramer A, Berencsi G, Weuffen W (Hres) Toxische und allergische Nebenwirkungen von Antiseptika. Bd !/5 Handbuch der Antiseptik, Stuttgart: Fischer, 1985; 113-210.
- Kramer A. Hygienische Risiken des Oderhochwassers 1997. Podiumsdiskussion des Deutschen Verbands für Wasserwirtschaft und Kulturbau "Qualitative Probleme des Hochwassergeschehens" Potsdam, 1.10.1998.
- Kroczek C, Thumulla J. Case report QAC. Proc 15th Fungi Conf June 2011 Bad Staffelstein, Germany, Fürth: *AnBUS*, 2011; 133-6.
- Lambert RJ. Comparative analysis of antibiotic and antimicrobial biocide susceptibility data in clinical isolates of methicillin-sensitive Staphylococcus aureus, methicillin-resistant Staphylococcus aureus and Pseudomonas aeruginosa between 1989 and 2000. *J Appl Microbiol.* 2004:97(4):699-711.
- Lee BH, Kim SH. Benzalkonium chloride induced bronchoconstriction in patients with stable bronchial asthma. *Korean J Intern Med*, 2007:22(4): 244-8.
- Li CS, Wang YC. Surface germicidal effects of ozone for microorganisms. AIHA J, 2003: 64(4): 533-7.
- Mickenhagen A Siefer O, Neugebauer P, Stennert E. The influence of different alpha-sympathomimetic drugs and benzalkoniumchlorid on the ciliary beat frequency of in vitro cultured human nasal mucosa cells. *Laryngorhinootol*, 2008:87(1): 30-8.
- Müller G, Kramer A. Biocompatibility index of antiseptic agents by parallel assessment of antimicrobial activity and cellular cytotoxicity. *J Antimicrob Chemother*, 2008:61(6) 1281-7.
- Neely AN, Maley MP. Survival of enterococci and staphylococci on hospital fabrics and plastic. *J Clin Microbiol.*, 2000:38(2):724-6.

- Neugebauer P, Stennert E. The influence of different alpha-sympathomimetic drugs and benzalkoniumchlorid on the ciliary beat frequency of in vitro cultured human nasal mucosa cells. Laryngorhinootol, 2008:87(1): 30-8.
- Price DL, Ahearn DG. Sanitation of wallboard colonized with *Stachybotrys chartarum. Curr Microbiol* ,1999:39(1): 21-6.
- Russell AD. Introduction of biocides into clinical practice and the impact on antibiotic-resistant bacteria. *Symp Ser Soc Appl Microbiol.*, 2002:31:121S--35S.
- Schnuch A. Benzalkoniumchlorid. Informationsverbund Dermatologischer Kliniken (IVDK). Universitäts-Hautklinik Göttingen. *Dermatosen.* 1997:45(4):179-80.
- Sharma M, Hudson JB. Ozone gas is an effective and practical antibacterial agent. *Am J Infect Control*, 2008:36(8): 559-63.
- Sidhu MS, Langsrud S, Holck A. Disinfectant and antibiotic resistance of lactic acid bacteria isolated from the food industry. *Microb Drug Resist.* 2001:7(1):73-83.
- Swiercz R, Halatek T, Wasowicz W, Kur B; Grzelinska Z; Majcherek W Pulmonary irritation after inhalation exposure to benzalkonium chloride in rats. *Int J Occup Med Environ Health*, 2008:21(2): 157-63.
- Thimm N, Feld H. Proof/identification of residues of disinfectants on surfaces. Report ULM01-1587, OFG - Analytik Münster, Germany, 09 March 2005.
- Trinetta V, Vaidya N, Linton R, et al. Evaluation of chlorine dioxide gas residues on selected food produce. *Food Sci*, 2011:76(1): T11-5.
- Tovalin H, Valverde M, Morandi MT, et al. DNA damage in outdoor workers occupationally exposed to environmental air pollutants. *Occup Environ Med*, 2006:63(4): 230-6.
- Uter W, Lessmann H, Geier J, Schnuch A. Is the irritant benzalkonium chloride a contact allergen? A contribution to the ongoing debate from a clinical perspective. *Contact Dermatitis* 2008:58(6) 359-63.
- Victorin K. Genotoxicity and carcinogenicity of ozone. *Scand J Work Environ Health,* 1996:22 (Suppl 3): 42-51.
- Vogelzang P FJ, van der Gulden JWJ, Tielen MJM, Folgering H, et al. Health-based selection for asthma, but not for chronic bronchitis, in pig farmers: an evidence-based hypothesis. *Eur Resp J,* 1999:13 (1): 187-9.
- Widulle H, Kramer A, Reichwagen S, Heldt P. Oberflächenaktive Verbindungen. In: Kramer A, Assadian O (Hrsg) Wallhäusers Praxis der Sterilisation Desinfektion, Antiseptik und Konservierung. Thieme: Stuttgart, 2008; 770-85.
- Wilcox MH, Fawley WN. Hospital disinfectants and spore formation by Clostridium difficile. *Lancet*, 2000:356(9238): 1324.

TOXICITY STUDY OF FIELD SAMPLES FROM WATER DAMAGED HOUSES IN FLOODED AREAS IN POLAND

Magdalena Twarużek, Jan Grajewski, Manfred Gareis

INTRODUCTION

This presentation summarizes the results of air quality investigations conducted in flooded homes in Poland. During the last decades there have been periods of significant draughts and floods in certain areas of Poland. These natural disasters may result in additional health threats to the inhabitants because of the major moisture damage and mold growth inside residential houses.

Flood simulations performed by the British Hadley Centre in Bracknell have shown that in the coming years large areas of Poland may be increasingly threatened by flooding. Other European countries may be also affected by this. The evidence for these threats is strengthened by the disaster caused by the flooding of the Oder and Vistula rivers, and recently also in Germany and the Czech Republic. The flooding of the Oder in 1997 resulted in the death of 114 people in 2592 villages and about half a million hectares of farm land were under water. The total loss was estimated at 6.5 billion euros (Hołdys, 2004).

The environmental conditions caused by flooding in these areas resulted in large scale mold growth and contamination in buildings and resulted in an immediate threat to the health of the residents. This fact is confirmed by the example of the Wroclaw - Brochów geographic area. Temporary residential housing had to be provided by the government. Many tenants could not return to their former homes because the interior mold growth and the likely adverse health effects in their homes prevented them from returning (Twarużek, 2005). The General Office of Building Control provided these temporary homes, however they were often in poor physical condition, and most importantly, many design defects were obvious: lack of proper heating, inadequate ventilation and damp soil, which all resulted in stains, moisture and interior mold growth (Janińska, 2002).

In 2002, another large scale flooding occurred in Poland and damaged buildings and houses in the north and south of Poland. Due to the increasing research interest in the fungal pathology of different species in the indoor environment, an effort was made to study the mycology and mycotoxins analysis of air and building partitions in the houses of the flooded areas. Earlier reports by other authors (Gareis 1994, Gareis et al., 1998; Johanning et al., 1998) were a motivation to conduct the following examinations.

MATERIAL AND METHODS

Field investigations were carried out over several years following widespread flooding in Poland in 2002. Air and wall samples were taken from the inside of buildings and houses in the previously flooded areas of northern Poland (Gdańsk, n=14) and southern Poland (Zembrzyce, Budzów, Baczyn, Maków Podhalański, n=15). A total of 58 samples (air samples n=29 and samples taken from walls by washing, n= 29) were mycologically analysed. Control air- and wall-samples were taken from non-contaminated rooms. Some of the houses were sampled again with the permission from the owners. For this purpose, samples of wallpaper or plaster were taken from selected homes in South Poland (n=8) and in North Poland (n=4). Samples were analyzed using a cell culture cytotoxicity assay method (MTT-bioassay, Hanelt et al., 1994) in order to screen for the presence of cytotoxic compounds and an ELISA for the determination of macrocyclic trichothecenes. Additionally, ergosterol and ochratoxin A were analyzed by HPLC method.

RESULTS

In the air samples, fungi of at least 13 different genera were found. One year after the flooding occurred, air samples from moldy rooms from homes in South and North Poland contained on average $4.1x10^3$ and 2.2×10^3 cfu/m^3, respectively (fig. 1, 2). Air samples from control rooms contained on average $1.2x10^3$ and $4.7x10^2$ cfu/m^3, respectively. In the north of Poland eight different fungi and in the south of Poland fifteen different fungi, including *Stachybotrys* were detected in wall samples taken from visibly moldy homes. These molds typically occur in water-damaged buildings, and mainly belong to the genera *Penicillium, Cladosporium, Acremonium* and *Aspergillus* (tab. 1,2). Contamination of walls with molds was found to be higher in the North (1.1×10^7 cfu/ 100 cm^2) than in the south of Poland (4.1×10^6 cfu/100 cm^2) (fig. 3,4). *Stachybotrys spp.* were found in four samples (3 wall paper samples and 1 plaster sample). A total of 12 samples originating from the north and south of Poland have been tested with cytotoxicity test. Testing results showed that 5 of these samples were cytotoxic. The cytotoxicity correlated with findings of *Stachybotrys* and data obtained with the trichothecene-ELISA. In the samples Roridin A-equivalents up to 7.8 ng/g have been detected. Ochratoxin A was detected in only 3 samples in low concentrations (< 1.0 µg/kg) (tab. 3,4).

Figure 1 Indoor air concentration of fungi detected in homes in North-Poland (cfu/m³).

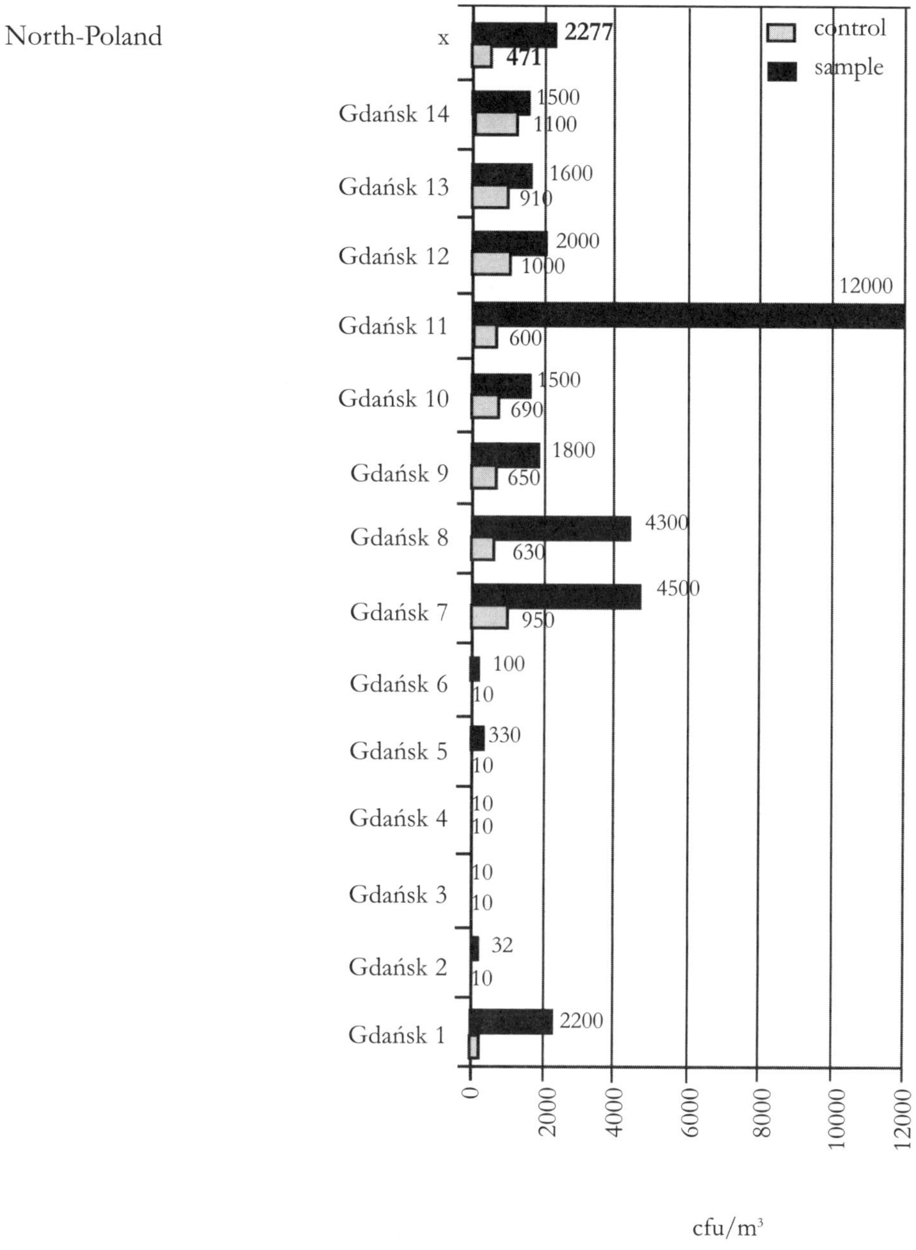

Figure 2 Indoor air concentration of fungi detected in homes in South-Poland (cfu/m³).

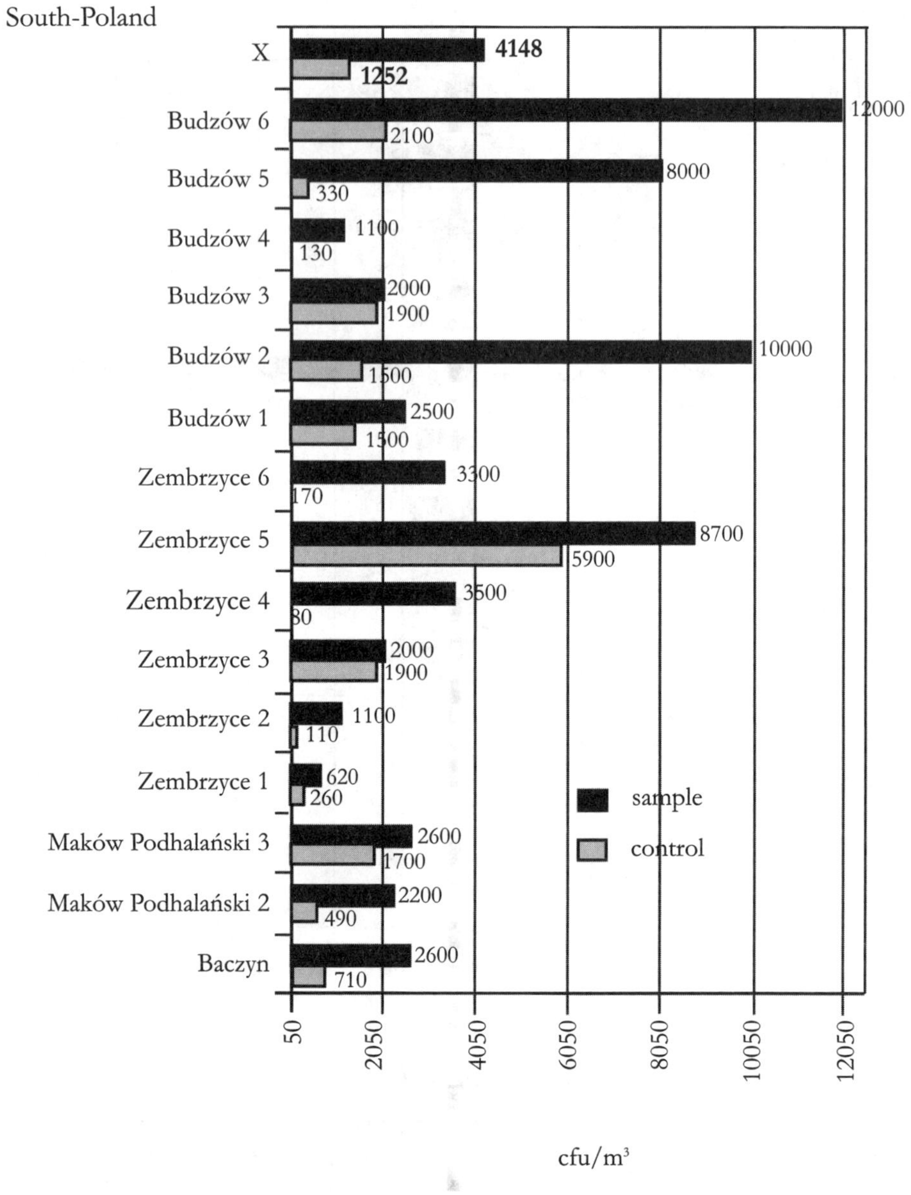

Figure 3 Indoor wall concentration of fungi detected in homes in North-Poland (cfu/100 cm²)

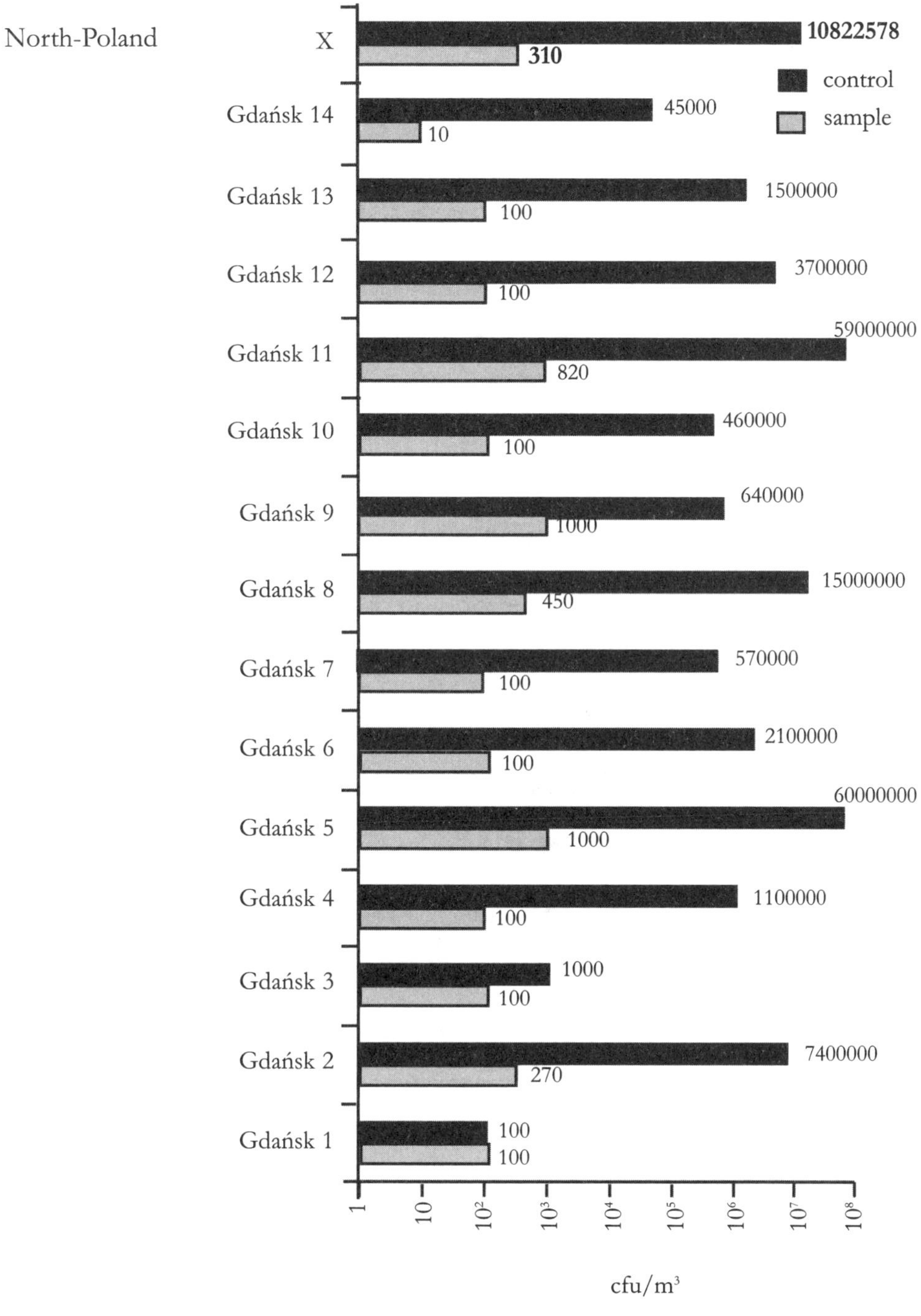

Figure 4: Indoor wall concentration of fungi detected in homes in South-Poland (cfu/100 cm²) (MP- Maków Podhalański).

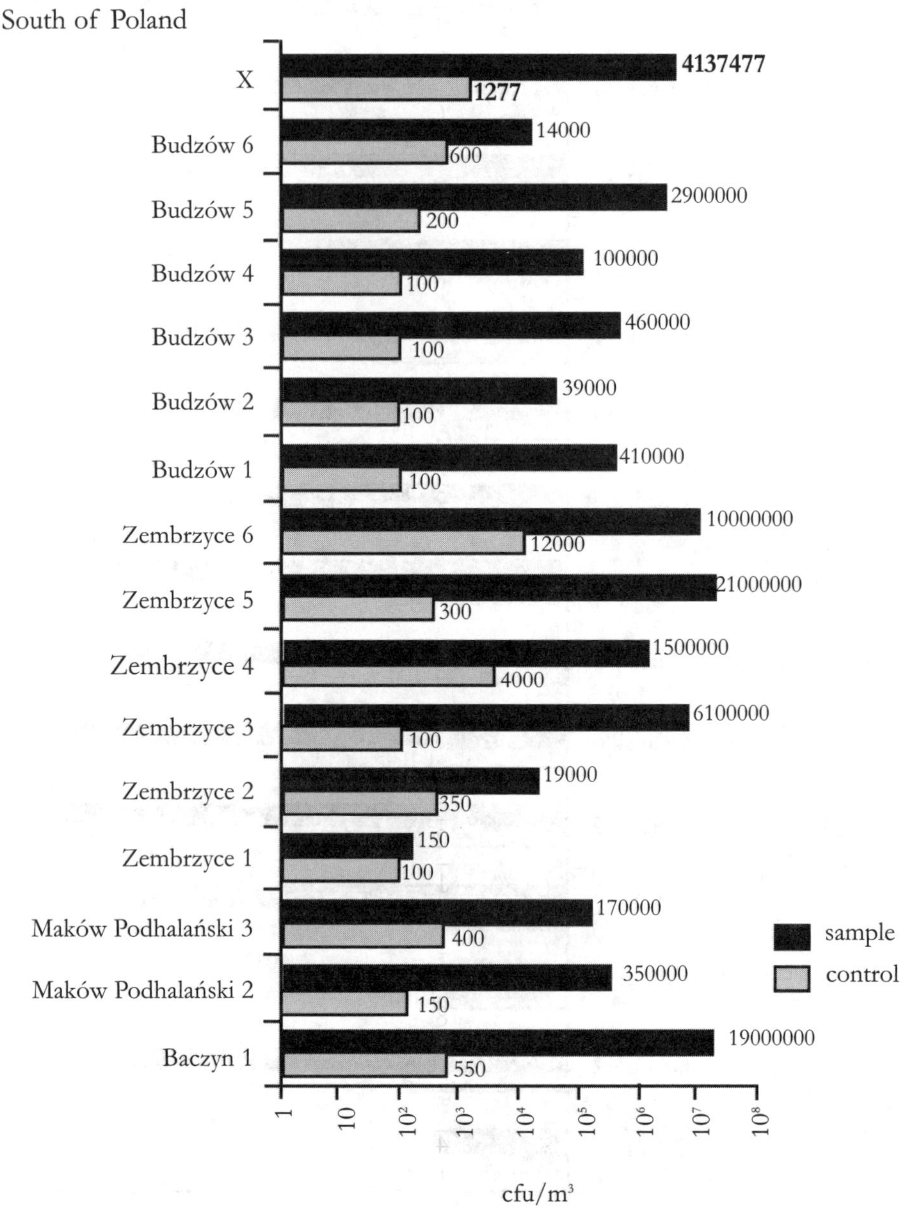

Table 1: Fungus genera identified in air samples from the inside homes.

North Poland		South Poland	
control	sample	control	sample
20% *Cladosporium*	21% *Penicillium*	21% *Penicillium*	22% *Penicillium*
18% *Penicillium*	17% *Alternaria*	20% *Cladosporium*	22% *Cladosporium*
16% *Alternaria*	17% *Cladosporium*	20% *Alternaria*	17% *Alternaria*
11% *Trichoderma*	13% *Fusarium*	15% *Aspergillus*	9% *Aspergillus*
11% *Fusarium*	8% *Aspergillus*	7% *Fusarium*	5% *Fusarium*
9% *Humicola*	6% *Chaetomium*	6% *Ulocladium*	5% *Trichoderma*
7% *Chaetomium*	6% *Trichoderma*	3% *Trichoderma*	5% *Epicocum*
4% *Ulocladium*	4% *Humicola*	3% *Humicola*	5% *Ulocladium*
4% *Verticillium*	4% *Ulocladium*	5% *Mucor*, *Paecilomyces*, *Trichothecium*, *Epicoccum*, *Chaetomium*	10% *Mucor*, *Paecilomyces*, *Acremonium*, *Verticillium*, *Rhizopus*
	4% *Acremonium*, *Verticillium*		

Table 2: Fungus genera identified on wall samples from the inside of homes.

North Poland		South Poland	
control	sample	control	sample
35 % *Cladosporium*	31% *Cladosporium*	26% *Cladosporium*	18 % *Penicillium*
25 % *Penicillium*	29% *Aspergillus*	23% *Penicillium*	18% *Cladosporium*
21% *Aspergillus*	23% *Penicillium*	13% *Acremonium*	17% *Acremonium*
7 % *Alternaria*	9% *Acremonium*	10% *Trichoderma*	15% *Aspergillus*
3% *Acremonium*	2% *Fusarium*	10% *Aspergillus*	6% *Ulocladium*
3% *Trichoderma*	2% *Alternaria*	8% *Alternaria*	5% *Alternaria*
3% *Myrothecium*	2% *Chaetomium*	3% *Stachybotrys*	5% *Chaetomium*
3% *Aureobasidium*	2% *Stachybotrys*	3% *Paecilomyces*	5% *Trichoderma*
		4% *Verticillium*, *Botrytis*	2% *Stachybotrys*
			9% *Verticillium*, *Pae-cilomyces*, *Epicocum*, *Fusarium*, *Chrysospori-um*, *Scopulariopsis*

Table 3 Results of bulk samples from North-Poland

City	Description of the sample		Cytotoxicity*	Roridin A-Equivalent (µg/kg)	OTA (µg/kg)	Ergosterol (mg/kg)
Gdańsk 1	Wall paper	control	-	n.d.	n.d.	1,02
		sample	++	1	n.d.	3,63
Gdańsk 4	Wall paper	control	-	n.d.	n.d.	1,80
		sample	-	<1	n.d.	16,00
Gdańsk 5	Wall paper	control	-	n.d.	n.d.	2,17
		sample	+++	3,1	n.d.	92,97
Gdańsk 14	Wall paper	control	-	n.d.	n.d.	1,16
		sample	+++	1,3	0,28	5,80

Samples marked bold: positive for *Stachybotrys* sp.
* : – (no cytotoxic effects) to +++ (high cytotoxicity)
n.d. - not detected

Table 4 Results of bulk samples from South-Poland

City	Description of the sample		Cytotoxicity*	Roridin A-Equivalent (µg/kg)	OTA (µg/kg)	Ergosterol (mg/kg)
Baczyn 1	Wall paper	control	-	n.d.	n.d.	0,72
		sample	+++	7,8	0,58	42,11
Maków Podhalański 3	plaster	control	-	n.d.	n.d.	1,03
		sample	-	<1	0,8	6,65
Zembrzyce 2	plaster	control	-	n.d.	n.d.	1,00
		sample	+	1	n.d.	8,02
Zembrzyce 6	plaster	control	-	n.d.	n.d.	1,80
		sample	-	<1	n.d.	26,35

Table 4 continued

City	Description of the sample		Cytotoxicity*	Roridin A-Equivalent (µg/kg)	OTA (µg/kg)	Ergosterol (mg/kg)
Budzów 1	plaster	control	-	n.d.	n.d.	0,48
		sample	+	1	n.d.	4,50
Budzów 2	plaster	control	-	n.d.	n.d.	0,76
		sample	-	<1	n.d.	1,82
Budzów 5	plaster	control	-	n.d.	n.d.	1,61
		sample	-	<1	n.d.	10,41
Budzów 6	plaster	control	-	n.d.	n.d.	1,32
		sample	++	1,2	n.d.	4,00

Samples marked bold: positive for *Stachybotrys* sp.
* : – (no cytotoxic effects) to +++ (high cytotoxicity)
n.d. - not detected

CONCLUSION

In conclusion, water-damaged homes in the flooded areas were found to be highly contaminated with different fungi known to occur in water-damaged building materials. Geographical differences regarding mold contamination problems were not found in this study. The finding of toxigenic *Stachybotrys* isolates in these indoor environments are proven for the first time in Poland.

REFERENCES:

- Gareis M. Cytotoxicity of samples originating from problem buildings. In: Johanning E., Yang Ch.S., Fungi and bacteria in indoor air environments: Health effects, detection and remediation. 1994:139-144.
- Gareis M., Johanning E., Dietrich R. Mycotoxin cytotoxicity screening of field samples. In Johanning E.: Bioaerosols, fungi and mycotoxins: health effects, assessment, prevention and control, New York. 1999:202-213.
- Hanelt M., Gareis M., Kollarczik B. Cytotoxicity of mycotoxins evaluated by the MTT-cell culture assay. Mycopathologia 1994:128: 167-174.

- Hołdys A. Rzeki się wściekną. *Gazeta Wyborcza* (in Polish). 2004:3-4.04, 9.
- Janińska B. O termomodernizacji i zmianach mikroklimatu sprzyjających rozwojowi grzybów pleśniowych. Inżynieria i Budownictwo, (in Polish). 2002:7: 360-363.
- Johanning E, Gareis M, Yang CH. S, Hintokka E-L, Nikulin M, Jarvis B, Dietrich R. Toxicity screening of materials from buildings with fungal indoor air quality problems (*Stachybotrys chartarum*). Mycotoxin Research 1998:4: 60-73.
- Twarużek M. Wykorzystanie biologicznych testów (MTT, PremiTest) w ocenie skażeń pomieszczeń mieszkalnych mikotoksynami grzybów pleśniowych (in Polish). PHD thesis, 2005:pp. 119.

MYCOTOXIN SCREENING OF INDOOR ENVIRONMENTS IN SENTINEL HEALTH INVESTIGATIONS

Eckardt Johanning, Manfred Gareis

ABSTRACT

In environmental and occupational studies of bioaerosols risk assessment is typically based solely on fungal identification and quantification. However, specific health outcomes are often poorly correlated with these parameters. This may be in part due to biological fungal air contaminants that have different toxic properties. Traditional in-vitro chemical laboratory analyses of fungi have practical limitations. In-vivo toxicity screening appears preferential to explore complex health reactions reported by exposed patients. In these sentinel health effect studies, we compared conventional fungal identification methods with a screening test for mycotoxins (cytotoxicity of mycotoxins evaluated by the MTT-cell culture assay) and EIA quantification of trichothecenes designed for detection of Roridin A and other macrocyclic trichothecenes. High (24h) volume air sampling to collect air-borne particles (n=225) was conducted in homes and work places of patients (n=115) with environmental symptomatology and visible fungal indoor growth. The crude extracts of approximately two thirds of the air samples showed mild to high toxicity in the MTT cytotoxicity assay and 19 % of n=214 samples had Roridin A results of >10 ng/g. Among all the fungi identified, there was only a weak association of viable Stachybotrys fungi and Roridin but not with other fungi. In conclusion, traditional fungal identification methods in bioaerosols exposure studies appear to be a poor predictor of toxicity without the use of the effect-based toxicity bioassay to assess and confirm toxicity. The MTT cell culture cleavage assay is an effect-based screening test which has appears to be quick and easy method. It facilitates evaluations of the biological activity of many different mycotoxins. It may provide a useful tool for the testing of a large variety of sample materials, including indoor air contaminants.

BACKGROUND

In conventional environmental indoor investigations, typically either viable or non-viable fungal sampling or identification methods are regularly used. These appear to only poorly correlate in some studies with known adverse health effects of wet and moldy indoor environments such as respiratory tract abnormalities (i.e., asthma and rhino-sinusitis) or especially with immunological and neurological symptoms. Some of the indoor fungi in damp and wet building are known to be able to produce mycotoxins, however the actual production of mycotoxins depends on specific field conditions and varies a lot. Most importantly past research by this investigator group and others has shown, that the presence of certain fungi (i.e., *Stachybotrys, Chaetomium, Paecilomyces, Penicillium, Aspergillus*) does not predict in-vivo toxicity. In the past the majorities of indoor dampness and mold field studies have either no or only limited airborne mycotoxins exposure assessed due to methodological and sampling limitations. Furthermore, it has been shown, that human health risks from mycotoxins exposures appear to be higher through inhalation than other routes of exposure in indoor settings.

OBJECTIVE

The objective of these clinically driven sentinel health studies was to detect airborne fungal toxins. In addition, the goal of this study was to compare conventional sampling methods and fungal identification methods (viable mycological identification) with an effect--based toxicity screening tests (MTT-test) and a macrocyclic trichothecene-specific enzyme-linked immunosorbent assay (Roridin A-ELISA) of.

METHODS

Indoor environments (i.e. homes, offices, etc.) of 115 patients evaluated for building related diseases and non-allergic health symptoms between 1999 and 2010 were studied with verified moisture-related building damage and indoor fungal growth. All studied buildings were located throughout North America, and varied in size, age and climate zones. In total, 225 high-volume air samples were systematically analyzed for trichothecene (Roridin A-equivalents) content by the ELISA method and fungi.

Fungal identification and counts:

Culture Method: All filter samples were cultured following standard laboratory practices. Stachybotrys was identified using light microscopy at a power of 400x. Non-viable fungal structure were identified and counted utilizing a volumetric portable air sampler (Mfg. Burkard Company™); sampling rate was 50 ml/min; spores, hyphal fragments and conidiophores were counted.

MTT-Bioassay

The colorimetric tetrazolium MTT cleavage test is based on the transformation of the yellow tetrazolium salt MTT by viable, living cells (via mitochondrial dehydrogenase) to purple formazans. At the end of the incubation period a volume of 20 µl of the MTT stock solution ([3(4,5-dimethylthiazol-2-yl)2,5-diphenyltetrazolium bromide; Sigma-Aldrich) in PBS at a final concentration of 5 mg/ml was added to each well and the plates incubated for another 4 hours at 37 °C in 5% CO_2. Supernatant was then removed using a multichannel micro pipette and 100 µl DMSO was added to each well in order to dissolve the dark formazan crystals. The optical density of each well was measured spectrometrically with an ELISA-Reader at a wavelength of 510 nm and data calculated with MicroWin 2000. Mean extinction values and standard deviations of each sample concentration were compared with those of the corresponding control and expressed as % cleavage activity in comparison to cell controls (100 %). The cytotoxic endpoints, i.e. the minimum concentration of the test reagents measured to cause toxic effects were determined on the basis of the IC_{50} value (Inhibitory Concentration $_{50}$= concentration resulting in 50% inhibition of the MTT cleavage activity).

Toxicity Analysis of Air Samples

Air sampling for 24 h was carried out with a high-volume sampler at a rate of 60 cf/min with a 8x11 inch micropore-paper filter. Two g of the filter paper material were crushed by cutting and soaked in 40 ml of methanol over night. The paper was then filtrated and extracted twice with 40 ml chloroform and 40 ml methanol for 30 minutes each. An aliquot of the raw extract from the sample was evaporated to dryness. The residue was dissolved in phosphate buffered saline (PBS) containing 10% methanol and directly assayed by EIA using the EnviroLogix QuantiTox kit for Trichothecenes (EnviroLogix Inc. 500 Riverside Industrial Parkway Portland, Maine USA). This is a 96-well plate kit designed for the quantitative laboratory detection of Tricothecenes, including Roridin A, E, H and L-2, Satratoxin G and H, Isosatratoxin F, Verrucarin A and J, and Verrucarol in bulk samples. The assay's quantitation range is from 0.2 to 18.0 parts per billion (ppb) of Roridin A in the sample extract. This Kit does not distinguish between various macrocyclic trichothecenes, but detects their presence to differing degrees. Results of this assay are expressed as RoA-equivalents. For the statistical analysis, *ELISA trichothence RorodinA-equivalents* (RoA)results were categorized in: < 2 ng/g; 2 < 5 ng/g; 5 < 10 ng/g; 10 < 50 ng/g; and => 50 ng/g.

RESULTS

Key points and results of MTT toxicity screening test of 24h high volume air-samples (n=225):

More than 78 % showed some detectable toxicity, with 19% in the very high category. More than two-thirds of the samples (n=214) analyzed for RoA showed a level of >2 ng/g, 5.5 % were > 50 ng/g. Among all the fungi identified, there was only a weak association of viable *Stachybotrys chartarum* fungi and RoA -equivalent, but not with other toxigenic fungi or any of the non-viable fungal identification methods. The MTT screening test correlated well with the EIA Roridin A.

CONCLUSIONS

In conclusion, the detection of airborne trichothecene mycotoxins in sentinel investigations was confirmed with high-volume air samples utilizing the MTT effect-based screening test and further by the Roridin-A ELISA test for trichothecenes. It could be shown that traditional sampling methods and identification for fungi appear to be insufficient predictors of the airborne mycotoxins/trichothecene toxicity. It would appear prudent not to base a health risk assessment of toxic mold exposure solely on traditional fungal quantification and identification methods. Furthermore, the MTT cell culture cleavage assay is an effect- based screening test which has been found to be quick and easy to perform evaluations of the biological activity of many different mycotoxins. It may also provide a useful tool for the testing of a large variety of sample materials, including indoor air contaminants.

ACKNOWLEDGMENT:

We wish to thank Ms. Renate Schneider (lab. tech.) (MRI, Kulmbach, Germany for her dedicated assistance in this study).

GRAPH 1: Methodology of mycotoxin analysis in indoor environments

<u>Inhalation Exposure - Logistics and Methodology</u>

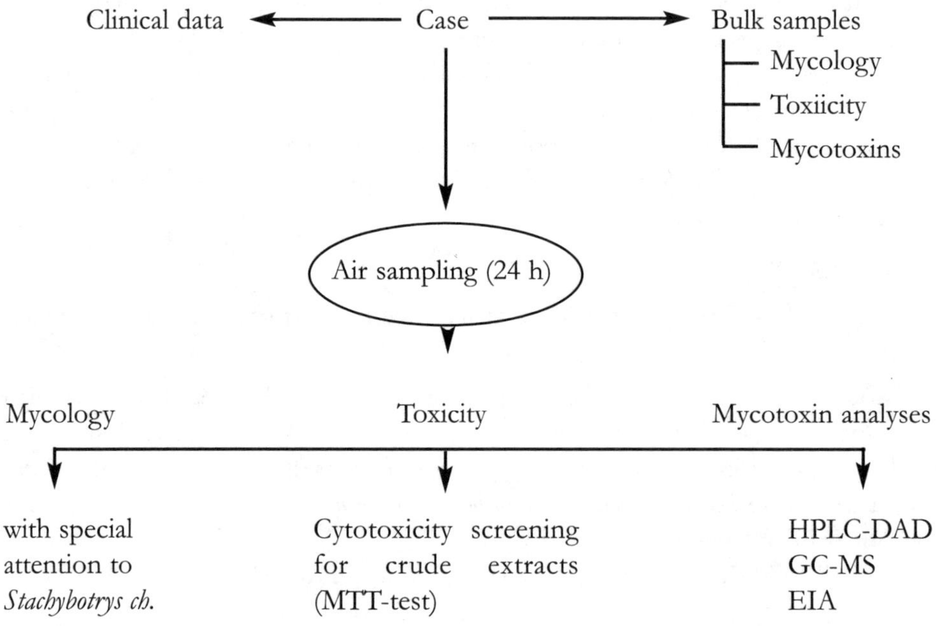

GRAPH 2: MTT cyto-toxicity screening test results: The results of the effect based toxicity screening test with the MTT assay showed that about 2/3 of the high-volume 24h air samples (n=225) showed some level of toxicity, with about 31 % showing moderate to high level toxicity.

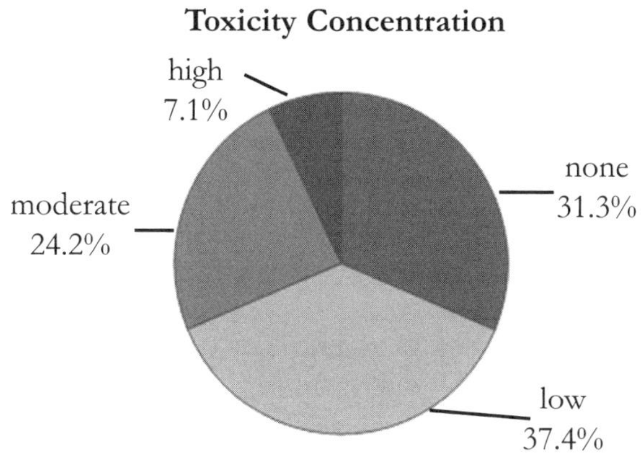

GRAPH 3: Results of Roridin A concentration. About two-third (58.8 %) of the samples (n= 214) analyzed for Roridin A mycotoxins showed levels >2 ng/g and 5.5 % were > 50 ng/g.

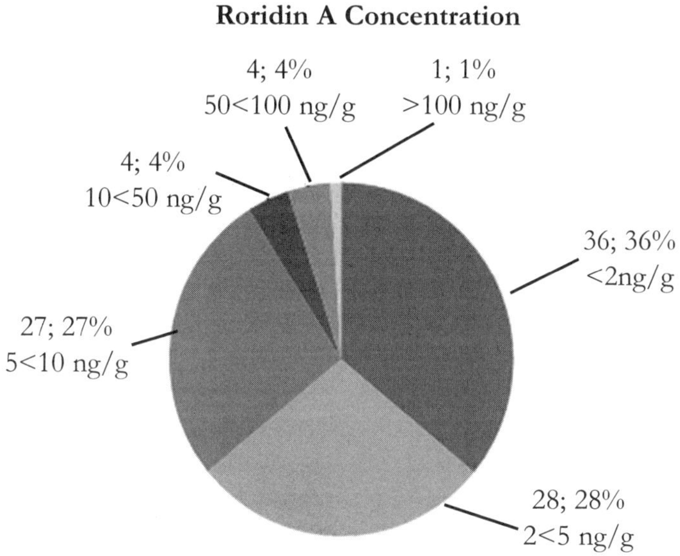

REFERENCES

- Gareis M, Johanning E, Dietrich R. Mycotoxin Cytotoxicity Screening of Field Samples. In: Johanning E, editor. Bioaerosols, fungi and Mycotoxins: Health Effects, Assessment, Prevention and Control. Albany, N.Y.: Eastern New York Occupational and Environmental Health Center, 1999: 202-213.
- Hanelt M, Gareis M, Komulainen H. Cytotoxicity of mycotoxins evaluated by the MTT cell culture test assay. *Mycopathologia.* 1995: 128, 167-174.
- Johanning E, Gareis M, Yang CS, Hintikka EL, Nikulin M, Jarvis B et al. Toxicity screening of materials from buildings with fungal indoor air quality problems (*Stachybotrys chartarum*). *Mycotoxin Research* 1999: 14(2):60-73.
- Johanning E.Mycotoxin air sampling in indoor environments with moisture related fungal contamination (*Stachybotrys chartarum*). Singapore: 26th International Congress on Occupational Health, 2000.
- Johanning E, Gareis M, Nielsen K, Dietrich R, Märtlbauer. Airborne mycotoxin sampling and screening analysis. Proceedings Indoor Air 2002, Hal Levin (Editor), 2002.
- Johanning E, Gareis M. Mycotoxin screening method using a MTT cleavage assay for environmental and occupational health studies. Medichem, Heidelberg, Germany. 2011.

A COMPARISON OF TWO SAMPLING MEDIA (MEA AND DG18) FOR ENVIRONMENTAL VIABLE MOLDS

Sirkku Häkkilä

ABSTRACT

Often the evaluation of fungal damage in buildings is carried out by culturing samples on different kind of growth media in laboratory conditions. This paper gives information about how choice of culture media does affect on the results you get, and what kind of differences are there between the two most often used mould culture media in indoor environmental studies. In the study a comparison of the culture results of 200 air samples and 200 material samples cultured both on Malt Extract agar and Dichloran Glyserol 18 agar was carried out. The mean total CFU counts and number of taxa on each media was almost the same. Although there were there were no significant differences between the two media in mean total CFU - counts /m^3 or /g in some samples the interpretation of the culture results would be different if only one type of culture medium was used. The biggest differences between the two media, was in the species composition: the xerophilic species were almost absent from MEA.

INTRODUCTION

There are many types of mycological analyses to estimate the degree of mold contamination in buildings - i.e. in building materials and in indoor air. Most often the evaluation of fungal damage in buildings is carried out by culturing the samples on different artificial media, by analysis of the composition of microbes in air or in material samples, and calculation of the microbial concentrations in samples (Wu et al., 2000). The ideal culture medium should allow the growth of most saprophytic fungi (Wu et al., 2000). The most frequent culture medium for detecting fungi growing in indoor environmental samples is MEA (malt extract agar). (Wu et al., 2000)) DG18 (dichloran glycerol-18 agar) is sometimes used for detecting fungi growing at lower water activity (Wu et al., 2000). DG18 was developed to detect fungi growing on low moisture foods (Hocking & Pitt, 1980; Samson, 2004).

The objective of this study was to compare the features of the two most often used culture media in indoor environmental studies: MEA (Malt extract agar, Blakeslee formula) and DG18 (dichloran glycerol-18 agar, Pitt). Does culturing samples also on DG18 -medium give information than culturing only on MEA doesn't when estimating the presence of mold and moisture damage in buildings? Are there differences between MEA and DG18 in species richness, in taxa growing on each media and/or in total cfu counts?

MATERIAL AND METHODS

In the comparison the growth results of two different kinds of samples were involved: 1) growth results of 200 air samples collected with a six-stage impactor and 2) growth results of 200 building material samples cultured by dilution method. All the samples were cultured both on DG18 and MEA. The samples were incubated for ten to fourteen days at 25°C. After 7 days incubation the colonies were counted and after 10 to fourteen days colonies were identified based on morphological features at genus or species level (some *Aspergillus* – species) and concentrations of fungi were expressed as CFU/m^3 (colony forming units/m^3) or CFU/g. The growth results of the two media were compared by the total CFU counts, by the mean number of taxa growing on each culture media and by the differences on species composition.

RESULTS

The comparison showed that there were no significant differences in mean total CFU/m^3 counts between MEA and DG18 in air samples and in the mean total CFU/g counts between MEA and DG18 in material samples (Fig 1. ans 2.). The mean number of taxa encountered in samples was approximately same on both culture media (Fig 3. and 4.).

Figure 1. and 2. Total CFU/m^3 in air samples and total CFU/g in material samples.

Fig. 1

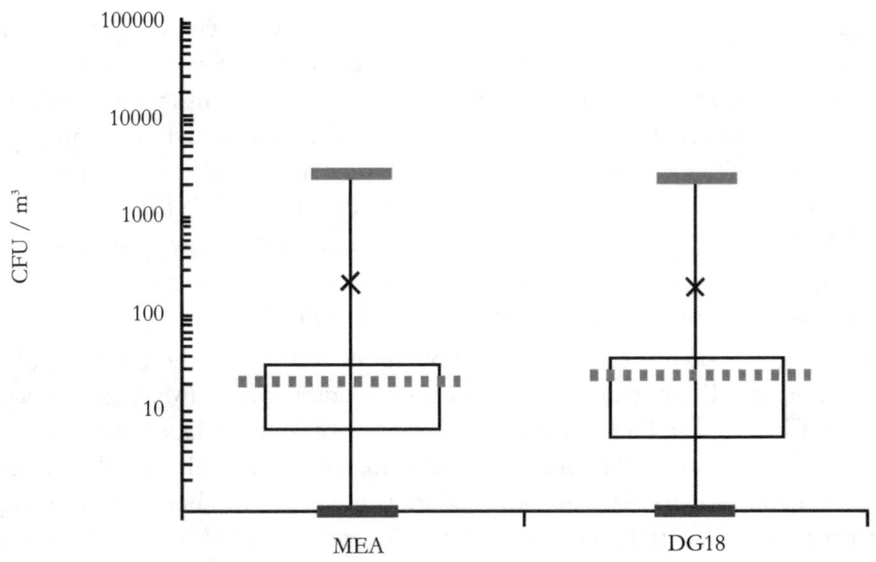

Fig. 2

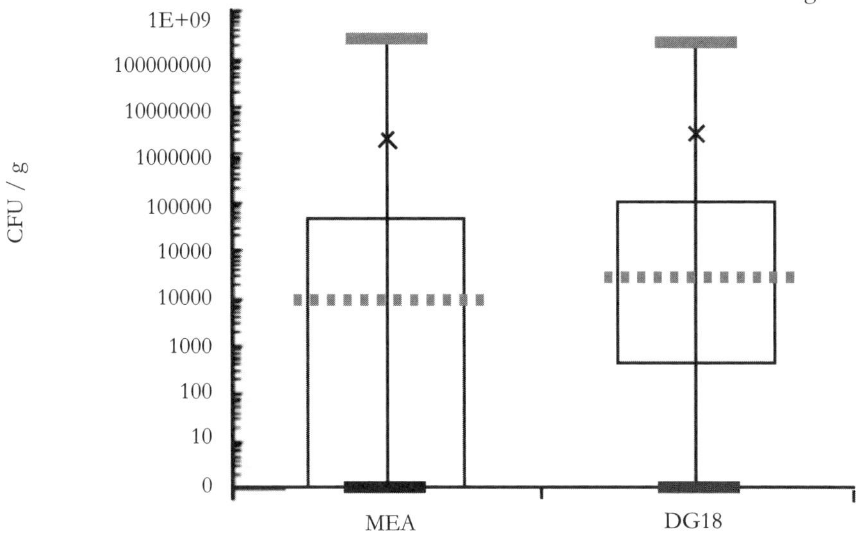

Figure 3 and 4. The number of taxa/sample in air samples and in material samples

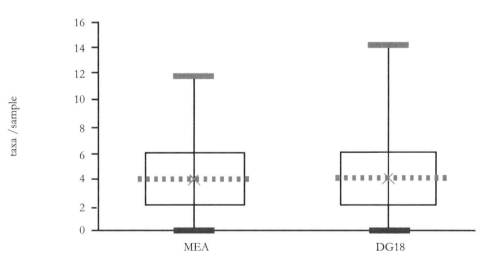

Fig. 3

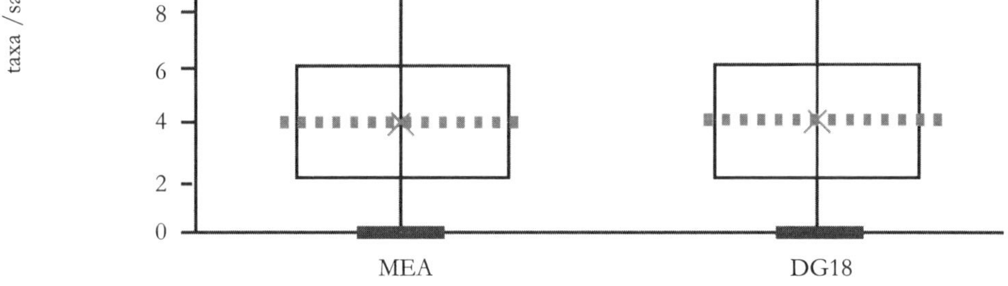

Fig. 4

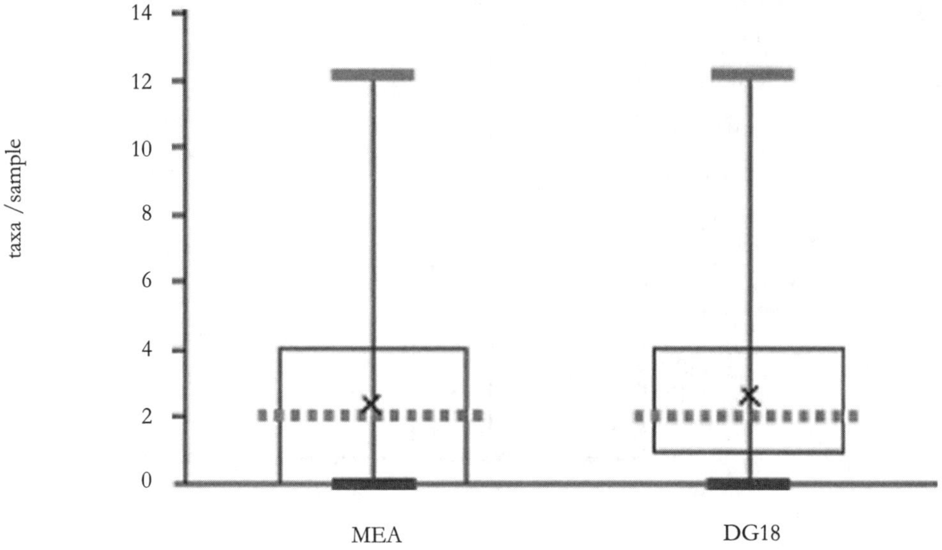

The biggest differences between the two media were on the taxa growing on the media. The xerophilic *Aspergillus restrictus* group was present in 50% of the air samples and in 30% of the material samples but was almost missing on MEA (present in 2 % of the air samples and 1 % of material samples) as well other xerophilic taxa such as *Wallemia* and *Eurotium* grew more often on DG18. Respectively *Engyodontium, Oidiodendron.* basidiomycetous fungi and mycelia sterilia (unidentified fungi) were encountered on malt extract agar more often than on DG18 in air samples. In material samples the genus *Fusarium* and *Acremonium* were growing on MEA more often than on DG18 but the differences in abundance were not as clear as with the xerophilic moulds (table 1).

Table 1. Taxon present in % of samples (all the taxa included that were present □ 10% of the samples on either one of the media).

	Air samles			Material samples	
	MEA	DG		MEA	DG 18
Penicillium	76	81	*Penicillium*	48	52
Cladosporium	47	54	*Aspergillus res.*	1	30
Aspergillus restrictus group	2	51	*Aspergillus vers.*	24	26
Mycelia sterilia	48	37	*Cladosporium*	13	18
Yeasts	30	31	Mycelia sterilia	17	17
Aspergillus versicolor	18	23	Yeasts	17	14
Eurotium	5	17	*Eurotium*	3	12
Wallemia	2	13	*Aspergillus* sp.	8	11
Engyodontium	10	9	*Fusarium*	11	6
Basidomycetous fungi	15	4	*Acremonium*	10	3
Oidiodendron	10	4			

Although the statistical values between the two culture media were almost the same, in some cases there were significant growth only on one culture media type. In this study were 18 material samples with over 10 000 cfu/g on DG18 and under 5000 cfu/g on MEA (Picture 4.) and 28 material samples with over 10 000 cfu/g on MEA and under 5000 on DG18.

Picture. Dilution plates of material sample: Petri dishes on left 360 000 cfu/g on DG18 (*Wallemia,Penicillium*), 1400 cfu/g on MEA (*Aspergillus restrictus*).

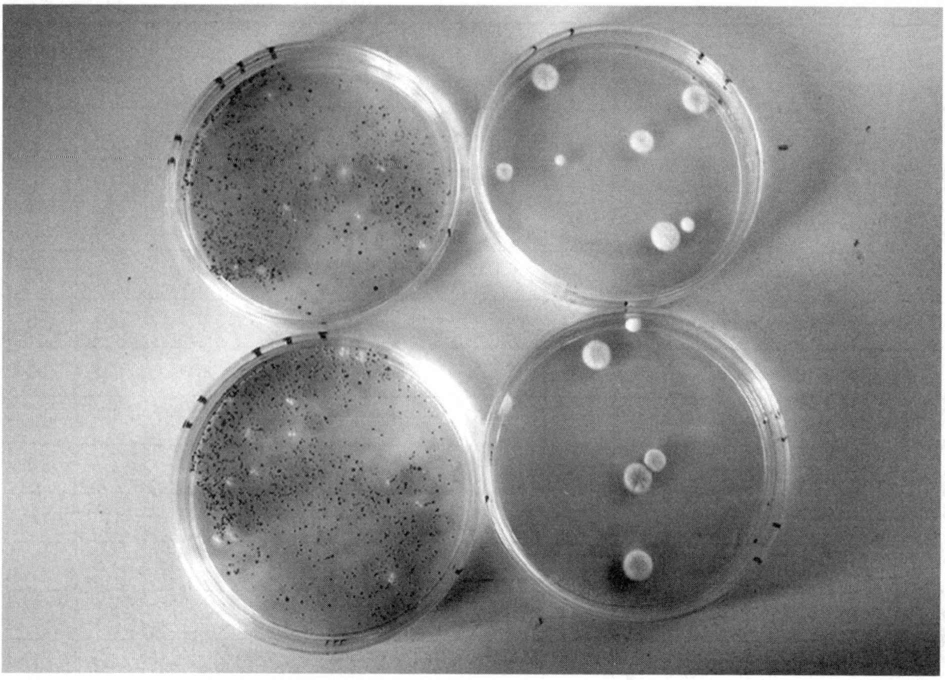

CONCLUSIONS

There were no significant differences between the total cfu counts on each media. Also number of taxa encountered on each media was approximately the same. Although the statistical values of the two media did not much differ, there were samples that would have had different interpretation if cultured only on one media. Use of different culture media resulted also in diverging species composition, judged to be more important than total cfu counts alone. These are the reasons why use of different media is encouraged.

REFERENCES

- Braendlin N. Enumeration of xerophilic yeasts in the presence of xerophilic moulds: a collaborative study. Int J Food Microbiol. 1996, 29(2-3):185-92..
- Hocking AD and Pitt JI. Dichloran-Glycerol Medium for Enumeration of Xerophilic Fungi from Low-Moisture Foods. Applied and Environmental Microbiology, 1980. p. 488-492 Vol. 39
- Samson R.A, Hoekstra ES and Frisvad JC. Introduction to Food-ând Airborne Fungi, 2004. 7th edn. Centraalbureau voor Schimmelcultures, Utrecht, 389pp.
- Wu PC, Su HJ and Ho HM. A comparison of sampling media for environmental viable fungi collected in a hospital environment. *Environ Res* 2000:82 , pp. 253–257.

EVALUATION STRATEGY IN DAMP BUILDINGS AND USE OF INFRARED THERMOGRAPHY

Yves Frenette

BACKGROUND

The Public Health Department in Montreal (PHDM) receives a hundred of notifications per year for health problems related to housing buildings. According to the Québec Public health law, physicians and municipalities for instance have to notify potential health threats related to biological, chemical and physical contaminants. Those mainly concern tenants living in multiplex buildings, but also schools, nursery homes, hospitals and other types of buildings.

Apart from these investigations, the PHDM supports city district inspectors in the application of the By-law concerning the sanitation and maintenance of dwelling, as well as local health professionals, provides training on indoor air quality problems and is implicated in research programs (Vg: Respiratory health of children, Effectiveness of home-based environmental interventions).

As part of its investigations, the environmental hygienist must quickly determine if the building is probably contaminated by molds and represents a risk to the health of the occupants. The aim of this article is to explain the strategy used to identify the conditions favouring molds growth and all the methods used in the most effective way, including the infrared thermography. We also present the results of the investigations we have performed on a sample of building apartments.

MATERIALS

GENERAL STRATEGY

It is very important to understand that molds need 3 conditions to grow: presence of organic matter (wood, cardboard, paper, dust), adequate temperature (~3-40° C) and water. The first 2 conditions are already present in every building. That is why water or humidity is the key element to look for outside and inside a building in any investigation. For that, we need to consider the history of the building (water damages, ...) and the health complaints of the occupants. Secondly, we proceed with visual inspection of the exterior of the building, looking for signs such as efflorescence, masonry deterioration or deformation, cracks, etc. Thirdly, we scan with an infrared camera all exterior walls of the building. Infrared and visible light photographs of suspected areas are taken.

Then, we proceed with the inspection of the interior of the building, looking for the various signs of past water damages, excess humidity, visible molds, etc. Secondly, the walls, floors and ceilings of the apartment(s) are visually and infrared inspected. The thermal camera can be adjusted based on the dew point measured in each room of the apartment. Infrared and visible light photographs of indoor suspected areas are also taken. Presence of excess moisture will be confirmed using a moisture meter. Air and surface sampling is used in addition in certain circumstances that will be detailed below.

INFRARED THERMOGRAPHY

Infrared thermography is a method used to measure, from a distance and without contact, the temperature emanating from an object. Given that the heat capacity of water is much greater than that of most dry materials, it is possible to detect thermal contrast on a building's internal envelope when an indoor/outdoor temperature differential of more than 5° C is observed. Use of an infrared camera enables us to visually identify water infiltration and thermal bridges that could cause condensation and mold growth. The thermal pattern is an excellent indicator of the presence of excess moisture. It helps to identify the origin of water infiltration. It targets sites where remedial and mitigation work must be carried out. It quickly allows assessment of the effectiveness of the corrective actions taken.

It allows the investigation of many apartments or buildings and enables to scan the whole side of a building in one thermographic photo. It also helps to show and easily explain the problem to all concerned parties.

The limits of thermography are:

- Metal objects in the building membrane as well as certain thermal reflections can erroneously generate suspect areas.
- Confirmation of excess moisture must be done using a moisture meter.
- Previous water infiltrations that have been repaired are difficult to identify once wet materials have dried It is very important to use an infrared camera with a sensitivity of 0.07° C or less.

MOLD SAMPLING

Molds sampling, particularly in the air, is mainly used to assess effectiveness of mitigation, when the validity of a report is doubtful or the data are in contradiction with other observations, in litigation, or in the case of a negative inspection but suggestive health effects. The analysis should be done at the species level in order to make a precise analysis. This allows the identification of indoor mold amplification, which is particularly useful when hidden mold contamination is suspected. Air sampling also allows for the verification of the effectiveness of control measures and to determine if the building can be occupied without health risk. It is also useful to obtain such data to support an expertise in court, but this is not always essential. In terms of disadvantages, it is time consuming and costly.

RESULTS

We have investigated more than 250 buildings of all kinds in the last 3 years. We report the results of an analysis of 87 studies of apartment buildings in Montreal, following health complaints related to indoor air quality (table 1).

It should be stressed that bad maintenance of flat roofs is a major cause of water infiltrations in these apartment buildings. For examples, roof drains can be plugged by leaves or insufficient in numbers or wrongly positioned, and roof flashing can become deteriorated due to stress during winter season. Unfortunately, roofs are rarely inspected and their refection is not always done according to the best practices.

As mentioned by others (Moularat, 2008), we have observed that in most buildings, the mold contamination is hidden in the walls, ceilings and floors. The strategy used allows to quickly identify the conditions that are known to promote the growth of molds.

We have also observed that the sampling methods of molds in the initial reports that were submitted to us were often not respecting the standards established by recognized organizations such as ACGIH and AIHA (Amman, 1999; D'Halewyn, 2002; Prezant, 2008). For instance, the volume of air sampled is often insufficient, the outdoor sample is too close to the building, the analysis is not performed at the species level, and the interpretation is often flawed because it does not respect the guidelines recommended by the above organizations. We should mention that there is no certification of indoor air consultants.

Table 1. Main findings from an analysis of 87 studies of apartment buildings in Montreal, 2010.

Characteristics	%	Comments
Previous water damage	25%	Reported by the tenant
Visible mold	20%	Observed on wall, ceiling or floor
Two floors	22%	
Three floors	35%	
More than 3 floors	43%	
Flat roof	100%	
Exterior walls made of bricks	96%	
Significantly deteriorated mortar joints	59%	

Characteristics	%	Comments
Cracks in brick walls	47%	
Significant traces of efflorescence visible on brick walls	60%	White crystal stains on the surface of masonry
Deteriorated lighters	22%	
Brick walls not vented at the base	11%	
Deteriorated roof flashing	30%	
Trees whose branches overlap the roof	14%	Leaves may plug the drains on the roof
Deteriorated roof soffits	10.5%	
Negative land slope	23%	Rain water accumulating at the foundations
Reported previous water damage, interior walls of buildings.	25%	Interior walls were made of gypsum (56%) or plaster (35%).
Medical data suggested the presence of illness or symptoms due to housing conditions.	80%	Not all persons had medical data confirming the relation with their initial health complaints.
Combination of visual inspection and moisture measurements was sufficient to identify the cause of the reported health complaints.	90%	
Confirmation by air samples of the presence of hidden contamination causing the reported health complaints.	10%	Air samples taken with a biocollector, with culture on Agar Petri dish and identification at the species level.

Figure 1. Example of a thermal photography Thermal anomaly that can be associated with excess humidity in an appartement in Montreal.

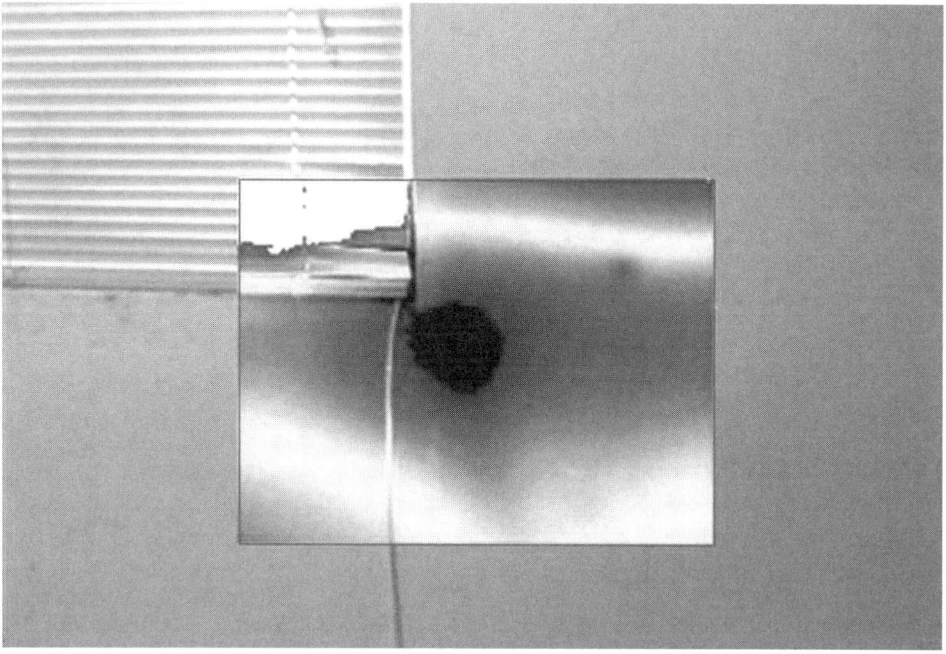

Figure 2. Photo of humidity detector

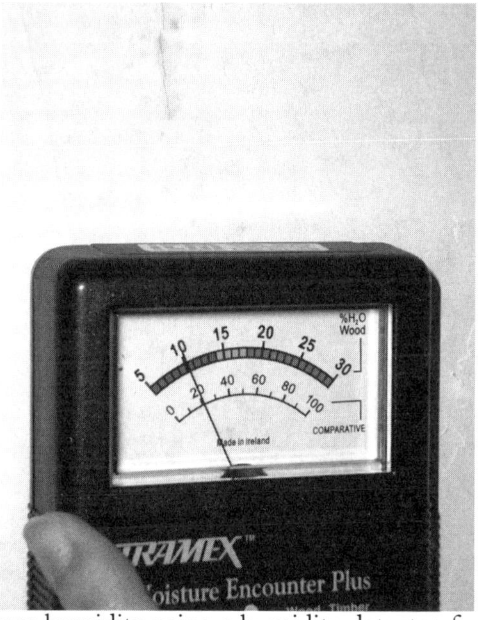

Confirmation of excess humidity using a humidity detector for material. Normally, a drywall should indicate a value of 0%.

Figure 3. Opening in the wall.

We cut the drywall where's humidity level is excessive confirm mold growth. It's important to understand that drywall don't need to be wet to allows mold growth.

Figure 4. Visible molds on a plaster wall

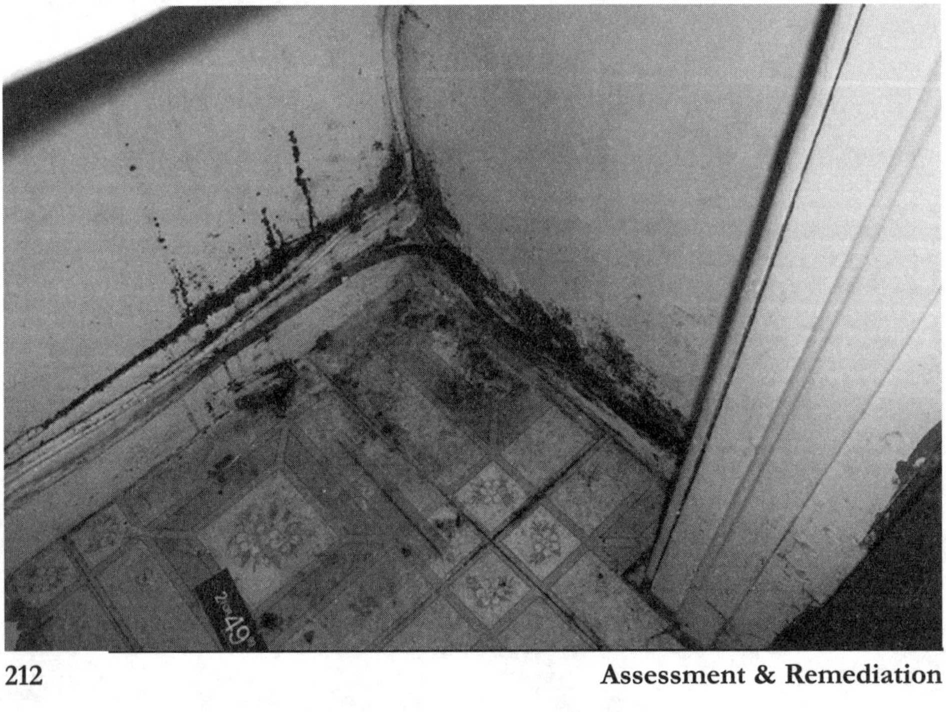

Figure 5. Example of a thermal photography

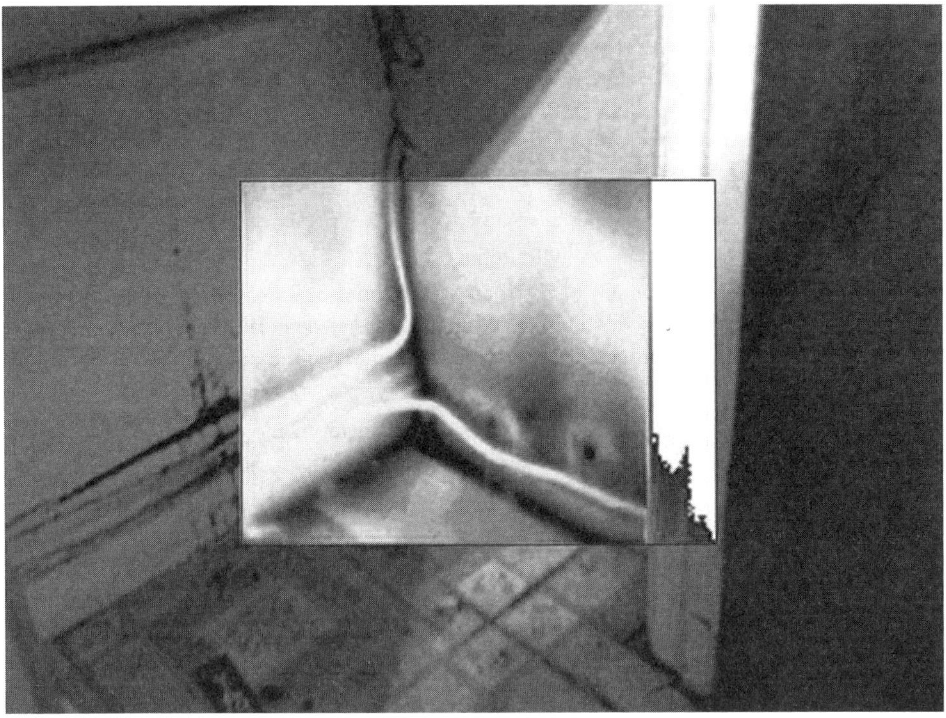

Pattern of a thermal anomaly corresponding to mold growth

CONCLUSION

Infrared thermography allows detection on large exterior and interior surfaces of thermal anomalies consistent with the presence of excessive moisture inside the material in order to target areas to be opened and to direct corrective actions precisely. Inspection with an infrared camera and moisture detector is an essential and efficient tool for the hygienist who has to support inspectors in cities, the workplace, and health care facilities, etc.

The strategy used by the PHDM maximized efficiency and effectiveness to assess the health risk associated with excessive moisture and mold on a large population scale.

Hygienists specializing in indoor air quality should develop expertise in building inspection. To determine whether or not there is a fungal contamination in a building is not sufficient. They should help to determine the cause of the problem, suggest how to correct it and give the green light for the occupancy at the end of the mitigation process.

ACKNOWLEDGMENT

Dr. Louis Jacques, Direction de la santé publique de Montreal Madame Carole Brizard, ISSBA, Université d'Angers, France

REFERENCES

- Ammann HM; Arlian LG; Burge HA; Cole EC; Foarde KK. Bioaerosols Assessment and control. ACGIH, ISBN: 882417-29-1, Cincinnati. 1999.
- D'Halewyn MA et al. Les risques à la santé associés à la présence de moisissures en milieu intérieur. Institut national de santé publique, ISBN: 2-550-40065-8, Montréal. 2002.
- Moularat S, Robine E, Draghi M, Derbez M, Kirchner S, Ramalho O. Moisissures dans les environnements intérieurs: état des connaissances et détermination de la contamination fongique des logements français par un indice chimique. *Pollution atmosphérique.* 2008:197: 33-46.
- Prezant B; Weekes DM; Miller JD. Recognition, Evaluation, and control of indoor mold. AIHA, ISBN: 978-1-931504-92-8, Fairfax. 2008.

PREVALENCE OF MOLD OBSERVATIONS IN EUROPEAN HOUSING STOCK

Ulla Haverinen-Shaughnessy

ABSTRACT

An assessment of the prevalence of mold problems in European housing stock was carried out. The assessment relies on recent studies, taking into account regional and climatic differences, as well as differences in study design, methodology, and definitions. It is based on general indicators of mold in dwellings, such as visible mold or mildew on surfaces. Similar indicators have been commonly used in an absence of more specific microbial markers of exposure. Epidemiological studies have also used similar indicators to estimate exposure-response relationships, associating presence of mold with health effects. Data were available from 22 European countries. Median prevalence of mold problems was 15.5%, and weighted mean prevalence was 10.3%. In addition to survey factors, climate characteristics (mainly temperature) appeared to influence on the prevalence values. Significant (up to 18%) differences were observed for prevalence of mold problems depending on survey factors and climate.

Learning objectives

The information gathered can be used to evaluate the quality of the housing stock, and the underlying mechanisms related to geographical differences in the prevalence of mold problems in Europe. Various survey factors have to be carefully considered in the interpretation of the results. Data on prevalence of mold problems are still not available from many European countries. Use of standardized definitions and survey methodologies should result in more reliable data and decrease uncertainties in the future assessments.

KEY WORDS: climate, microbial growth, region, survey factors

INTRODUCTION

In 2004 the Institute of Medicine (IOM, 2004) concluded there was sufficient evidence to link indoor exposure to mold with upper respiratory tract symptoms, cough, and wheeze in otherwise healthy people; with asthma symptoms in people with asthma; and with hypersensitivity pneumonitis in individuals susceptible to that immune-mediated condition. The IOM also found limited or suggestive evidence linking indoor mold exposure and respiratory illness in otherwise healthy children. In 2009, the World Health Organization came to similar conclusions and issued additional guidance, the WHO Guidelines for Indoor Air Quality: Dampness and Mold (WHO, 2009). A subsequent assessment of environmental burden of disease (EBoD) related to dampness and mold in Europe was carried out as a part of a WHO project on quantifying the burden with disease with housing inadequacy

(Jaakkola et al., 2011). Based on the assessment, about 12% of new childhood asthma in the European Region (WHO EUR) could be attributed to exposure to mold in home environments, where a crude prevalence of 10 (5-25)% was used as an exposure estimate.

Survey based prevalence estimates of dampness and mold indicators in residential buildings have varied broadly, from approximately 2 to 85%, depending on the climate, study design, and definition used (Bornehag et al., 2001, 2004). However, regional or climatic differences have not been widely studied.

Epidemiological studies have used similar indicators to estimate exposure-response relationships between dampness and mold indicators and health effects. This information, together with data on related health outcomes amongst the population, can be further utilized to conduct quantitative assessments of health impacts, such as presented in recent meta-analyses (Fisk et al., 2007; Mudarri and Fisk, 2007). Nevertheless, knowing the prevalence of dampness and/or mold problems is fundamental for any such assessment.

It is also likely that the prevalence of dampness and mold problems in the housing stock changes over time depending on the economical situation and other factors. Increasing public awareness about the association between dampness, mold, and poor health may prompt preventive and corrective actions. Environmental factors such as climate change and increasing demands towards more energy efficient buildings may also result in changes in the housing stock. Therefore, any assessment should rely on relatively recent studies, taking into account differences in study designs, methodologies, and definitions.

In this assessment, studies are considered recent if they were published in the first decade of the 21th century. This paper focuses on mold problems: only such mold observations that were reported being present in the dwellings either at the time of the study or in the past 12 months, were included in the assessment. Mold growth in indoor environments is strongly related to excessive moisture. However, observations of dampness, water damage, and other indicators of excessive moisture (without signs of mold growth), were considered separately from mold observations and excluded from this paper (Haverinen-Shaughnessy, 2012). The aim was to estimate the prevalence of mold problems in European housing stock and factors affecting the estimates.

MATERIAL AND METHODS

A literature search was performed with search terms 'dampness or mold or microbial growth'. The studies providing relevant information for the assessment were mainly large population based studies published in the last ten years. A total of 16 individual (country specific) surveys were identified. In addition, three large data sources consisting of data from multiple countries were included: the LARES survey (WHO, 2007), the European Community Respiratory Health Survey (Burney,

1994), and Eurostat Statistics on Income and Living Conditions (Lelkes and Zolyomi, 2010).

Country specific prevalence values were input into a common database. Some of the studies reported prevalence values for different types of observations separately. These values were classified into four variables based on type of observation reported: Damp (13), Mold (50), Water damage (34), and Combination (28), i.e. any one or more of mold, damp and water damage observations. Mold observations were defined, for example, as "mold or mildew on any surface inside the home in the last 12 months" (Zock et al., 2002) or "any visible presence of mold in the household" (Antova, 2007). Related to each survey and country specific data point, background information collected included: 1) year when the data were collected, 2) type of the sample, 3) target population, 4) study protocol, and 5) method (questionnaire or on-site visit).

Various survey factors, including type of sample, target population, study protocol, and method may have an influence on the prevalence values. Because many of the survey factors in these data were related to each other, they could not be assessed separately. Therefore, a composite Survey factor was formed, including three categories: on-site visits in random samples of homes, questionnaires among random samples of adults, and questionnaires among children. Further classifications were made based on 6) Time frame (only "current signs of mold" or "mold within the past 12 months" were included), 7) Region (Northern, Western, Eastern, or Southern European country of origin), and 8) Climate/temperature (cold, temperate, or warm/hot).

Statistical analyses were performed using SPSS Statistical Software Version 17.0. First, a pooled assessment was performed by further exploring prevalence values. In order to estimate standard errors, area population statistics were collected for each study. The data were mainly based on the country of origin, or more specific geographic area (e.g. city) if applicable. Population statistics were mainly based on European statistics, Census data, or other information sources available on the Internet (e.g. Wikipedia), where the time point closest available to the corresponding survey was selected. Utilizing the sample and population sizes, margins of errors and 95% Confidence Intervals for the prevalence values were estimated. While the prevalence estimate plus or minus 1.96 times standard error (SE) is a 95% Confidence Interval, this information was used to calculate SEs. Finally, weighted mean prevalence was obtained using the inverse of the squared SE as the weight (w) for each study (Neely et al., 2009).

Next, Kruskall Wallis non-parametric test for several independent samples procedure was used to compare crude differences in the prevalence of mold problems between different time frames, survey factors, regions, and climates. Finally, linear regression analysis was performed, where Time frame and Survey factor were first entered in the model as independent variables, after which Region and Climate/temperature were introduced in the models using step-wise selection procedure, where P-to-enter was specified as 0.05 and P-to-remove as 0.10. Weighted least-squares (WLS) models were obtained, using (w) as the weight variable.

RESULTS AND DISCUSSION

Large data sources were available from multi-national surveys that used the same protocol and definition of the mold problems throughout the survey and therefore provided comparable estimates between countries and regions. The LARES survey was undertaken in eight European cities relying on on-site home visits (WHO, 2007). It consisted of data on 3,373 dwellings. Country specific data were reported by Nicol et al. (2006). The European Community Respiratory Health Survey (ECRHS) investigated self-reported dampness and mold exposure in 38 study centers in 18 countries (Zock et al., 2002). Centers were located both in Europe (14 countries), and outside Europe (four countries, not included in the assessment). Both LARES and ECRHS provided similar overall estimates for indoor mold problems (22-25%).

Rest of the studies identified were country specific. PATY study (Antova et al., 2008) reported results of ten original studies, seven of which reported recent mold exposures in European countries. Other studies included Brasche et al. (2003), Bornehag et al. (2005), and Skorge et al. (2010).

In all, data reporting prevalence of mold problems were available from 22 European countries (i.e. countries with territory located in Europe). In pooled analyses, the median prevalence of mold problems was 15.5% (estimated margin of error ranging from 0.2 to 2.4%) and the weighted mean was 10.3%. With respect to the pooled analyses, it should be noticed that the Confidence Interval calculations assume a genuine random sample of the relevant population. If the sample is not truly random, the intervals may not be reliable. This would have an effect on the weighted mean values.

Another issue increasing uncertainty is related to the samples drawn not necessarily being representative for the entire population, e.g. in case the sample only included homes of children. There is relatively little information about the prevalence of mold problems among population sub-groups. Many socio-economic factors are assumed to have an influence on housing conditions (Braubach and Savelsberg, 2009; Braubach and Fairburn, 2010), and it appears that the prevalence of mold problems may be higher among certain sub-groups, e.g. homes of children, even after excluding studies on symptomatic children.

Based on this assessment, the median prevalence of mold was reported higher among populations based on children than those based on adults. In addition, the prevalence of mold was commonly estimated lower based on on-site home visits than by questionnaires. By the composite Survey factor, the median prevalence of mold (SE) was 8.5% (5.7%) for random samples of homes, 17.3% (7.4%) for questionnaires among random samples of adults, and 23.8% (10.1%) for questionnaires among children, respectively.

By Time frame, the median prevalence of mold observed in the past 12 months was significantly higher (18.1%, SD 5.7%) than prevalence of current mold problems (8.5%, SD 5.7%). The longer the time frame, the higher the prevalence can be expected to be due to the fluctuating nature of dampness and mold problems in buildings. Whereas some studies report longer time frames (e.g. mold problems in the past 5 years or mold problems ever), only current problems and the problems occurred in the past 12 months were included in this assessment, in order to provide a cross sectional evaluation of the extent of the mold problems as best possible.

In addition, significant differences ($p<0.05$) were observed between different climates, where median prevalence values for mold problems were higher for warm/hot (13.3%) and temperate climates (19.9%) as compared to cold climate (8.6%). Whereas the prevalence by region ranged from 13.6% to 20.7%, regional differences were not statistically significant.

There were apparent correlations between Time frame, Survey factor, Region, and Climate/temperature. Thus, linear regression analysis was performed in order to see which factors may independently associate with mold observations. After adjusting for Time frame and Survey factor, temperate and warm/hot climates had significantly higher prevalence of mold (+18%, and +11%, respectively) as compared to cold climate. The model R square was 0.77.

Region was factored in the assessment to evaluate the possibility of cultural or building related differences between Northern, Western, Eastern, and Southern European regions that could be attributed to mold problems in dwellings. Different countries have naturally adjusted their building stock according to their climate and temperature, so in that sense these factors may not be easily distinguished. However, only differences by climate appeared statistically significant.

There may also be differences considering the location of mold observations within the dwellings. Most of the studies do not differentiate between locations, but some studies emphasize observations in the bedroom or other living spaces. Some studies report the extent and/or severity of mold problems, but most are based on a dichotomous rating. Overall, there exists a considerable variation in how the questions on mold growth are framed, and prevalence estimates may therefore range widely dependent on the type of questions used, the level of detail asked for, and the judgment of those analyzing the data. These kinds of uncertainties could not be effectively controlled for in this study due to limitations related to reporting of the original data and a relatively small number of studies.

CONCLUSIONS

Based on this assessment, the weighted mean prevalence of mold problems in European housing stock is about 10%. Significant (up to 18%) differences were observed for the prevalence of mold problems depending on survey factors and temperature.

ACKNOWLEDGEMENTS

The assessment was performed as a part of the INTARESE project (project number 018385), co-funded by the European Commission under the Sixth Framework Programme. It includes some work performed for WHO project on quantifying disease from inadequate housing in Europe.

REFERENCES

- Antova T, Pattenden S, Brunekreef B et al. Exposure to indoor mould and children's respiratory health in the PATY study. *J Epidemiol Community Health.* 2008; 62(8): 708-14.
- Bornehag CG, Blomquist G, Gyntelberg F et al. Dampness in buildings and health. Nordic interdisciplinary review of the scientific evidence on associations between exposure to "dampness" in buildings and health effects (NORDDAMP). *Indoor Air* 2001; 11:72-86.
- Bornehag CG, Sundell J, Bonini S et al. EUROEXPO. Dampness in buildings as a risk factor for health effects, EUROEXPO: a multidisciplinary review of literature (1998-2000) on dampness and mite exposure in buildings and health effects. *Indoor Air* 2004; 14: 243-57.
- Bornehag CG, Sundell J, Hagerhed-Engman L et al. 'Dampness' at home and its association with airway, nose, and skin symptoms among 10,851 preschool children in Sweden: a cross-sectional study. *Indoor Air.* 2005;15 Suppl 10:48-55.
- Brasche S, Heinz E, Hartmann T, Richter W, Bischof W. Prevalence, causes, and health aspects of dampness and mold in homes. Results of a population-based study in Germany; Bundesgesundheitsblatt - Gesundheitsforschung - Gesundheitsschutz 2003; 46: 683-693.
- Braubach M, Savelsberg J. Social Inequities and their influence on housing risk factors and health. A data report based on the WHO LARES database. WHO Regional Office for Europe. Copenhagen. 2009.
- Braubach M, Fairburn J. Social inequities in environmental risks associated with housing and residential location – a review of evidence. *European Journal of Public Health* 2010; 20(1):36-42.
- Burney PG, Luczynska C, Chinn S, Jarvis D. The European Community Respiratory Health Survey. *Eur Respir J.* 1994;7(5):954-60.
- Fisk WJ, Lei-Gomez Q, Mendell MJ. Meta-analyses of the associations of respiratory health effects with dampness and mold in homes. *Indoor Air* 2007;17:284-296.
- Haverinen-Shaughnessy U. Prevalence of dampness and mould in European housing stock, *Journal of Exposure Science and Environmental Epidemiology* . 2012. (In Press).

- IOM (Institute of Medicine). Damp Indoor Spaces and Health. The National Academies Press, Washington D.C. 2004.
- Jaakkola M, Haverinen-Shaughnessy U, Douwes J, and Nevalainen A. Indoor dampness and mould problems in homes and asthma onset in children. WHO Housing Programme. Draft final report.. 2011.
- Lelkes O and Zolyomi E. Housing Quality Deficiencies and the Link to Income in the EU. European Centre. Policy Brief March. 2010.
- Mudarri D, Fisk WJ. Public health and economic impact of dampness and mold. *Indoor Air* 2007;17:226-235
- Neely J, Magit A, Rich J et al. A practical guide to understanding systematic reviews and meta-analyses. Otolaryngology - Head and Neck Surgery 2010;142 (6).
- Nicol S. The Relationship Between Housing Conditions and Health-Some Findings from the WHO LARES Survey of 8 European Cities. In: ENHR conference "Housing in an expanding Europe: theory, policy, participation and implementation", Ljubljana, Slovenia 2 - 5 July 2006.
- WHO. Large Analysis and Review of European housing and health Status. Preliminary overview. WHO Regional Office for Europe. Copenhagen. June 2007.
- WHO Guidelines for Indoor Air Quality: Dampness and Mould. WHO Regional Office for Europe. Copenhagen. 2009.
- Zock J-P, Jarvis D, Luczynska C, et al. Housing characteristics, reported mold exposure, and asthma in the European Community Respiratory Health Survey, *Journal of Allergy and Clinical Immunology,* 2002;110 (2), 285-292.

INDOOR MOLDS AND RESPIRATORY HYPERSENSITIVITY: A COMPARISON OF SELECTED MOLDS AND HOUSE DUST MITE INDUCED RESPONSES IN A MOUSE MODEL

Marsha D W Ward, Yong Joo Chung, Lisa B Copeland, Don Doerfler and Stephen Vesper

Learning Objectives:

Recognize a) the complex nature of molds and thus the complexity of responses to mold exposure;

b) that there are exposure threshold doses that lead to allergic and asthmatic responses.

ABSTRACT

Molds are ubiquitous in the environment and exposures to molds contribute to various human diseases. Damp/moldy environments have been associated with asthma exacerbation, but mold's role in allergic asthma induction is less clear. The molds selected for these studies are commonly found indoors, associated with water damaged buildings and/or sick building syndrome. The studies objectives were to 1) elucidate the association between specific molds and allergy/asthma and 2) assess the relative allergenicity of these molds by comparing responses to those induced by house dust mite (HDM) using a mouse model.

Female BALB/c mice received 1 or 4 exposures by intratracheal aspiration of 0-80 µg of mold extract or HDM. Airway responses (PenH) to methacholine (Mch) challenge were measured on day 1. Serum and bronchoaveolar lavage fluid (BALF) were collected on day 2 after the final exposure. Serum extract-specific IgE and BALF inflammatory cell counts are presented.

Responses to mold extract exposure varied among the molds but multiple exposures were required to induce significant increases in extract-specific IgE and elevated levels of BALF eosinophil counts. To achieve similar results to those induced by HDM in the extract-specific IgE assay required 1.5X more *Scopulariopsis brevicaulis* and 2.25X more *Epicoccum nigrum*. However, *Penicillium crustosum* group did not induce a significant extract-specific IgE response at any dose level. Multiple extract exposures also induced significant change in airflow (PenH) following Mch challenge.

The data suggest the capacity of molds to induce allergic responses varies. It also suggests there are threshold doses for allergic sensitization.

INTRODUCTION/GOAL OF STUDY

Molds are ubiquitous in the environment and exposures to molds contribute to various human diseases. There are more than 200 different types of fungi to which people are routinely exposed. The growth of molds in homes, schools, offices, and other public buildings has been implicated as the cause of a wide variety of adverse health effects. Respiratory exposures to fungi have been associated with adverse health effects including frank infection, toxic effects and respiratory tract conditions ranging from irritation to hypersensitivity responses such as allergy and asthma, and pulmonary hemosiderosis. Headlines resulting from moldy, water-damaged homes, particularly "toxic molds" and the aftermath of hurricanes on the East and Gulf coasts of the U.S. have raised public awareness and concern regarding molds as potential health hazards, particularly in indoor environments.

Fungal infections include superficial (skin and hair) and cutaneous (epidermis, invasive hair and nail) infections. Although systemic infections are considered rare compared to bacterial infections, they are caused by primary pathogens (lungs) and opportunistic organisms. In most cases these opportunistic infections occur in people with a comprised immune system such as those undergoing chemotherapy or those with AIDS. Generally ingestion has been considered the primary route for both acute and chronic mycotoxins poisoning but mycotoxins have also been associated with health effects resulting from inhalation exposures. Adverse health effects resulting from mycotoxin exposure range from immuno-modulating to carcinogenic effects. These agents have the potential to cause neurotoxic effects, although this effect is still controversial.

IgE-mediated allergic diseases occur in a genetically predisposed population in response to foreign substances (proteins) that should be innocuous. Allergic diseases include: asthma, allergic rhinitis, conjunctivitis, and allergic sinusitis. Asthma prevalence has increased over the last several decades for reasons that are thought to have an environmental component. It is estimated that up to 90% of asthmatics are atopic and have an allergy trigger for asthmatic episodes (Pearce, Pekkanen, Beasley, 1999; Holt et al., 1999).

Additionally, it has been estimated that up to 10% of people have fungal allergies (Horner et al., 1999). The Institute of Medicine reports and WHO guidelines concluded that the role of molds in asthma induction is not clear but an association exists between damp buildings, the presence of mold and asthma exacerbation (NAS, 2000; NAS, 2004; WHO, 2009). It has been estimated that 21% of current asthma in the US may be attributed to dampness and molds (Mudarri, Fisk, 2007).

A review by Denning et al. (2006) described epidemiological evidence of an association between fungal exposure and asthma severity and animal studies have demonstrated a cause-effect relationship for allergy sensitization and asthma-like responses (Ward, Sailstad, Selgrade, 1998; Ward et al., 2010; Chung et al., 2010).

The studies objectives were to 1) elucidate the association between specific molds and allergy/asthma and 2) assess the relative allergenicity of these molds by comparing responses to those induced by the well-characterized, indoor allergen house dust mite (HDM), a known inducer of allergic asthma, using a mouse model. The selected molds are commonly found indoors, associated with water damaged buildings and/or sick building syndrome.

METHODS

Female BALB/c mice (8-10 weeks old) were maintained in an Association for Assessment and Accreditation of Laboratory Animal Care International accredited vivarium. All animal procedures were reviewed and approved by the Institutional Animal Care and Use Committee of NHEERL, US EPA. In a series of studies mice received either 1 or 4 exposures to 0-80 μg of extract in HBSS (total volume of 50 μl) by intratracheal aspiration (IA) over a four-week period (as previously described in 12). The three fungi (*Epicoccum nigrum* (ENE), *Scopulariopsis brevicaulis* (SBE), and *Penicillium crustosum* group (PCG)) were grown on a nylon filter resting on plates of potato dextrose agar at 23°C for 3-5 weeks (Vesper et al., 2006). The fungal mass (mycelium and spores) was extracted as described in Ward et al. (1998) except polytron disruption of the fungal mass was added. The endotoxin level at the highest dose (80 μg extract protein) of each extract was measured using a Limulus amoebocyte lysate test kit. The endotoxin levels were below those demonstrated to have an impact in allergy induction. The fungal source and endotoxin levels are presented in Table 1.

Table 1.[a]Abbreviations used as mold extract identifiers for multiple (4X) and single (1X) exposures. [b] Endotoxin was assessed at the highest dose level (80 μg extract protein).

FUNGI	EXPOSURE PROTOCOL[a]		SOURCE	ENDOTOXIN LEVEL[c] (in 80 μg extract dose)	
	4X	1X			
Epicoccum nigrum	ENE	HB/EN	University of Alberta Microfungus Collection & Herbarium (UAMH) #3247	0.133 EU	0.01-0.03 ng
Penicillium crustosum	PCG	HB/PC	Australian Culture Collection (FRR) #1669	0.116 EU	0.01-0.02 ng
Scopulariopsis brevicaulis	SBE	HB/SB	UAMH #7771	0.089 EU	0.01-0.02 ng

Lyophilized whole body extracts of house dust mites (HDM: Dermatophagoides farinae and D. pteronyssinuss; Greer Laboratories, Lenoir, NC), were rehydrated in HBSS to 3 mg protein/ml, mixed equally and stored in aliquots at –80 °C until use. Endotoxin level of the highest dose (80 μg extract protein) was 0.152 EU (0.02-0.03 ng).

One day after the final exposure, airway responsiveness was assessed by methacholine (Mch; 0, 6.25, 12.5, and 25 mg/ml) challenge using whole body plethysmography (Viana et al., 2002). Serum and bronchoalveolar lavage fluid (BALF) were collected 2 days after the final exposure. BALF total and differential cell counts were performed. Serum antigen-specific IgE was assayed as previously described (Chung et al., 2006).

RESULTS

The allergic impact of the mold extract exposures was different for each mold. Multiple exposures were required to induce increased BALF eosinophil counts and serum antigen-specific IgE.

One day following the final exposure, airway responsiveness (Figure 1.A) indicated that multiple exposures (20 μg lowest significant dose level) and a single exposure (40 μg) to E. nigrum (ENE) and multiple exposures to HDM (40 μg) induced a significant increase in airflow restriction compared to controls. On day 2 (D2) BALF cells counts demonstrated that the neutrophil counts were significant compared to controls at a lower dose with multiple exposures (5 μg) than a single exposure to ENE (20 μg) (Figure 1. B). HDM exposed mice displayed a dose-dependent response that was more robust than that induced by either multiple or single exposures to ENE. Although the difference in neutrophil counts was not significant, the single exposure to ENE was higher than the multiple exposure response. Multiple exposures to ENE induced significant increases in both BALF eosinophil counts and serum antigen (extract)-specific IgE compared to controls (Figure 1. C & D). Additionally, eosinophil counts were significantly higher than those of HDM-exposed mice at the 40 and 80 μg dose levels. In order to compare ENE and HDM antigen-specific IgE responses, the dose that would result in 10% of total mediator release was calculated to be 14.5 μg and 6.5 μg, respectively. This suggests that 2.25X more ENE is required to achieve the same response as HDM. A single exposure to ENE did not induce a significant response in either of these endpoints compared to control mice. In the higher dose range ENE induced more robust responses than HDM with the exception of BALF neutrophil counts. However, the response to higher doses of HDM in this mouse cohort was generally lower than we have seen in other studies. Additionally, a single ENE exposure induced a significant increase in airway responsiveness (40 μg) and BALF neutrophils (20 μg) suggesting that exposure to ENE might exacerbate asthmatic responses.

Figure 1. A) Airway response to methacholine challenge presented as area under the curve (AUC). BALF inflammatory cell counts B) neutrophils and C) eosinophils following either 1 (HB/EN) or 4 (ENE) exposures to *E. nigrum* extract or 4 exposures to HDM. D) Extract-specific IgE: shaded box indicates the calculated dose that results in 10% of total mediator release stated in the legend. Symbols indicate (§) the lowest significant dose compared to Control (0 μg dose), (a) indicates significant vs. other treatments within dose, (b) indicates significant vs. single exposure within dose ($p<0.05$). Error bars represent standard error of the mean. Control *n* = 12; HDM and ENE *n* =5-6; HB/EN *n* =4.

A

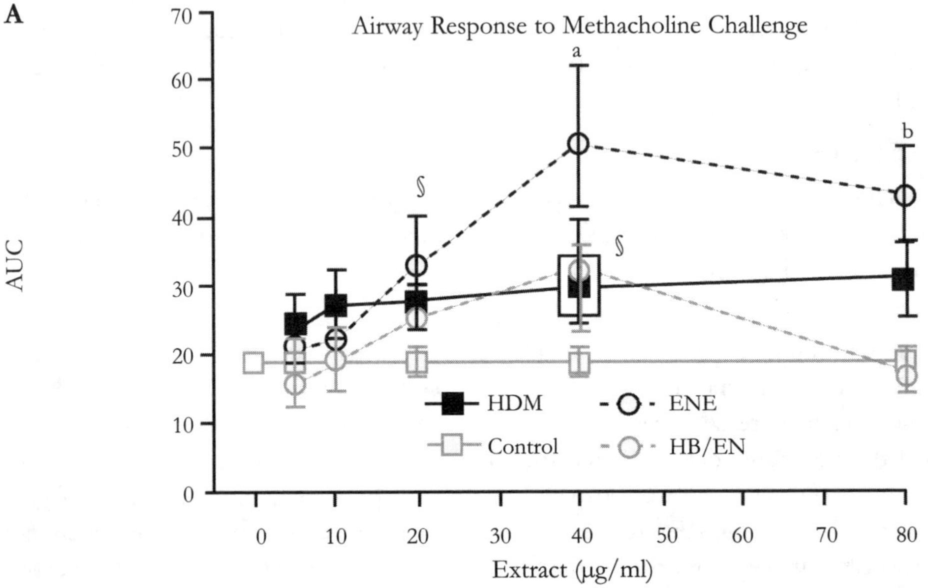

B

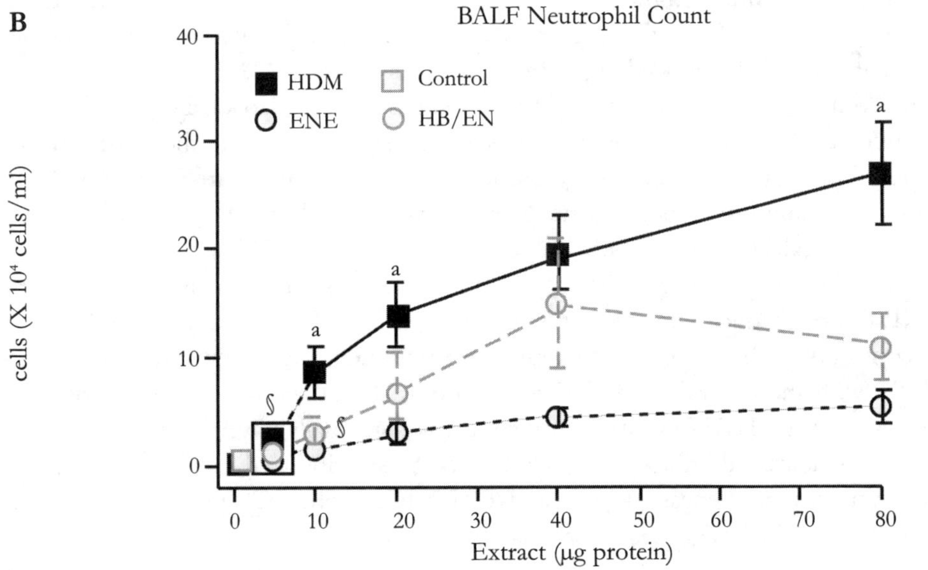

Figure 1 continued

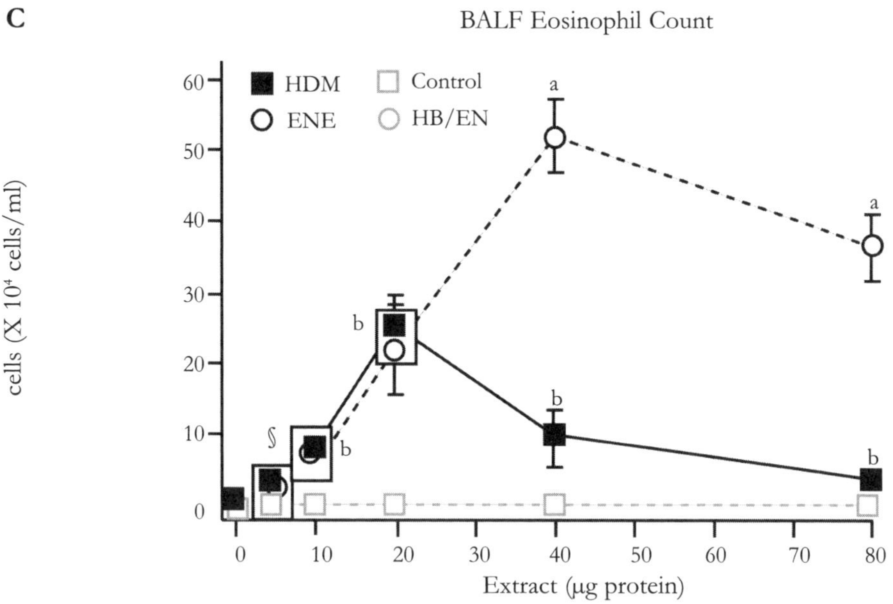

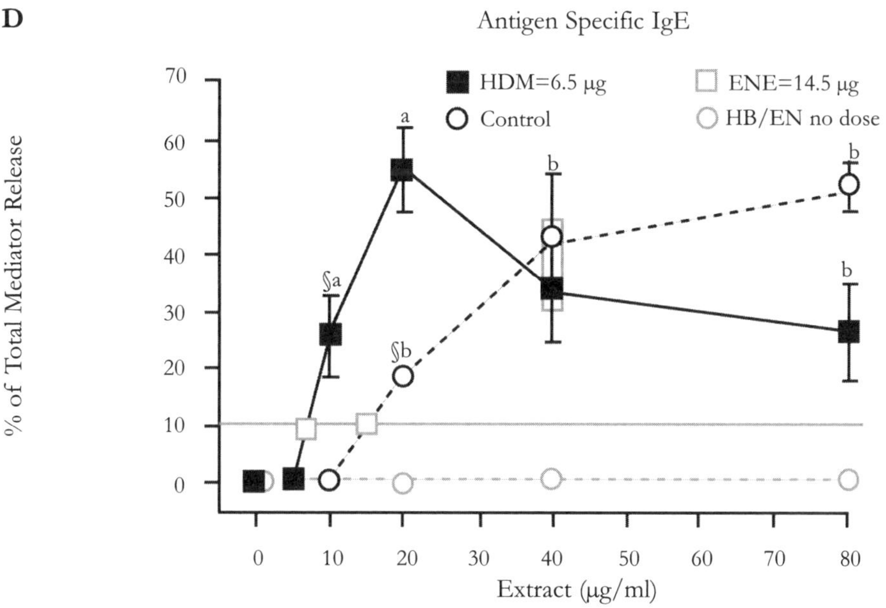

The airway responsiveness to Mch challenge indicated that only multiple exposures to *P. crustosum* (PCG; 40 μg) and HDM (40 μg) induced a significant increase in airflow restriction compared to controls (Figure 2.A). At the 80 μg dose level BALF neutrophil cell counts (multiple and single exposure) were significant compared to controls with the singly exposed mice displaying a slightly more robust and significant response (Figure 2.B). However, HDM exposed mice displayed a dose-dependent response that was more robust and significant compared to multiple- or singly-exposed PCG mice across the dose range. PCG induced a significant increase in BALF eosinophil counts (5 μg) compared to controls (Figure 2.C). However, the response tended to level off above the 10 μg dose level. Additionally, the HDM-exposed mice eosinophilic response was significantly higher than that of PCG exposed mice at the 20-80 μg dose levels. Neither multiple nor single exposures to PCG at any dose tested induced a significant increase in serum IgE quantified as % of total mediator release level (Figure 2.D).

Only multiple exposures to *S. brevicaulis* (SBE) at 80 μg dose level induced a significant increase in airflow restriction compared to controls during Mch challenge (Figure 4.A). On the other hand multiple exposures to HDM induced a significant increase at 40 μg dose level. HDM exposed mice displayed a dose-dependent and more robust BALF neutrophil response compared to multiple- or singly-exposed SBE mice. Neither SBE dosing protocol resulted in significant neutrophil counts compared to controls (Figure 4.B). Although, multiple exposures to SBE induced a significant increase in BALF eosinophil counts (40 μg) compared to controls (Figure 3.C) the HDM-exposed mice eosinophilic response was significantly higher than that of SBE exposed mice across the dose range. The calculated dose of SBE and HDM that would result in 10% of total mediator release (antigen-specific IgE response) was calculated to be 28.0 μg and 19.0 μg, respectively (Figure 3.D). This suggests that 1.5X more SBE is required to achieve the same response as HDM. A single exposure to SBE did not induce a significant eosinophil or antigen-specific IgE response compared to control mice.

Figure 2. A) Airway response to methacholine challenge presented as area under the curve (AUC). BALF inflammatory cell counts B) neutrophils and C) eosinophils following either 1 (HB/PC) or 4 (PCG) exposures to *P. crustosum* group extract or 4 exposures to HDM. D) Extract-specific IgE shaded box indicates the calculated dose that results in 10% of total mediator release stated in the legend. Symbols indicate (§) the lowest significant dose compared to Control (0 μg dose), (a) indicates significant vs. other treatments within dose, (b) indicates significant vs. single exposure within dose ($p<0.05$). Error bars represent standard error of the mean. Control $n = 12$; HDM and PCG $n = 5\text{-}6$; HB/PC $n = 4$.

Figure 2

A

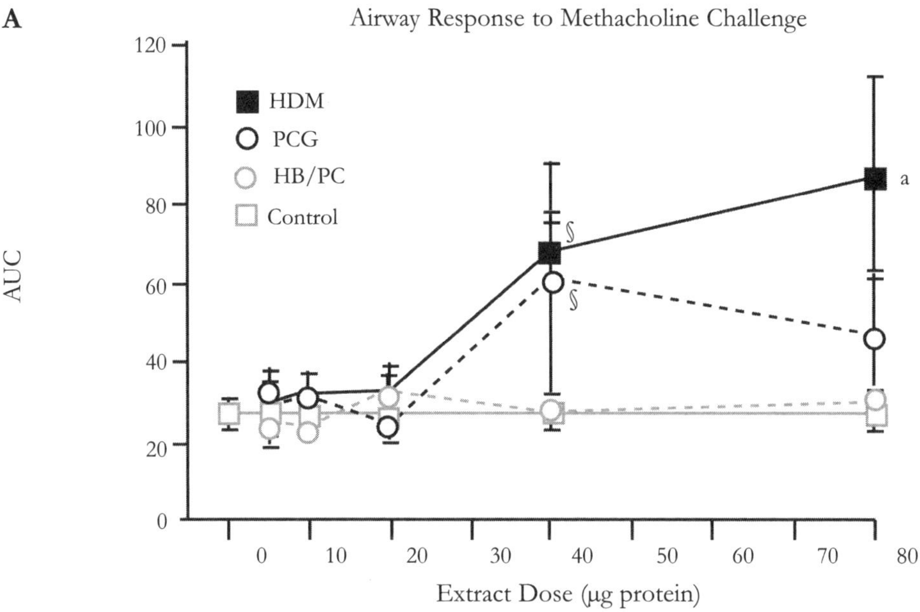

Airway Response to Methacholine Challenge
120
100
80
60
40
20
0
AUC
HDM
PCG
HB/PC
Control
§
§
a
0
10
20
30
40
50
60
70
80
Extract Dose (μg protein)

B

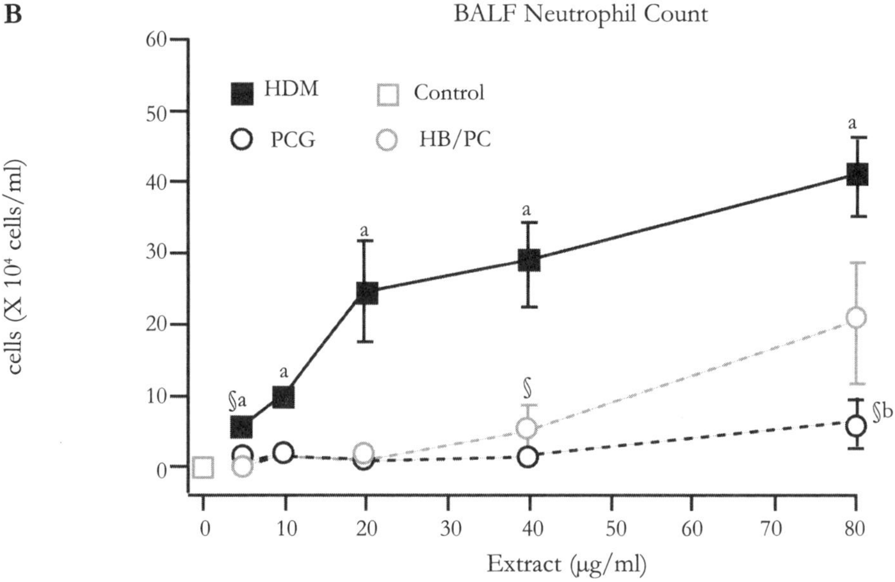

BALF Neutrophil Count
60
50
40
30
20
10
0
cells (X 10⁴ cells/ml)
HDM
PCG
Control
HB/PC
§a
a
a
a
a
§
§b
0
10
20
30
40
50
60
70
80
Extract (μg/ml)

Figure 2 continued

C

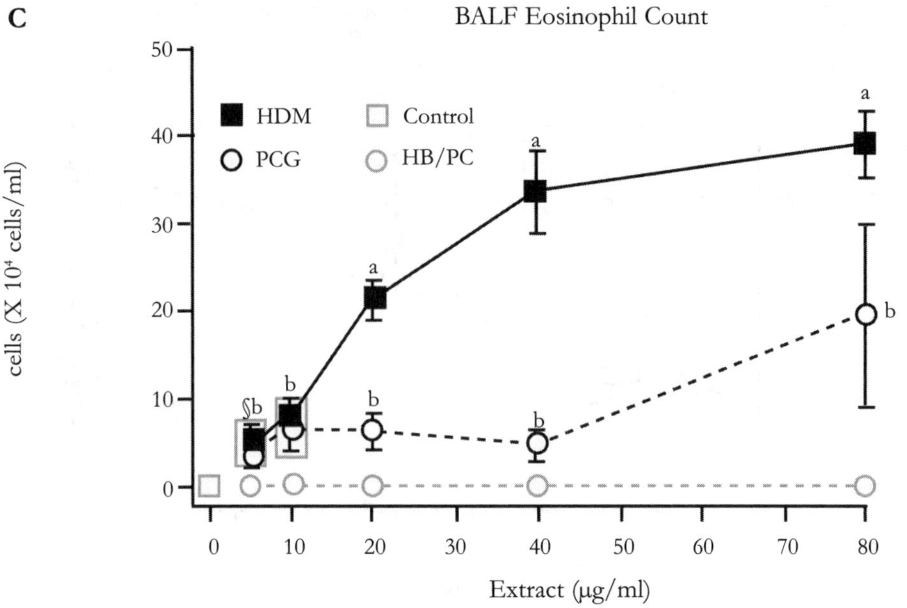

D

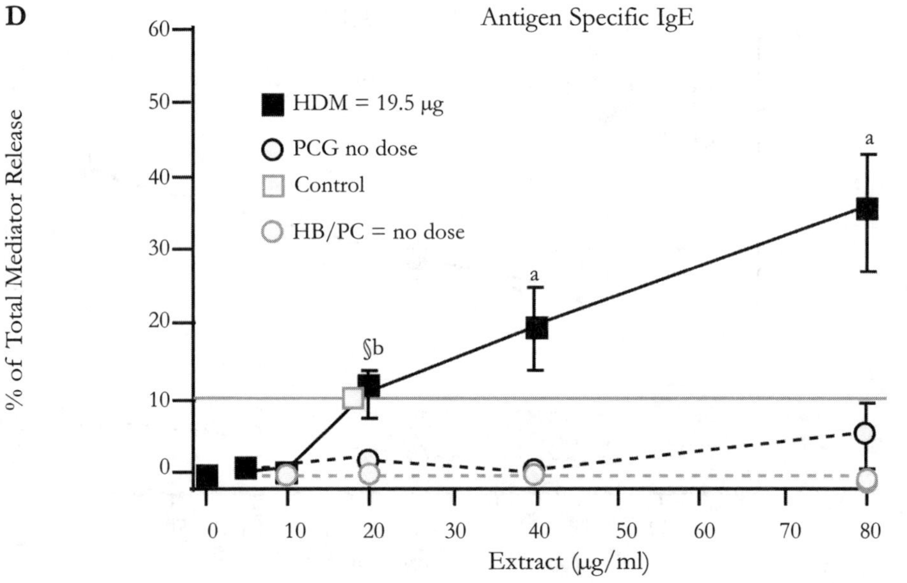

Conclusions

In the dose-response animal studies presented here as well as earlier studies (*Stachybotrys chartarum* (Chung et al., 2010), *Penicillium chrysogenum* (Ward et al., 2010) and the biopesticide *Metarhizium anisopliae* (Ward et al., 2011)) comparing responses induced by mold extract exposures to those induced by house dust mite (a well characterized indoor allergen) the molds demonstrated differential capacities (potencies) to induce allergic/asthma-like responses. Multiple respiratory exposures to certain molds cause immune, inflammatory, and respiratory physiological responses in BALB/c mice characteristic of human allergic lung disease. However, not all molds were able to cause allergic responses in the animal model but some of these molds may have components that exacerbate asthmatic responses. Our studies indicate that molds should be assessed individually for their allergenic potential.

Multiple exposures were required to induce increased BALF eosinophil counts and serum extract-specific IgE. SBE did induce these allergic responses as well as increased airway responsiveness but higher doses were required to induce significant increases in these endpoints compared to HDM. Of the three molds evaluated in these studies ENE appears to induce a more robust response than HDM based on the lowest significant dose compared to controls and/or dose-response profile for airway responses and BALF eosinophil influx. However, ENE was less potent than HDM based on the calculated dose required for a similar response in the antigen-specific IgE assay. The significant increase in airway responsiveness following a single exposure to ENE (40 μg) also suggests the potential for asthma exacerbation. Although, PCG did not induce significant levels of antigen-specific IgE there was a low but significant increase in BALF eosinophils induced and a significant airway response to Mch challenge (40μg). These data suggest that PCG is not a robust allergen but may have the capacity to exacerbate asthma based on airway response and BALF eosinophil increases.

Figure 3. A) Airway response to methacholine challenge presented as area under the curve (AUC). BALF inflammatory cell counts B) neutrophils and C) eosinophils following either 1 (HB/SB) or 4 (SBE) exposures to S. brevicaulis extract or 4 exposures to HDM. D) Extract-specific IgE shaded box indicates the calculated dose that results in 10% of total mediator release stated in the legend. Symbols indicate (§) the lowest significant dose compared to Control (0 μg dose), (a) indicates significant vs. other treatments within dose, (b) indicates significant vs. single exposure within dose ($p<0.05$). Error bars represent standard error of the mean. Control *n* = 12; HDM and SBE *n* =5-6; HB/SB *n* =4.

Figure 3.

A

Airway Response to Methacholine Challenge

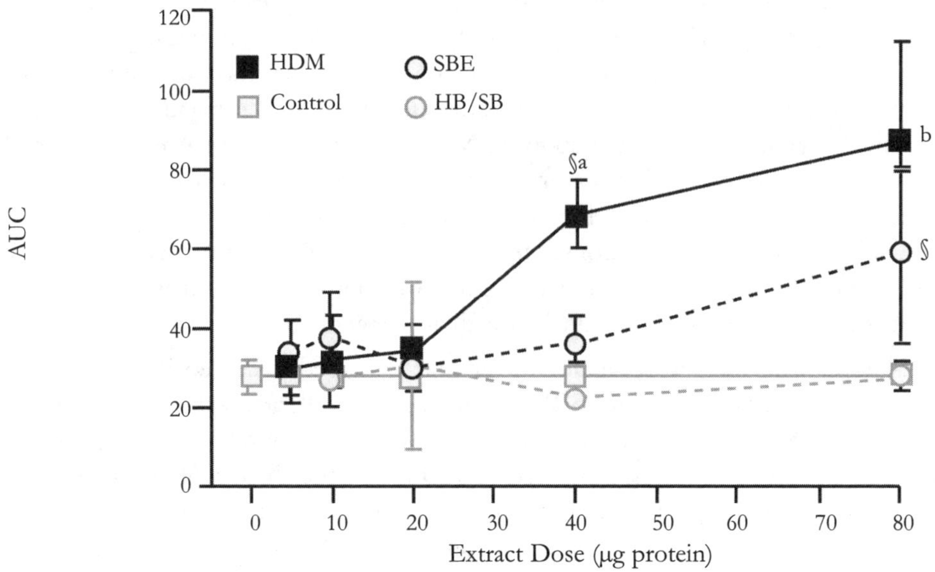

B

BALF Neutrophil Count

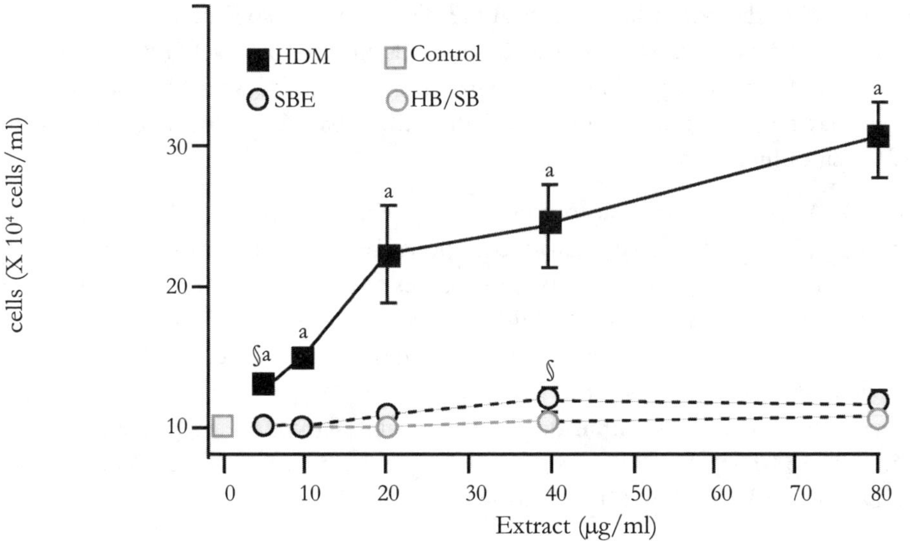

Figure 3. continued

C

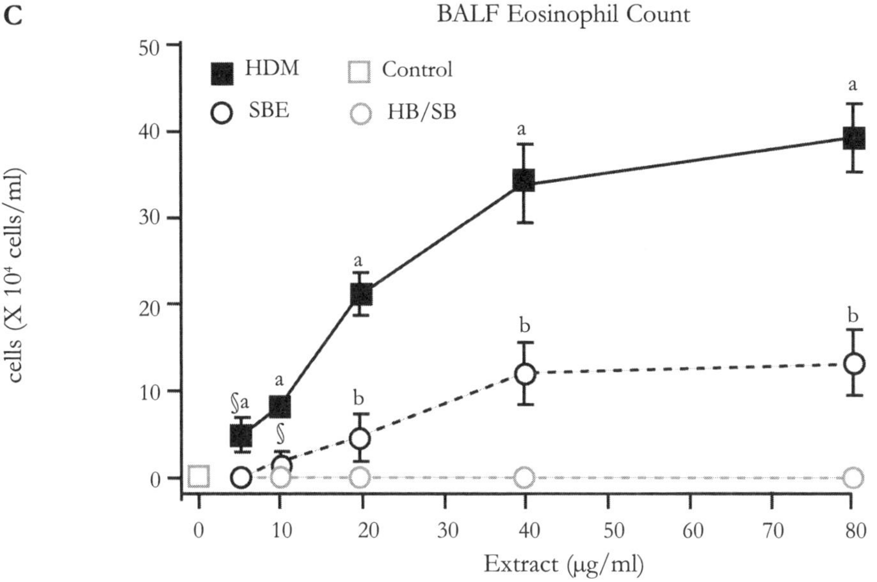

D

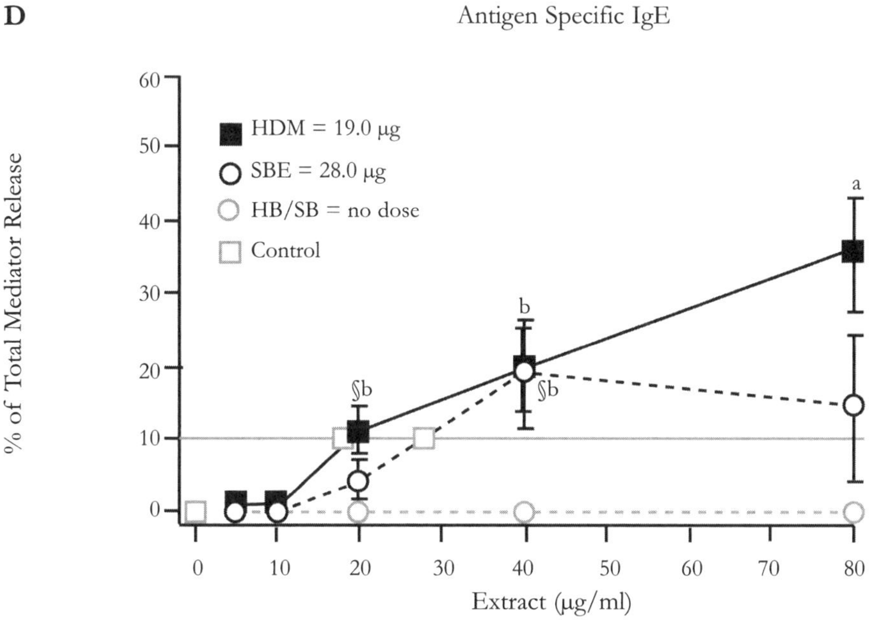

In previous studies we demonstrated that *P. chrysogenum* (Ward, et al., 2010) is a robust allergen source and concluded that *P. chrysogenum* is a more potent allergen than HDM. This is in contrast to our conclusions based on data presented here for *P. crustosum* (PCG), a member of the same genus. Many building mold assessments do not identify the molds, particularly *Penicillium* and *Aspergillus*, at the species level. Although a direct comparison cannot be made between these two studies, differences in allergic responsiveness induced by the two species is noteworthy because it suggests that the species, not just the genera, makes a difference in the health outcome with regard to allergy and possibly asthma induction.

These dose-response studies demonstrate the complex nature of molds and their differing capacities to induce allergy and by extension allergic asthma. Having said this, it must be stated that regardless of the allergenic potential of an agent, exposure levels and sensitization thresholds are critical factors in allergic disease development. At this time human exposure levels and sensitization thresholds are unknown for most allergens including molds/fungi. Importantly, these data suggest a threshold dose for the induction of allergic responsiveness and a role for molds in allergic asthma.

Acknowledgements

The authors would like to thanks Debora Andrews, Elizabeth Boykin, James Lehmann and Judy Richards of US EPA for expert technical assistance. Additionally, the authors would like to thank Drs. Christal Bowman and Doris Betancourt for their review of this manuscript.

(This abstract has been approved by the NHEERL, U.S. EPA. The contents may not reflect the views and policies of the agency.)

References

- Chung YJ, Coates NH, Viana ME, Copeland L, Vesper SJ, Selgrade MK and Ward MD. *Toxicology,* 2005:209, 77-89.
- Chung YJ, Copeland LB, Doerfler DL and Ward MD.W. *Inhalation Toxicology,* 2010:22(6), 460-468.
- Denning DW et al. *Eur Respir J* 2006:27, 615-26.
- Holt PG, Macaubas C, Stumbles PA and Sly PD. *Nature*, 1999:402, E12–E17 .
- Horner WE, Helbling A, Salvaggio J E and Lehrer SB. *Clin Microbiol Rev,* 1995:2, 161-179.
- Mudarri D and Fisk WJ. *Indoor Air*, 2007:17, 226–235.
- NAS. Clearing the Air: Asthma and Indoor Air Exposures. National Academy of Sciences, Institute of Medicine, Board on Health Promotion and Disease Prevention. Washington, DC: National Academies Press. 2000.
- NAS. Damp Indoor Spaces and Health. National Academy of Sciences, Institute of Medicine, Board on Health Promotion and Disease Prevention. Washington, DC: National Academies Press. 2004.

- Pearce N, Pekkanen J, Beasley R. *Thorax.* 1999:54, 268-272.
- Vesper SJ, McKinstry C, Yang C, Haugland RA, Kercsmar CM, Yike I, Schluchter MD, Kirchner HL, Sobolewski J, Allan TM and Dearborn DG. *J Occup Environ Med,* 2006:48, 852-858.
- Viana ME, Coates NH, Gavett SH, Selgrade MK, Vesper SJ and Ward MD. *Toxicol Sci,* 2002:70, 98-109.
- Ward MDW, Chung YJ, Copeland LB, and Doerfler DL. *Indoor Air,* 2010:20, 380–391.
- Ward MDW, Chung YJ, Copeland LB and Doerfler DL. *Journal of Toxicology,* .Vol 2011, Article ID 360805.
- Ward MDW, Sailstad DM and Selgrade MJK. *Toxicol. Sci.* 1998:45: 195-203.
- WHO. WHO guidelines for indoor air quality: dampness and moulds. WHO Regional Office for Europe, Copenhagen. 2009.

WHAT DOES THE DEVELOPMENT OF FUNGAL SYSTEMATICS MEAN TO DNA-BASED METHODS FOR INDOOR MOLD INVESTIGATIONS?

De-Wei Li and Chin S. Yang

Learning objectives:

1. To understand the development of fungal systematics and its effect on indoor mold studies and investigation.
2. To understand the pros and cons of DNA-based methods for indoor mold detection.

ABSTRACT

In the last decade, DNA sequence technology has greatly advanced our understanding of phylogenetic relationships of fungi. This article reviewed the literature on the latest development of fungal systematic and its impact on DNA-based analytical methods currently used in indoor mold investigation. Among many studies, *Cladosporium cladosporioides* s.l. was redelineated and 22 species had been newly described from it. This study will have a significant impact on indoor fungi investigations using either morphological methods or DNA-based methods. *Stachybotrys chartarum* collected indoors was found to have two distinct phylogenic clades which represent two different species. Current Quantiative Polymerase Chain Reaction or QPCR method is not able to differentiate *Stachybotrys chartarum, S. chlorohalonata,* and *S. yunnanensis,* and *S. cylindrospora* (non *S. eucylindrospora*). The biggest challenge of DNA sequence technology in mycological research is that up to 27% DNA sequence data deposited in GenBank are questionable or erroneous due to incorrect identification of the fungal materials. The conclusion is that DNA-based analytical methods, similar to morphology based methods, have showed great advantages but also have their limitations. Morphology and DNA-based analytical methods are supplementary. Indoor fungi investigators should understand the pros and cons of each test method and carefully choose the method of lab analysis which will best accomplish their objectives.

INTRODUCTION

Fungal systematics has made significant developments in the last decade. DNA sequence technology has greatly advanced our understanding of phylogenetic relationships of fungi, including those of indoor significance. DNA-based methods are an important and powerful part of taxonomy and systematics. It allows us to study, understand, and determine phylogenetic relationships among fungi and fungal taxa from molecular level. It assists us to better define fungal taxa at all ranks and to clarify confusions and to develop new methodology for fungal detection, diagnosis, and quantification. It has resulted in changes at almost all ranks of fungal classification and in placement of some fungal taxa including several important indoor fungi (Celio et al., 2006; James et al., 2006; Liu et al., 2006). The advancement of molecular technology has shaped the future development of fungal systematics.

The objective of this article is to review current literature and to evaluate the significance of these studies to indoor mold research and investigations using the DNA-based analytical methods.

Number of Fungi and development of fungal systematics

According to the Dictionary of the Fungi, ca 97,861 species were cataloged (Kirk et al., 2008). In the last decade, 1196 new species of fungi in average were described each year (Hibbett et al., 2011). Among these newly described species, a large number of fungi (74.4%, 8895 species) was described based on morphological characters without DNA sequence data from 1999 to 2009 and remainders were described based on both morphological and DNA sequence data (Hibbett et al., 2011).

Hawksworth (1991) estimated that there are 1.5 million fungal species in the world. O'Brien et al. (2005) opined that 3.5-5.1 million species are present. Schmit and Mueller (2007) suggested a conservative minimum estimate of 712,000 species. Other estimates published since 1990 ranged from 0.5 million to 9.9 million (Hawksworth, 2001; Mora et al., 2011). Hawksworth's estimate has been widely accepted by mycological community. A huge number of fungi (>93%) remain undescribed. At the current rate, it will take >1000 years to catalogue all fungal species. Clearly mycologists are facing a long journey ahead in cataloguing and describing all fungal species. Expectation is that this process will be accelerated with assistance from the development of new molecular technology (Blackwell, 2011). DNA sequence data from the samples taken from environmental, ecological, and indoor fungal studies have shown that many fungal taxa may be new to science (Pitkaranta et al., 2008; Schadt et al., 2003). For example, phylogenetic analysis of tundra soil fungi revealed a high diversity of fungi and three novel clades that constitute major new groups of fungi and an abundance of previously unknown fungi that are active beneath the snow (Schadt et al., 2003).

According to The International Code of Nomenclature for algae, fungi, and plants (ICN) (formerly the International Code of Botanical Nomenclature (ICBN)) passed at The XVIII International Botanical Congress (IBC) held in Melbourne, Australia, 24–30 July 2011, a new fungal taxon can be described with morphological characters alone, or morphological characters and DNA sequence data, but not only with molecular data. The diagnosis or description of a new fungal taxon can be in Latin or English effective on January 1, 2012. The most significant changes has been published in September 2011 (Knapp et al., 2011; Hawksworth, 2011). Recent molecular taxonomic approaches also have changed the old taxonomies which were based primarily on morphology (Samson, 2011).

"Systematics or biosystematics is the study of the relationship and classification of organisms and the processes by which they have evolved and are maintained" (Kirk et al., 2008). Whenever taxonomic placement change for fungal species or new combinations (name change) were proposed, the conclusion was drawn based on the evidences of phylogenetic relationships and/or better understanding of morphological characters of these fungi. To understand the effect of fungal growth indoors, it is essential to have a correct identification (Samson, 2011). Fungal taxonomy has changed rapidly in the last decade. Consequently and unfortunately some fungal names which have been used for many decades sometimes change. It is often difficult for a non-mycologist to understand the naming and name changing of fungi and sometimes leads to confusion when the identification of molds is made. The correct nomenclature name of a fungal taxon reflects the current state of taxonomic knowledge (Samson, 2011).

We should keep in mind that there is a major drawback associated with DNA-based analytical methods. A significant number of sequences deposited in GenBank, possibly as many as 20-27% of the ITS sequences, have been found to be from erroneously identified specimens or cultures (Nilsson et al., 2006, Rossman and Palm-Hernández, 2008). Up to 14% of the GenBank sequences for the genera *Alternaria* and *Ulocladium* were incorrect entries (De Hoog and Horré, 2002). These problems no doubt will compromise the reliability of the results when using BLAST to identify a fungus or analyzing phylogenetic relationships of a group of fungal taxa using sequence data deposited in Genbank is conducted. One way may avoid the problem is to examine origin and identifiers of the fungal isolates used to sequence DNA data to determine whether these sequence data are reliable or verifiable. It is crucial for molecular biologists to work with mycologists to avoid or exacerbate this problem. The current problem needs to be recognized and corrected. Bar coding of fungi (to link or DNA sequence data to a voucher specimen or a culture) is the first step towards this direction. Whenever the GenBank data are utilized, caution is warranted. We should be diligent to examine the data to be used so as to avoid the pitfalls.

Fungal systematics and indoor fungi research

Systematics of *Acremonium, Aspergillus, Cladosporium, Penicillium*, and *Stachybotrys* were intensively studied using DNA sequence data in the last several years and many species in these genera were redelineated (Bensch et al., 2010; Summerbell et al., 2011; Zalar et al., 2007). Some species in these genera are common fungi in indoor environments, especially the one with water damage or dampness.

Cladosporium cladosporioides s.l. was redelineated in 2010 and 22 species had been newly described from it based on phylogenetic analysis using three gene/region sequences and cryptic morphological differences (Bensch et al., 2010). *Cladosporium sphaerospermum* s.l. was redefined also. *Cladosporium dominicanum, C. halotolerans, C. psy-*

chrotolerans, C. spinulosum and *C. velox*, all with globoid conidia (Zalar et al., 2007). Since some common indoor fungi belong to these genera, the results from these studies have a significant effect on our indoor mold studies and investigations. The subtle morphological differences among these newly delineated species pose a significant challenge to identify them collected from indoor environments. At the same time these developments raise a major question: which and how many species of these genera are we facing indoors? Can we differentiate or detect these species using morphological characters or QPCR (sometimes called Mold-Specific PCR, MSPCR) probes and primers we are using at present? To answer these questions, it is necessary to conduct further studies on indoor fungi of these genera from both morphology and molecular aspects on the samples.

At least five species of *Stachybotrys* have been identified from indoor environments (per. obser.). The exact number of species of *Stachybotrys* present indoors is not clear. Koster et al. (2003) found that *Stachybotrys chartarum* s.l. collected indoors had two distinct phylogenic clades which represent two different species based on multi-gene sequence data. However, they could not draw a conclusion whether one of them is a new taxon due to the fact that type species of Corda for the genus *Stachybotrys* and the type specimen of *Stachybotrys chartarum* (Ehrenb.) Hughes are either in a poor condition or no longer available. Clearly, *Stachybotrys chartarum* indoors is a species complex. The current qPCR method for detecting *Stachybotrys chartarum* is not able to differentiate *Stachybotrys chartarum, S. chlorohalonata*, and *S. yunnanensis*, and *S. cylindrospora* (non *S. eucylindrospora* (Li, 2007)). The reason is that the primers and probes were developed prior to the publication of *S. chlorohalonata* and *S. yunnanensis* was not included in their study (Andersen et al., 2003; Haugland et al., 2001; Kong, 1997). It is beneficial to improve or redesign the current primers and probes so that these *Stachybotrys* species can be differentiated. Before the new primers and probes are available, the *Stachybotrys* detected by current methods should be treated as species complex. A thorough monographic study of *Stachybotrys* is urgently necessary to clarify the confusion about *Stachybotrys* species indoors.

Penicillium chrysogenum is the most common fungus indoors (Scott et al., 2004) and in water damaged buildings in Denmark and Greenland (Andersen et al., 2011). Scott et al. (2004) conducted a study on indoor *P. chysogenum* using 4 genes/region in Ontario, Canada. The isolates collected during this study formed four clades. The first clade contained more than 5.6% of house isolates clustered with the ex-type strains of *P. chysogenum* and *P. notatum*. The second clade was 1%. The third clade was 3%. The 4th clade clustered with Fleming's strain was the dominant one which included more than 90% of house isolates. Follow-up sampling of outdoor air failed to reveal *P. chrysogenum*, confirming the rarity of this fungus in outdoor air (Scott et al., 2004).

Houbraken et al. (2011) further studied *P. chrysogenum* using partial β-tubulin, calmodulin and RPB2 datasets. They concluded that Fleming's strain is not

Penicillium chrysogenum, but *P. rubens* (Houbraken et al., 2011). They also opined that the clades 1 and 2 belong to *P. chrysogenum* in Scott et al. study and the clade 3 is probably an undescribed species. The significance of the study by Houbraken et al. to indoor mold field is that one of the most common fungi (clade 4 in Scott et al. study) in indoor environments is *P. rubens,* not *P. chrysogenum.*

QPCR for detecting indoor fungi and ERMI to screen indoor environments for molds were DNA-based methods developed in late 1990's and 2000's, respectively (Haugland et al., 1999; Vesper et al., 2007). These methods provided alternative tools of morphology-based methods for indoor mold research and investigation. Based on the recent development, we cannot be sure which species within the *P. chrysogenum* complex is actually detected by QPCR or ERMI methods?

Isolates of "*Penicillium chrysogenum svar.* I" were used to develop PenGrp3 including: ATCC 9179 (deposited as *P. notatum*), ATCC 10106, FRR 807 (Type Strain) as part of isolate components (Haugland et al., 2004). Among these isolates, both ATCC 10106 and FRR 807 are the same type culture of *P. chrysogenum* with the same origin. PenGrp3 was developed to detect a group of *Penicillium* species (EPA 2011). It is known that two isolates are *P. chrysogenum,* but it is not clear which species ATCC 9179 belong to. "*Penicillium chrysogenum svar.* II" (Pchry) is the one used in QPCR and ERMI (group 2). Among the isolates (ATCC 9480, EPA 467 (Control), EPA 534, EPA 535, EPA 578, EPA 579, NRRL 832) used for developing Pchry (Haugland et al., 2004), ATCC 9179 and NRRL 832 are the same isolate and NRRL 832 and ATCC 9480 (= NRRL 1951) belong to *P. rubens* (Houbraken et al., 2011). Since the taxonomic identities of EPA 467, EPA 534, EPA 535, EPA 578, and EPA 579 are unknown within the species complex of *P. chrysogenum,* it is not clear which species that Pchry detects at present. These isolates should be further studied and compared with the data used in the study by Houbraken et al. (2011) to determine the taxonomic identities of these isolates. Further studies are necessary to verify the specificities of the primers and probes of qPCR for detecting indoor fungi based on the latest development of fungal systematics. It will advance our understanding of indoor molds and their effects on public health.

A recent study found that the biodiversity of indoor fungi is greater in the temperate regions than in the tropics based on analyses of dust samples collected around the globe using pyrosequencing data (Amend et al., 2010). The result was rather unexpected. It raises a set of questions. Do fungal species indoors evolve independently from their outdoor counterparts and if so at what evolutionary rate? Can we still claim that all indoor fungi are originated from outdoor sources? Since no samples were collected from the central and northern Africa and South America as well as the continental Asia and Japan in the study by Amend et al. (2010), it is necessary to conduct additional studies to confirm their results and to answer the aforementioned questions.

PCR inhibitors

The advantages of using QPCR and ERMI are well reviewed by Vesper (2011). However, no method is perfect. It is no exception for DNA-based analytical methods. One of the drawbacks of DNA-based methods is the presence of inhibitors of PCR in environmental samples. Gypsum dust originated from dry walls is very common in buildings where repair work or mold remediation has been conducted or undergoing. Gypsum dust and cement dust are PCR inhibitors (Vesper, 2011). Inhibitory effects from gypsum dust and other factors on qPRC results had been observed since 2001 (Cruz-Perez et al., 2001). The presence of these dusts will lead to a reduced the detection sensitivity or even false negative results of qPCR (Cruz-Perez et al., 2001; Halstensen, 2008). However, no research has been conducted to determine the limits of amount of gypsum dust in the samples at which will show significant negative effect on the reliability of the QPCR results.

Cruz-Perez et al. (2001) found that several *S. chartarum* isolates were found to contain highly potent PCR inhibitors coextracting with the DNA and had led to false negative results. Unfortunately, these results were overlooked in the last decade. The mechanism of such inhibitory effects has not been studied either. Diluting the extracted DNA sample may alleviate the problem, but it is not clear if it is the best approach to solve this problem.

Additional PCR inhibitors reported are: humic acids, polysaccharides, tannins, other biological particles, organic compounds, phenols, fulvic acids in soil, polyphosphates in fungi, heavy metals, non-target DNA, and high concentrations of non-target DNA (Cruz-Perez et al., 2001; Halstensen, 2008; Mitchell and Zuccaro, 2006; Wu et al., 2002). Some of these factors are present in indoor environments. Inhibitory factors cannot and should not be overlooked. Opel et al. (2010) indicated that potential mechanisms of PCR inhibitors include binding to the polymerase, interaction with the DNA, and interaction with the polymerase during primer extension.

The significance of PCR inhibitors has been recognized by forensic and food scientists (Opel et al., 2010; Maurer, 2011). At present major studies are conducted in these fields. The lack of research to PCR inhibitors in indoor mold industry is troublesome. It is important to know the pros and cons of molecular methods and to optimize its application so that the cons are minimized, whenever it is possible.

Intrinsic limitations

Paralogous (duplicate) copies of ITS region exist in some genera, such as *Fusarium* (O'Donnell and Cigelnik, 1997). Several fungi are not amenable to PCR amplification and/or cycle sequencing (Summerbell et al., 2005). These intrinsic characters are the hurdles for developing methods for detecting fungi and also may interfere with test results of DNA-based fungal detection methods.

CONCLUSIONS

Pitkaranta et al. (2008) compared DNA sequence, cultivation, and quantitative PCR methods and found that the three methods complemented each other and generated a more comprehensive picture of mycobiota than any of the methods would give alone. "DNA sequence data are an important and powerful part of taxonomy and systematics. Molecular data have an indisputable role in the analysis of biodiversity. However, DNA-based data should not be seen as a substitute for understanding and studying whole organisms when determining identities or systematic relationships" (Will and Rubinoff, 2004). It is imperative to have the knowledge of the latest development of fungal taxonomy and the advantages and shortcomings of all the methods available for mold investigation and research. Such knowledge will assist us in choosing methods suitable to our objectives and to generate new hypotheses for our future research.

DNA-based methods, similar to morphology based methods, have showed great advantages but also have their limitations. Morphology and DNA-based methods are supplementary. Indoor fungi investigators should understand the pros and cons of test methods and carefully choose the methods to fit their objectives.

REFERENCES

- Amend AS, Seifert KA., Samson R & Bruns TD. Indoor fungal composition is geographically patterned and more diverse in temperate zones than in the tropics. *Proceedings of the National Academy of Sciences*, 2010:107:13748-13753.
- Andersen B, Frisvad JC, Sondergaard I, Rasmussen IS & Larsen LS. Associations between fungal species and water damaged building materials. *Appl. Environ. Microbiol.:AEM*, 2011:02513-02510.
- Andersen B, Nielsen KF, Thrane U, Szaro T, Taylor JW & Jarvis BB. Molecular and phenotypic descriptions of *Stachybotrys chlorohalonata* sp. nov. and two chemotypes of *Stachybotrys chartarum* found in water-damaged buildings. *Mycologia*, 2003:95:1227-1238.
- Bensch K, Groenewald JZ, Dijksterhuis J, Starink-Willemse M, Andersen B, Summerell BA, Shin, H-D, Dugan FM, Schroers H-J, Braun U & Crous PW. Species and ecological diversity within the *Cladosporium cladosporioides* complex (Davidiellaceae, Capnodiales). *Studies in Mycology*, 2010:67:1-94.
- Blackwell M. The Fungi: 1, 2, 3 … 5.1 million species? *American Journal of Botany*, 2011:98:426-438.
- Celio G, Padamsee M, Dentinger B, Bauer R & McLaughlin D. Assembling the fungal tree of life: constructing the structural and biochemical database. *Mycologia*, 2006:98:850.
- Cruz-Perez P, Buttner MP & Stetzenbach LD. Specific detection of *Stachybotrys chartarum* in pure culture using quantitative polymerase chain reaction. *Molecular and Cellular Probes,* 2001:15:129-138.
- De Hoog GS & Horré R. Molecular taxonomy of the *Alternaria* and *Ulocladium* species from humans and their identification in the routine laboratory. *Mycoses*, 2002:45:259-276.

- Halstensen A. Species-specific fungal DNA in airborne dust as surrogate for occupational mycotoxin exposure? *International Journal of Molecular Sciences*, 2008:9:2543-2558.
- Haugland RA., Heckman JL & Wymer LJ. Evaluation of different methods for the extraction of DNA from fungal conidia by quantitative competitive PCR analysis. *Journal of microbiological methods*, 1999:37:165-176.
- Haugland RA, Varma M, Wymer LJ & Vesper SJ. Quantitative PCR analysis of selected *Aspergillus, Penicillium* and *Paecilomyces* species. *Systematic and Applied Microbiology*, 2004:27:198-210.
- Haugland RA, Vesper SJ & Harmon SM. Phylogenetic relationships of *Memnoniella* and *Stachybotrys* species and evaluation of morphological features for Memnoniella species identification. *Mycologia*: 2001:54-65.
- Hawksworth D. A new dawn for the naming of fungi: impacts of decisions made in Melbourne in July 2011 on the future publication and regulation of fungal names. *MycoKeys,* 2011:1:7-20.
- Hawksworth DL. The fungal dimension of biodiversity: magnitude, significance, and conservation. *Mycological Research*, 1991:95:641-655.
- Hawksworth DL. The magnitude of fungal diversity: the 1.5 million species estimate revisited. *Mycological Research*, 2001:105:1422-1432.
- Hibbett DS, Ohman A, Glotzer D, Nuhn M, Kirk P & Nilsson RH. Progress in molecular and morphological taxon discovery in Fungi and options for formal classification of environmental sequences. *Fungal Biology Reviews*, 2011:25:38-47.
- Houbraken J, Frisvad JC & Samson RA. Fleming's penicillin producing strain is not *Penicillium chrysogenum* but *P. rubens. IMA Fungus*, 2011:2:87-95.
- James TY, Letcher PM, Longcore JE, Mozley-Standridge SE, Porter D, Powell MJ, Griffith GW & Vilgalys RA molecular phylogeny of the flagellated fungi (Chytridiomycota) and description of a new phylum (Blastocladiomycota). *Mycologia*, 2006:98:860.
- Kirk P, Cannon PF, Minter D & Stalpers J. Dictionary of the Fungi. 2008.
- Knapp S, McNeill J & Turland NJ. Changes to publication requirements made at the XVIII International Botanical Congress in Melbourne—what does e-publication mean for you? *Cladistics*:no-no. 2011.
- Kong H-Z. *Stachybotrys yunnanensis* sp. nov. and Neosartorya delicata sp. nov. isolated from Yunnan, China. *Mycotaxon*, 1997:62:427-433.
- Koster B, Scott J, Wong B, Malloch D & Straus N. A geographically diverse set of isolates indicates two phylogenetic lineages within *Stachybotrys chartarum. Canadian Journal of Botany*, 2003:81:633-643.
- Li D-W. S*tachybotrys eucylindrospora*, sp. nov. resulting from a re-examination of *Stachybotrys cylindrospora. Mycologia*, 2007:99:332-339.
- Liu Y, Hodson M & Hall B. Loss of the flagellum happened only once in the fungal lineage: phylogenetic structure of kingdom Fungi inferred from RNA polymerase II subunit genes. *BMC evolutionary biology,* 2006:6:74.

- Maurer JJ. Rapid Detection and Limitations of Molecular Techniques. *Annual Review of Food Science and Technology,* 2011:2:259-279.
- Mitchell JI & Zuccaro A. Sequences, the environment and fungi. *Mycologist,* 2006:20:62-74.
- Mora C, Tittensor DP, Adl S, Simpson AGB & Worm B. How many species are there on earth and in the ocean? *PLoS Biol,* 2011:9:e1001127.
- Nilsson RH, Ryberg M, Kristiansson E, Abarenkov K, Larsson K-H & Kõljalg U. Taxonomic reliability of DNA sequences in public sequence databases: a fungal perspective. *PLoS ONE,* 2006:1:e59.
- O'Brien HE, Parrent JL, Jackson JA, Moncalvo J-M & Vilgalys R. Fungal community analysis by large-scale sequencing of environmental samples. *Appl. Environ. Microbiol.*, 2005:71:5544-5550.
- O'Donnell K & Cigelnik E. Two divergent intragenomic rDNA its2 types within a monophyletic lineage of the fungus *Fusarium* are nonorthologous. *Molecular Phylogenetics and Evolution,* 1997:7:103-116.
- Opel KL, Chung D & McCord BR. A study of PCR inhibition mechanisms using real time PCR. *Journal of Forensic Sciences,* 2010:55:25-33.
- Pitkaranta M, Meklin T, Hyvarinen A, Paulin L, Auvinen P, Nevalainen A & Rintala H. Analysis of fungal flora in indoor dust by ribosomal DNA sequence analysis, quantitative PCR, and culture. *Appl. Environ. Microbiol.*, 2008:74:233-244.
- Rossman AY & Palm-Hernández ME. Systematics of plant pathogenic fungi: Why it matters. *Plant Disease,* 2008:92:1376-1386.
- Samson RA. Ecology and general characteristics of indoor fungi In: Adan OCG & Samson RA (eds), Fundamentals of mold growth in indoor environments and strategies for healthy living. Wageningen Academic Publishers, 2011, pp. 101-116.
- Schadt CW, Martin AP, Lipson DA & Schmidt SK. Seasonal dynamics of previously unknown fungal lineages in tundra soils. *Science,* 2003:301:1359-1361.
- Schmit J & Mueller G. An estimate of the lower limit of global fungal diversity. *Biodiversity and Conservation,* 2007:16:99-111.
- Scott J, Untereiner WA, Wong B, Straus NA & Malloch D. Genotypic variation in *Penicillium chysogenum* from indoor environments. *Mycologia,* 2004:96:1095-1105.
- Summerbell RC, Gueidan C, Schroers H-J, de Hoog GS, Starink M, Rosete YA, Guarro J & Scott JA. *Acremonium* phylogenetic overview and revision of *Gliomastix, Sarocladium,* and *Trichothecium. Studies in Mycology,* 2011:68:139-162.
- Summerbell RC, Lévesque CA, Seifert KA, Bovers M, Fell JW, Diaz MR, Boekhout T, de Hoog GS, Stalpers J & Crous PW. Microcoding: the second step in DNA barcoding. *Philosophical Transactions of the Royal Society B: Biological Sciences,* 2005: 360:1897-1903.
- Vesper S. Traditional mould analysis compared to a DNA-based method of mould analysis. *Critical Reviews in Microbiology,* 2011:37:15-24.
- Vesper S, McKinstry C, Haugland R, Wymer L, Bradham K, Ashley P, Cox D, Dewalt G & Friedman W. Development of an Environmental Relative Moldiness Index for US

homes. *Journal of Occupational and Environmental Medicine,* 2007:49:829-833 810.1097/JOM.1090b1013e3181255e3181298.

- Will KW & Rubinoff D. Myth of the molecule: DNA barcodes for species cannot replace morphology for identification and classification. *Cladistics,* 2004:20:47-55.
- Wu Z, Blomquist G, Westermark S-O & Wang X-R. Application of PCR and probe hybridization techniques in detection of airborne fungal spores in environmental samples. *Journal of Environmental Monitoring,* 2002:4:673-678.
- Zalar P, de Hoog GS, Schroers H-J, Crous PW, Groenewald JZ & Gunde-Cimerman N. Phylogeny and ecology of the ubiquitous saprobe *Cladosporium sphaerospermum*, with descriptions of seven new species from hypersaline environments. *Studies in Mycology*, 2007:58:157-183.

CASE STUDY: DETERMINATION OF MOISTURE DAMAGES ON ITEMS OF ART IN EXHIBITIONS BY THE USE OF MICROBIAL ANALYSIS

Judith Mueller and Urban Palmgren

ABSTRACT

The described case studies concerning the distinctive features and demands of investigations of mold growth on art and other objects of exhibition. The surfaces and materials used in this objects, pose a special challenge for sampling and analysis. In the microbiological assessment of damages of art objects, the conservation of the value of the object, the search for the cause and the age determination of the damage and the minimization of the damage were the primary targets, as well as attending to the interests of museums, artists and insurances.

KEYWORDS: mold, art, object of exhibition, microbiologic analysis

INTRODUCTION

Inaccurate storage and transportation can cause moisture damage on art objects.

As soon as microbiological damage occurs, many interest groups, like artists and museums, but also insurances, are concerned about the conservation of the value of the damaged objects. The often very delicate surfaces of these objects pose the particular difficulty and only allow for limited sampling and restoration. In some cases, only a minimization of the damage, instead of an elimination of the damage, can be achieved. Microbiologists, restorers and artists have to work together closely to decide on and carry out the options of sampling and removal of the damage. The preservation of art and other objects of exhibition as well as safety and health of restorers, employers and visitors should always be the crucial factor of decision, even if financial considerations have a strong influence on all work carried out.

MATERIALS AND METHODS

The investigations were carried out together with a German laboratory which is specialized on mold and bacterial analysis of moisture damaged building materials, but lately investigations were carried out in museums and exhibitions. After a case of water leakage in a storage building microbiologists worked as team workers closely together with restorers, conservators, directors and insurances experts. With the accurate microbial analysis, correct estimations of the damage could be accomplished, further microbial growth prevented and the objects with mold growth restored.

Before items are sent to other museums and after they are returned it is of great importance to investigate their microbial status. To detect the microbial status of items, the age and activity of microbial growth or contamination on surfaces are evaluated. The investigation of the microbial status of items can also confirm the storage standards at museums and let them prevented self made microbial damages depending on incorrect storage.

The samples were analyzed by using different methods:

Adhesive tape direct microscopy, sterile cotton swab samples, material samples, i.e. wood or paper, if possible without destroying objects, and air sampling were accomplished in the building to estimate microbial health hazards. Adhesive tape samples are analyzed using a microscope. Mold spores and mycelia are examined and counted and the number per cm^2 was calculated.

All material samples, including swab samples, were analyzed using a three-level method of analysis in order to get as much information as possible about the microbiological infestation. Following, the three levels (Palmgren, 2004) are defined:

1) Determination of the total number of cells (Tnc)
Using the Camnea method[2], all mold and bacteria are made visible with a fluorescent dye and an epifluorescence microscope. The microorganisms are microscopically detected, counted, categorized and defined as the total number of mold and bacteria cells.

2) Determination of the number of biochemical active cells (Nbac)
Using the fluorescent dye and an epifluorescence microscope, the microorganisms, with active metabolisms, are made visible. The active microorganisms can be divided into bacteria and mold. They are counted separately and therefore deliver crucial data for the age determination of the microbiological damage.

3) Determination of the number of colony forming units (NCFU)
Colony formed units (CFU) are assessed on three standard media (Holah, Betts, Thorpe, 1988) (TGE for bacteria, DG18 and Malt agar for mold), incubated at 25 °C and counted, as well as differentiated, after seven days.

Air samples are made with the filtration method (Camnea method, (Palmgren et al. 1986)). Polycarbonate filters with a pore size of 0.4µm and a diameter of 37mm are placed on support pads in sterilized filter cassettes (Millipore). With an air flow of 2 liters per minute over 4 hours, air is sucked through the filter holder with an air sampler (SKC). After the sampling, the filter is removed out of the cartridge and washed with a sterile washing liquid. 100 mµ of the sample is dyed with acridine-orange (fluorescent dye) and filtrated through a black polycarbonate filter. Thereafter, the microorganisms on the filter surface are counted under a fluorescence microscope. In the next step, colony formed units (cfu) are plated out on three standard media, incubated at 25 °C and counted, as well as differentiated, after seven days. Total number of cells (tnc) of microorganisms stained with fluorochromes, such as acridine orange, has been used for a long time in different environments, such as marine research (Zimmerman, Meyer-Reil, 1974), rapid determination in food samples (Holah, Betts, Thorpe, 1988) and counting of airborne microorganisms in highly contaminated environments (Palmgren et al., 1986).

A large percentage of bacteria and mold in buildings do not growth under laboratory condition. The total amount of microorganisms can be detected by the total cell count analysis. In regard to air samples the amount of the total number of cells can deviate from the number of cfu up to 100 times1.

Additional staining of material samples with flourochromes shows the metabolic activity of microorganisms and gives the researcher information that makes a determination of the age of the mold growth on building materials possible. Labor Urbanus GmbH, Duesseldorf Germany for example, invented this extended method of investigation to confirm if more than just one water damages have caused particular mold damage to a building.

The microbial damages on surfaces and materials of art were estimated with rating of Labor Urbanus GmbH publicized 20041. The evaluation and the rating of the results by these extended microbial material investigations depend on the experience and classification of > 200.000 microbial samples in 25 years. The classification of the degree of microbial damage is based on normal back ground concentrations found on materials without additional water damage. The critical values are not defined as sterile material samples; on the contrary a certain contamination with microorganisms on all surfaces is ubiquitary existent. Background values state that no abnormal numbers of microorganisms were detected in the samples.

TABLE 1: background or critical value of bacteria depending on reference parameter.

Bacterial classification Microbial method of investigation	Background values in cm^2	Background values in gram
Total number of bacteria (Tnb)	< 100.000	< 1.000.000
Number of biochemical active bacteria (Nbab)	< 10.000	< 100.000
Share of Nbab in Tnb	< 10%	< 10%
Number of colony forming units (NCFUb)	< 10.000	< 100.000
Share of NCFUb in Tnb	< 10%	< 10%

TABLE 2: Background or critical value of mold depending on reference parameter

Mold classification	Background values in cm^2	Background values in gram
Microbial method of investigation		
Total number of molds (Tnm)	<10.000	< 100.000
Number of biochemical active molds (Nbam)	<1.000	< 10.000
Share of Nbam in Tnm	< 10%	< 10%
Number of colony forming units (NCFUm)	< 1.000	< 10.000
Share of NCFUm in Tbn	< 10%	< 10%

TABLE 3: Classification of microbial damage

Classification of microbial damage	Assessment
Up to background values	No abnormal microbial growth
Slightly increased numbers	Up to 10 times higher than background
Increased numbers	Up to 100 times higher than background
Strongly increased numbers	More than 1.000 times higher than background

If one of the described microbial methods of investigations show an increased number or strongly increased number of microorganisms in the sample the sample is assessed as microbial damaged.

The age determination of microbial damages by microbiological analyses is based on microbe's lifecycles and the metabolic activity of bacteria and mold. The expanded analysis including total cell count, biochemical activity and cfu has to be carried out. The rates of tcc / ba and cfu / ba serve as a basis of decision. Table 4 sums up the simplified basis of this method. For the final age determination the overall picture of the analysis, the biochemical activity, the composition of the microorganisms and experiences play a major role.

TABLE 4: Basis of a simplified age determination of microbial damages

Rate of CFU/Tnc	Approximate age
~ 1%	> 1 year
~10%	> 6 month
~25%-50%	about 3 month
>50%	< 3 month

RESULTS AND DISCUSSION

In these case studies, different cases the laboratory worked on will be examined and presented. Each case is distinct, regarding to varied questions, protagonists and difficulties. For the case studies objects and examinations were chosen to illustrate the diversity and the specific characteristics of mold infestation of art objects. The delicate structures, sensitive materials and the uniqueness of many objects are some of the challenges that need to be met at these appointments. The diverse interest groups also have to be regarded and accounted for. Often, the artist, who wants to preserve his art object and conserve its value, is in opposition to the insurance experts, who want to keep the cost of the damage low. Initially, it is important to thoroughly debate the question in each case and then to plan and execute the sampling and analysis as well as the possible measures with all experts involved.

Case 1: Water damage in a museum

In a museum, there was water entry due to a leaking roof. The entire wall structure, composed of plastered and painted drywall as well as mineral rock wool insulation, was completely saturated with moisture. Material samples of the drywall and the mineral rock wool showed a strong, microbiological infestation with a high fraction of the toxic fungus *Stachybotrys chartarum*. The wall was part of an exhibition hall that contained several pieces of art. However, the artwork was not directly affected. The goal was to remove the wet wall material, to ascertain whether or not the artwork had microbiological damage and, if possible, to continue the exhibition during the clean-up operations without endangering the employees and visitors of the exhibition. The extent of the damage was defined using material samples of the wall and then the clean-up area was shielded with dust-tight walls, behind which the work was performed in under negative pressure. All contaminated material was removed and transported outside through air-locks. The artwork was checked for contamination by using swab samples and, if necessary, was cleaned with Hepa filter protected vacuum cleaner. While the clean-up operations were going on, the air in the exhibition hall was regularly checked via air sampling to ensure that there was no hazardous exposure to microbial contamination. The negatively pressurized dust tight walls were a successful strategy: the hazard for the employees and visitors in the exhibition hall was controlled by air samples. No higher microbial concentration could be found. In this way, the museum could carry out the clean-up operations without endangering the artwork and without closing the exhibition.

Case 2: Water damage in the storage area of a museum

The origin of the damage was the failure of the climate control unit in the storage area of a museum. The storage area did not have an alarm function and therefore there was a high temperature and humidity in the area for two weeks before the malfunction was detected. Most of the objects were stored in especially fabricated wooden boxes, but some of the objects were only covered with plastic foil sealing. The objects showed varyingly potent visible microbiological infestation. A new storage area was rented to assess the extent of the damage. The area was divided in three zones: a black area for infested objects, a gray area for examination and a white area for unsoiled and cleaned objects. Each object and each box was examined and assessed using adhesive tape and direct microscopy. At first, the boxes were divided into „no infestation", „to be cleaned" and „to be reassessed" groups and placed in the black or white area. The boxes that were to be cleaned, were sanded and then checked again and were then either discarded or placed in the white area, depending on the new result. The examination of the art work was performed in collaboration with the restorers. Depending on the material and the surface, swab samples or adhesive tape samples were used to decide which adequate methods for cleaning and conservation to choose.

With all the additional security measures and the extensive analysis, more serious damage to the objects could be prevented. Despite the additional cost for specialists and measures, the insurance companies saved millions of Euros in compensation claims from the artists.

Case 3: Mold damage on borrowed objects

A private collector lent his icons collection to a museum for 12 months and regulated the correct storage of the objects by contract. Icons are religious art work most commonly a painting on wooden board. After the return of the items, the collector examined the art work and detected a microbiological infestation in the form of a white coating. The allegation towards the museum was of microbial damages depending on incorrect storage resulting in a loss in value of the icons. The storage rooms were professionally examined and neither differences in room temperature or humidity nor structural damage of the rooms could be detected. The icons were investigated, using adhesive tape samples and swab samples. The adhesive tape samples confirmed the microbiological infestation. The wipe samples were used to determine the age of the microbiological infestation. Some icons had been chemically treated and were therefore not suited for the age determination. Consequently, an untreated infestation was chosen and examined in the laboratory for the total number of cells, the biochemical activity and the CFU. The analysis cleared showed, that the infestation had occurred more than one year ago. Therefore, the museum was cleared of all compensation claims. In the future, the condition of borrowed objects is to be checked before these objects are accepted into the museum.

Every case in this study and every problem to solve were different, therefore a standard recommendation cannot be given. However, in many cases the microbiology can help to understand the damage and provide solutions for decontamination. If the microbiological status of museum objects is assessed before the items go for loan the claim of damages can be evaluated. In another case the analysis could help to define compensation for an artist after water damage affected his objects of art in a storage room. There was also a case where a total damage of an entire exhibition could be prevented because the spread of the microorganisms in an entire museum was prevented. The microbiological analysis can help to understand and deal with microbiological damages on items of art or the museum itself. The items are sensitive and microbiology is just one part of a close collaboration between artists, restorers and microbiologists that is essential to solve the damage of art in a professional way.

REFERENCES

- Holah JT, Betts RP, Thorpe RH. The use of direct epifluorescent microscopy (DEM) and the direct epifluorescent filter technique (DEFT) to assess microbial populations on food contact surfaces. *J. Appl.* 1988 Vol.65. 61, 215-221.
- Palmgren U. Gesamtzellzahl, KBE und biochemische Aktivität von Bakterien und Schimmelpilzen in "Baumaterialien". VIII Lübecker Fachtagung für Umwelthygiene. Verlag: Schmidt-Römhild, Lübeck, 69-93. 5. ... Environments", In: *Indoor Air.* 2004:15:11:87-98.
- Palmgren U, Ström G, Blomquist G, Malmberg P. Collection of airborne microorganisms on Nuclepore filters, estimation and analysis, CAMNEA-Method. *J. Appl. Bact.* 1986:61:401-406.
- Zimmerman R, Meyer-Reil L. A new method for fluorescence staining of bacterial populations on membrane filters. Kiel. *Meeresforsch.* 1974:30:24- 27.

INVESTIGATION OF EFFECTIVENESS OF MOLD DISINFECTANTS AND CHEMICALS ON TOTAL CELL NUMBERS OF MOLD ON BUILDING MATERIALS

Judith Mueller and Urban Palmgren

ABSTRACT

In this investigation mold disinfectants H_20 (30%), H_20_2 (5%) with hydroxy-acids (fruit acids) and isopropanol (70%) were tested on *Aspergillus versicor* growth on wall paper pieces. The control treatment was H_20. The aim of the study was to examine if the chemicals have an effect on the total cell count. The disinfectants had four incubation times: 0h, 2h, 24h and one week. The results have shown that the amount of reduction of the total cell count is not sufficient to replace the remediation of infected material with disinfection. In addition, the biochemical activity and the CFU have shown a habit of recovering after only one week, without nutrients and fluids available. This emphasizes the ineffectiveness of disinfectants to remove the biomass. Furthermore, the long term effect of mold disinfectants has not been investigated. In regard to preventive health protection, indoor mold contamination should be removed and replaced rather than treated by disinfectants.

KEYWORDS: mold, total cell numbers, disinfection, mold disinfectants

INTRODUCTION

Disinfection is the inactivation of microorganisms and inhabition of the ability of reproduction and division. The disinfectant varies in the effect and action. The destruction of the cell does not always occur but even airborne cell components can have an allergic and toxic potential (WHO, 2009). Disinfection of mold and bacteria on building material is a common alternative to a costly remediation. Guidelines in the USA and Europe recommend the removal of microbiologic contaminated material. Depending on the parts and the size of the contamination the remediation is costly and complex. Disinfection seems to be an easy and fast alternative. Tests shows, that chemicals could minimize the presence of colony forming units of microorganisms. However, within these studies, the biomass was not taken into consideration. The aim of our investigation was to examine if the total cell numbers (biomass) of mold on samples was affected by mold disinfectants. Furthermore the metabolism activity of the cells and the CFU were investigated. As a result we tested the hypothesis "mold disinfectants and chemicals do have an effect on total cell numbers of mold on building materials".

MATERIALS AND METHODS

In our assay, we tested the effectiveness of mold disinfectants as well as other chemical substances for the purpose of mold disinfection. We want to prove that disinfection cannot replace decontamination of microbiological contaminated material while the chemical substances fail to destroy the biomass. The disinfection only affects the CFU; therefore, without testing the total cell count, the measured success of chemicals as mold disinfectants may be based on false positive results. Our research tests the effects on CFU as well as the total cell number and biological activity, thereby giving us a more accurate result of the effectiveness chemicals has towards mold.

The studies were based on mold contamination of wall paper. A pure culture of *Aspergillus versicolor* was transferred in 10 ml of sterile buffered solution and worked as the mold suspension. The mold suspension was spread on 2x2 cm wall paper and incubated on DG18 Agar for one week. The colonies of the suspension and the wallpaper were analyzed in preexperiments to ensure that samples were homogenous. The pieces of wallpaper were treated with three different substances: H_2O_2 (30 %), H_2O_2 (5%) with fruit acid and Isoporpanol. For every treatment and incubation time, three replicates have been done. Further, the samples with these substances are tested in different incubation times: immediately, after 2 hours, after 24 hours and after one week in regard to the reduction of the biomass and the reactivation of the biological activity and the effect on the CFU. The contaminated samples were washed with 10 ml of sterile buffered solution to achieve a sample solution. The control group was treated with pure water and analyzed in the same way. The total number of microorganism/ cm^2 was investigated by the Camnea method (Palmgren et al., 1986) dyed with acredin orange and counted with an epifluorescence microscope. The biochemical or metabolism activity of the mold was investigated with the dye "fluorescein diacetate" (Palmgren, 1984) and counted with an epifluorescence microscope. Colony formed units (CFU) were assessed on three standard media (Samson, 2010) TGE for bacteria, DG18 and Malt agar for mold, incubated at 25 °C counted as well as differentiated after seven days.

RESULTS AND DISCUSSION

In the preparation of the mold suspension the amount of *Aspergillus versicolor* was counted with an epifluorescence microscope. The suspension was cultivated on wall paper and washed out after a week to compare the concentration of the mold between the origin suspension and the contamination of the wall paper samples. The results are stated in figure 1. The pre-tests have shown that the chosen method were successful to prepare homogenous samples.

Figure1: Total cell numbers of *Aspergillus* suspension and Pre-tests

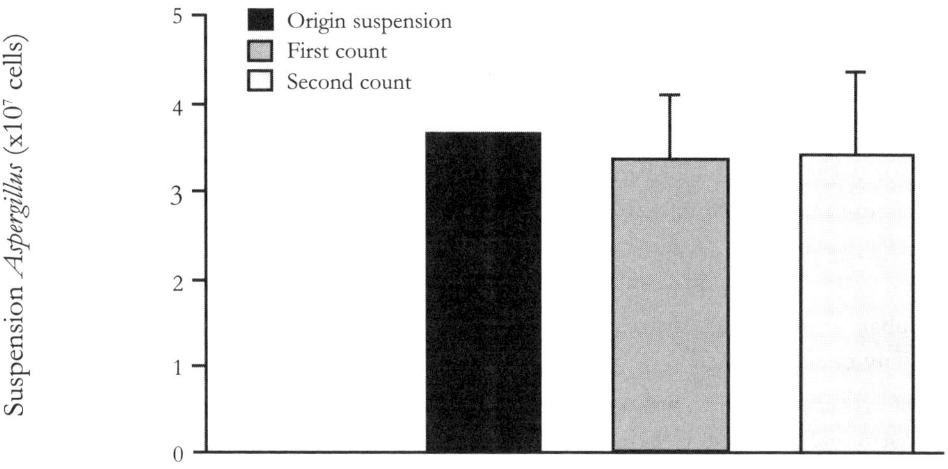

The prepared samples were treated with the chemicals and the total cell numbers were counted after an incubation time of 0, 2, 24 hours and one week.

Figure 2: total cell count after different incubation times of disinfectants.

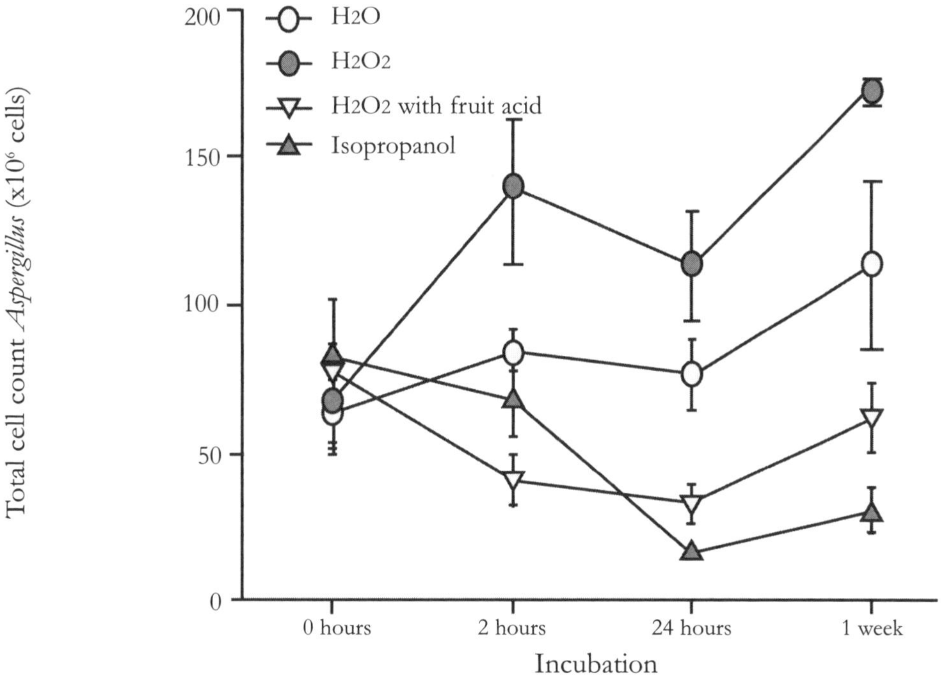

Figure 2 shows the results of the different disinfectants. Water was used as the control group and has shown only a slightly difference between the four cell counts. Hydrogen peroxide has shown an increase of microorganisms, hydrogen peroxide with fruit acid a slightly decrease and isopropanol has shown the highest reduction. The *Aspergillus* suspension started with about $3.6x10^7$ microorganism per ml. The highest reduction was detected after the incubation time of 24 hours to a cell number of $1.4x10^7$. With all disinfectants, a recovery and an increase of total cell numbers could be seen after one week of incubation.

All samples and replications were also investigated for metabolism or biochemical activity with the dye "fluorescein diacetate" and were counted after the same incubation times. The decrease of activity was significant. After 2 hours almost no activity could detected.

Furthermore, the water treated samples hardly show biochemical activity. Surprisingly after one week incubation all samples could recover and show metabolism activity. Only the samples treated with isopropanol alcohol had a slightly lesser recovery.

Figure 3a: biochemical activity after different incubation times of disinfectants

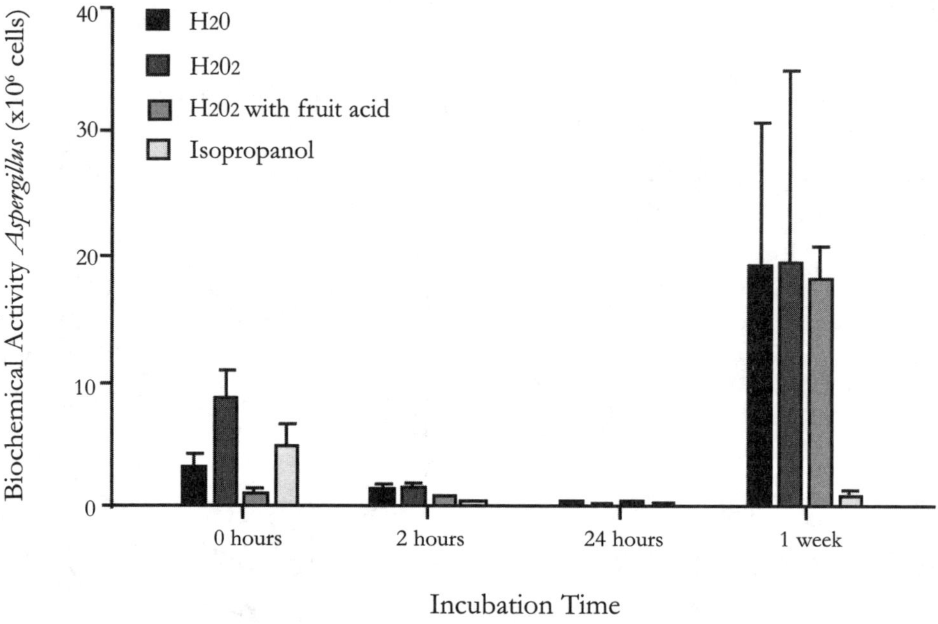

After counting the total cell number and the biochemical activity all samples were incubated for one week to investigate the CFU. The control showed a rapid increase until 24h incubation and a slight decrease after one week. Hydrogen peroxide could not reduce the CFU, however hydrogen peroxide with fruit acid has shown a reduction but also an increase after one week incubation. The alcohol only had an extreme effect on CFU, where even after one week incubation, the CFU were underneath the limit of detection.

Figure 3b: CFU after different incubation times of disinfectants.

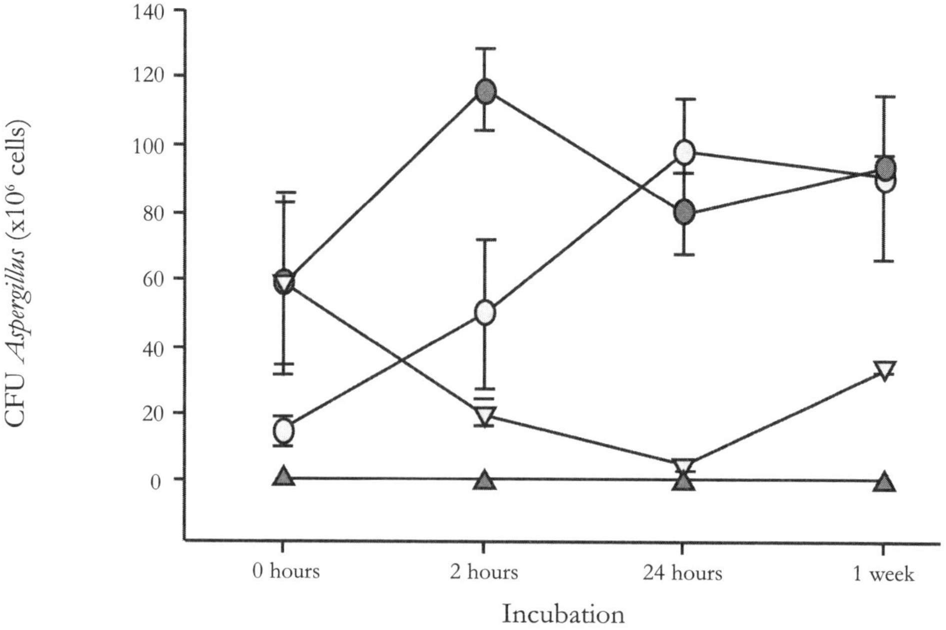

In the German mold remediation guidlines (Tischer, Chen, Heinrich, 2011) the removal of contaminated material is endorsed in regard to preventive health protection. Further, dead and dried microorganisms can produce allergic and toxin components. Due to this publication mold disinfectants produces praise their products as not only to kill but also to destroy the biomass. After disinfection, the components are still in the porous indoor material and emit these components to the indoor air. If the products can lead to a lyses of the microorganisms, the allergen potential is still present on the surface of the cells and can also emit to the indoor air. These investigations and the results show the reduction of total cell number of *Aspergillus versicolor* occur, but only in a marginal percentage. The number of 10^7 per ml could not be reduced to tolerable microbial contamination. The camnea method dye intact spores and mycelia and a prediction about cell components cannot be given after this investigation. Within the biochemical activity a reduction could be seen but also a recovery after one week. This leads to the assumption that the treatment did shock rather than kill the microorganisms. However, the recovering was visible and in the further experiments with a longer period of recovery should be recovery investigated. Normally one expects a recovery if the availability of water is still present. In our investigation we have stored the samples dry and no nutrients or fluids were available. Hence we can confirm that the trend of recovering take place without the aid of nutrients of fluids.

On this stage of the study we conclude that the tested chemicals could not reduce the biomass of contaminated building materials and could not substitute the remediation.

Disinfections cannot replace the decontamination of microorganism in building material and hence our null hypothesis is rejected.

In further studying an increase of replication is necessary. Furthermore, at the present time the same investigation is carried out with real samples and the results will be published in the near future.

References:

- Palmgren U, Ström G, Blomquist G and Malmberg P. Collection of airborne microorganisms on Nuclepore filters, estimation and analysis, CAMNEA-Method. *J. Appl. Bact.* 1986:61, p. 401-406.
- Palmgren U. Microbial and biochemical changes occurring during deterioration of Hay and Preservative effect of urea, *Swedish J. agric Res.* 1984:14 p.127-133
- Samson RA et al. Food and indoor fungi, CBS-KNAW Fungal Biodiversity Centre Utrecht, Netherlands, 2010.
- Tischer, Chen and Heinrich. Association between Domestic Mould and Mould Components, and Asthma and Allergy in Children: a systematic Rewiew., 2011 in Press.
- WHO, Dampness and Mould. Guidelines for Indoor Air Quality, 2009.

MICROBIOLOGICAL CHARACTERIZATION OF AEROSOLS ISOLATED FROM REMOTE LAKES IN THE CHILEAN PATAGONIA

Escalante G, León C, Campos V, Urrutia R and Mondaca MA.

INTRODUCTION

The long range mobilization of biological material mainly bacteria and fungi (bioaerosols) has increased in recent years due to the effects of Climate Change (Moulin, Chiapello, 2006; Neff et al., 2008). The bioaerosols transport occurs from warm regions and lower latitudes of the planet to colder regions and higher latitudes, such as Patagonia area. According to this mechanism, most bioaerosols may accumulate in the Polar Regions, where low temperature would allow them to be deposited (De la Rosa, Mosso, Ullan, 2002).

The role of bioaerosols depositation into remote lake ecosystems is a potentially important process but not yet totally explored. The flow of airborne bacteria could be an important route of colonization of remote and pristine environments, with ecosystems that have no local anthropogenic influence, such as Patagonian lakes (Catalan et al., 2006). The risk of microorganisms dispersion to remote areas may have effects on the ecosystem of Patagonian lakes such as alteration of biogeochemical cycles, food webs and also can produce different diseases in plants, animals and even to the human beings (Griffin, 2007).

In Chile, there are few studies related to bioaerosols, so it is important to study the effects of the presence of allochthonous microorganisms in lake ecosystems, allowing a better understanding of the response of these ecosystems to the influence of external agents.

The aim of this study was to detect and characterize microorganisms in bioaerosols of remote lakes in the Chilean Patagonia (Alto Lake, Esponja Lake and Verde Lake).

MATERIALS AND METHODS

Location

The lakes: Alto, Esponja and Verde Lake, were selected according to their altitudinal location and rate of precipitation.

Sample

Bioaerosol samples were taken using simple filtering technique with a Tactical Air Sampler with an airflow of 5 liters per minute. through nitrocellulose or polycarbonate filters of 47mm. The suspended particles that impacted on the filter were trapped in the pores, forming a deposit.

Fig.1 Determination of the presence of microorganisms in bioaerosols

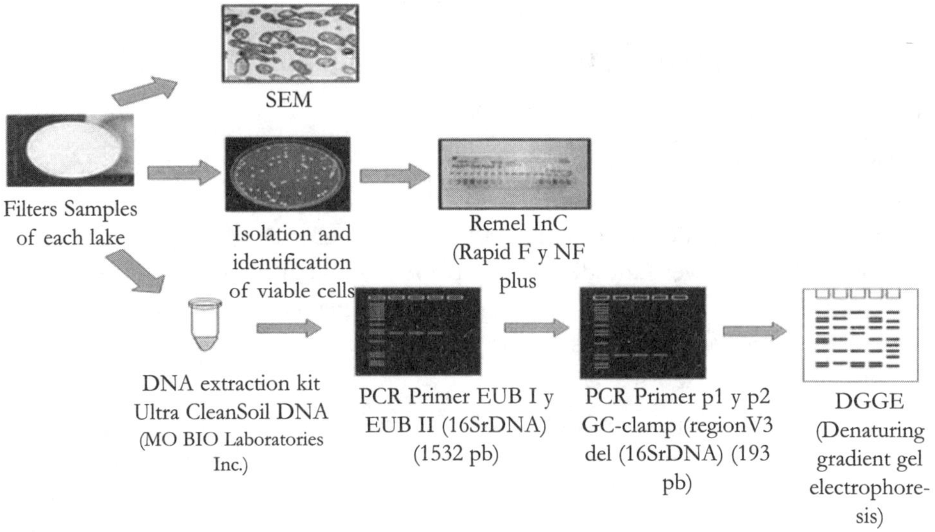

RESULTS

The scanning electron microscop (SEM) show the presence of bacterial cells with different morphological characteristics, adhered to the nitrocellulose filters.

Figure 2. DGGE analysis of DNA 16S from samples of Chilean Patagonia lakes, F1: Alto lake; F2: Esponja lake; F3: Verde lake.

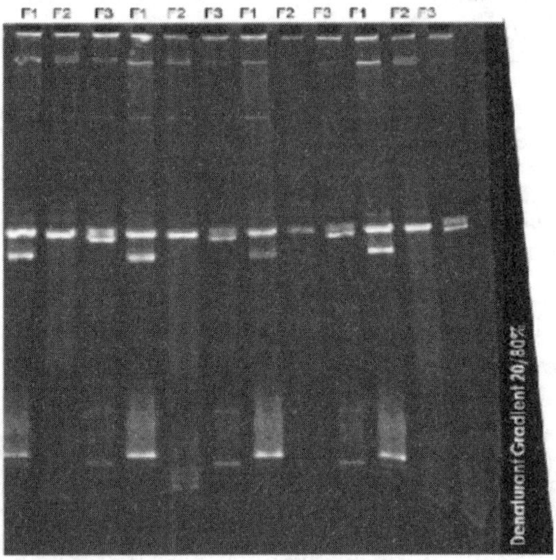

DGGE, shows the bacterial communities associated with filters and these are different for each lake.

Table 1. Identification of bacteria isolated of bioaerosols from Chilean Patagonia lakes

Alto Lake	Esponja Lake	Verde Lake
Acinetobacter (3)	*Burk. cepacia*	*Acinetobacter spp.*
Alc. piechaudii	*Morax. osloensis Acinetobacter*	*Ac. calcoaceticus*
CDC NO-1	*Ac. calcoaceticus*	CDC NO-1
		Ps. stutzeri
		Burk. cepacia
		Morax. osloensis
		Sph. paucimobilis

Each lake has a characteristic bacterial flora. The CDC NO-1 is an unusual bacterium, recently identified associated with dog and cat bite wounds. designated as CDC nonoxidizer 1 group (NO-1) and was found in two of the lakes under study.

CONCLUSIONS

- These results are the first report that demonstrate the presence of bacteria in air samples of Patagonian lakes.
- The isolation of CDC NO-1 strain is the first in Chile and because of that, the etiologic role and the pathogenicity of NO-1 are unclear. At present no surveillance system exists for reporting dog or cat bite wound-associated infections and, therefore, the incidence of NO-1 isolated from bite wound infections is not known.
- This is the first study in Chile and in Patagonia Lakes, since there is no information, the proposed study would be a pioneer in our country and serve as a baseline for future studies.

REFERENCES

- Catalan J, Camarero L, Felip M, Pla S, Ventura M, Buchaca T, Bartumeus F, Mendoza G, Miro A. Casamayor EO, Medina-Sanchez JM, Bacardit M, Altuna M, Bartrons M and Diaz de Quijano D. High mountain lakes: extreme habitats and witnesses of environmental changes. *Limnetica.* 2006:25:551–584.
- De la Rosa MC, Mosso MA and Ullan C. El aire: habitat y medio de transmision de microorganismos. Observatorio Medioambiental. 2002:5:375-402.
- Griffin DW. Atmospheric Movement of Microorganisms in Clouds of Desert Dust and Implications for Human Health.*Clinical Microbiology Reviews,* July 2007:20:3;459–477.

- Hollis DG, Moss CW, Daneshvar MI, Meadows L, Jordan J, Hill B. Characterization of Centers for Disease Control group NO-1, a fastidious, non-oxidative, gram-negative organism associated with dog and cat bites. *J Clin Microbiol* 1993:31:746-8.
- Moulin C, Chiapello I. Impact of human-induced desertification on the intensification of Sahel dust emission and export over the last decades. *Geophysical Research letters*, vol. 33. 2006
- Neff JC, Ballantyne AP, Farmer GL, Mahowalda NM, Conroy JL, Landry CC, Verpeck JT, Painterio TH, Lawrence CR, Reynolds RL. Increasing eolian dust deposition in the Western United States linked to human activity. *Nature Geoscience*, 2008:1(3): 189-195.

USE OF CULTURE AND PCR ANALYSIS IN MOLD ASSESSMENTS

Philip R. Morey

ABSTRACT

Both PCR and culture analysis of collected samples can result in similar building assessment conclusions. However, both methodologies are subject to limitations. For example, the utility of PCR data in a mold assessment is limited by missing target taxa. Interpretation of PCR data is also problematic, as in the ERMI approach, when a taxon is misclassified as to its water indicator or phyloplane status. The use of culture data is limited by the longevity of spore culturability among species in aged samples. However, the most important aspect of a mold assessment is the physical inspection of the building and its HVAC system for evidence of dampness and biological growth. This study shows that the greatest limitation in mold assessments occurs when there is an over reliance placed on data obtained from sampling.

KEYWORDS: Culture Analysis, Mold, PCR Analysis, ERMI, Data Interpretation

INTRODUCTION

Air and dust sampling and culture analysis of those samples to the species level by an experienced mycologist has been considered to be the gold standard in mycological investigations in buildings (Miller, 2011). Rank order comparison of taxa identified in samples collected in damp buildings including control locations has been used for many years for purposes of data interpretation. Quantitative polymerase chain reaction (PCR) for mold species identified has become a popular analytical methodology used in mold assessments (Summerbell et al., 2011; Vesper, 2011). This study was undertaken to determine (a) if culture and PCR analysis of samples to the species level can result in similar conclusions during mold assessments and (b) to identify limitations associated with both analytical methodologies.

METHODS

Building A. Dust samples were collected in a northeastern USA schoolroom during the winter and then again during the following summer. Filter cassette minivacuum samplers were used to collect settled dust primarily from various above floor surfaces (Morey, 1995). The dust sample from the winter evaluation was analyzed exclusively by PCR methodology with a target list of 10 "normal or outdoor sourced" and 26 "atypical or water sourced" molds (Vesper, 2011). During the subsequent summertime evaluation dust samples were collected from various locations in the same room including its HVAC system and these samples were analyzed exclusively by culture methodology (direct and dilution plating on malt extract agar [MEA]; AIHA, 2005). Both the wintertime and summertime evaluations in Building A identified the presence of a moldy crawlspace beneath the room where sampling had taken place. The airstream surface of the supply air duct supplying ventilation air to the schoolroom was also inspected during the summertime evaluation and was found to be mold colonized.

Building B. A water loss with subsequent mold growth (approximately 10M2 of visible colonization) on wallboard and wood surfaces occurred in one room in a large house in the USA southwest. Paecilomyces variotii was identified as a predominant mold in air and dust samples from the water loss area. Subsequent clean-up efforts were problematic because of repeated breaches in critical barriers that resulted in dissemination of remediation dusts throughout the house. Approximately 12 to 16 months after the water loss sampling of settled dust throughout the house was carried out to determine if cleaning had removed cross-contaminating, water indicator molds such as Paecilomyces variotii. Some of the collected dust samples were analyzed by both PCR and culture methodologies.

RESULTS AND DISCUSSION

Building A. Table 1 presents the rank order concentrations (PCR spore equivalents per milligram of dust; SE/MG) of the seven most abundant molds found in the settled dust in the schoolroom in Building A during wintertime heating conditions. Based on the concentrations of all the 36 molds (data not shown) an environmental relative moldiness index (ERMI) value of 9.6 was calculated according to the method described by Vesper (2011). An ERMI value of 9.6 suggests a moldy condition in the room, although visible mold growth was not found on material surfaces. However, the investigators did find a moldy condition in the crawlspace beneath the room and that the crawlspace was positively pressured relative to air in the classroom.

Among the seven taxa in Table 1, four (*Aureobasidium pullulans, Wallemia sebi, Penicillium variable*, and *Cladosporium sphaerospermum*) are considered by the ERMI concept (Vesper, 2011) to be water indicator molds, and three of the taxa (*Cladosporium cladosporioides, Epicoccum nigrum*, and *Cladosporium herbarum*) are grouped as outdoor sourced molds. It should be noted that *Aureobasidium pullulans* is considered by taxonomic authorities (Gravesen et al., 1994; Samson et al., 2011) to be primarily a phyloplane (outdoor sourced) mold and not a water indicator taxon. Thus, the moldy condition indicated by the ERMI analysis value of 9.6 (Table 1) is likely due to the incorrect assignment of *Aureobasidium pullulans* to the water indicator mold group.

TABLE 1. PCR Analysis of Settled Dust From Building A January Heating Conditions

Taxa	SE/MG
Aureo. pullulans[a]	5,500
Clad. cladosporioides[b]	3,100
Epico. nigrum[b]	1,000
Wallemia sebi[a]	330
Clad. herbarum[b]	310
Pen. variable[a]	190
Clad. sphaerospermum[a]	150

[a] Water indicator molds (Vesper, 2011)
[b] Outdoor – sourced molds (Vesper, 2011), ERMI Value = 9.6

Table 2 presents the rank order concentrations of culturable molds found in settled dust in the HVAC system serving the same schoolroom in the building six months after the first evaluation. It had been noted during the summertime inspection that microfungal colonization (*Cladosporium* conidiophores) had occurred on the duct liner in the variable air volume (VAV) box which provided conditioned air to the subject classroom. *Cladosporium halotolerans* was the predominant mold found by direct plating of the dust from the VAV duct liner as well as dust from the upper surface of the diffuser downstream of the VAV box. *Cladosporium halotolerans* was not on the target list of the 36 molds analyzed by PCR in the dust collection in the subject schoolroom six months earlier.

TABLE 2. Direct Plating Analysis of Settled Dust from HVAC in Building A; June Air-Conditioning

Direct Plating Taxa, VAV Dust	**Colonies on MEA**
Clad. halotolerans	163
Clad. sphaerospermum	29
Phoma herbarum	2
Direct Plating Taxa, Diffuser Dust	**Colonies on MEA**
Clad. halotolerans	13
Clad. sphaerospermum	6
Pithomyces chartarum	6
Aureo. pullulans	2

[a] *Cladosporium* fruiting structures found on VAV box liner by tape slide analysis. MEA = Malt Extract Agar.

An additional dust sample was collected during summertime conditions from a cabinet in the subject schoolroom directly above the moldy crawl space. *Aspergillus sydowii* (a water indicator mold) was the predominant taxon cultured from this dust sample.

The collective wintertime and summertime evaluations in Building A showed that physical inspection for mold colonization in both the crawlspace and the VAV box liner was more important in building assessment than sampling results. Important limitations of the ERMI method used in the first evaluation in the schoolroom were the absence of *Cladosporium halotolerans* from the PCR target list as well as the misclassification of *Aurobasidium pullulans* as a water indicator mold.

Building B. The objective of sampling was to determine if clean-up efforts had been effective in removing *Paecilomyces variotii* from surfaces in the non-water loss areas of Building B. Dust samples were collected 12 to 16 months after the water loss. One dust sample was obtained from the canister of a central vacuum cleaner, which had remained unused since shortly after the time of the water loss. Some samples were of sufficient size that dust was analyzed by both PCR and culture methods. Samples of very small size (about 2MG) were analyzed by PCR only. The data from only a few representative samples is presented herein. Physical inspection throughout Building B showed that considerable dust was still present on room surfaces 12 to 16 months after the water loss.

Table 3 presents PCR and culture analysis of above-the-floor settled dust obtained from non-porous surfaces from a room about three meters from the water loss area. Rank order examination of the PCR data showed that *Cladosporium cladosporioides* was the predominant taxa recovered with *Paecilomyces variotii* being a distant second. By contrast, *Paecilomyces variotii* was the most abundant mold recovered by culture with *Cladosporium cladosporioides* being fifth on the rank order list of taxa (Table 3). One explanation for the lack of dominance of *Cladosporium cladosporioides* recovered by dilution plating is its short culture halflife (≤0.1 year; AIHA, 2005, pp. 99) in the aged sample (house had been vacant and closed-up for eleven months).

TABLE 3. PCR and Culture Analysis of Dust from Non-Water Loss Room in Building B[a]

Taxa	SE/MG
Clad. cladosporioides	13,000
Paec. variotii	2,300
Pen. brevicompactum	1,500
Eur. amstelodami	830
Taxa	**Colony Counts on MEA, Dilution**
Paec. variotii	14
Yeasts	10
Alt. alternata	8
Pen. chrysogenum	5
Clad. cladosporioides	4

[a] Samples collected and analyzed 12 months after water loss.

The same pattern of taxa recovery was evident in dust from a central vacuum cleaner that had not been used since about the first month after the water loss (Table 4). *Cladosporium cladosporioides* dominated the PCR analysis and *Paecilomyces variotii* dominated the culture data.

TABLE 4. PCR and Culture Analysis of Dust from Central Vacuum Cleaner in Building B[a]

Taxa	SE/MG
Clad. cladosporioides	4,500
Paec. variotii	1,500
Pen. brevicompactum	290
Asp. ustus	250
Pen. chrysogenum	180
Taxa	**Colony Counts on MEA, Dilution**
Paecilomyces spp.	17
Yeasts	14
Pen. chrysogenum	13
Asp. niger	6
Alt. alternata	3

[a] Samples collected and analyzed 12 months after water loss.

Some upholstered furnishings had been wrapped in polyethylene sheeting, removed from the house about one month after the water loss and placed in storage for 15 months. A small amount (about 2MG) of dust extracted from an upholstered chair was analyzed by PCR (insufficient dust for culture analysis). PCR analysis of the dust showed that *Paecilomyces variotii* was the dominant taxon recovered with *Cladosporium cladosporioides* being second in rank (Table 5).

TABLE 5. PCR Analysis of Dust From Upholstered Chair from Building B[a]

Taxa	SE/MG
Paec. variotii	11,000
Clad. cladosporioides	6,100
Eur. amstelodami	1,000
Asp. niger	610
Pen. brevicompactum	480
Asp. fumigatus	320

[a] Sample collected and analyzed 16 months after water loss

In Building B analysis by both PCR and culture methods, showed that *Paecilomyces variotii* was still abundant in dusts 12 to 16 months after the water loss

(Tables 3-5). HEPA vacuum cleaning was required to reduce residual dust to acceptable levels (See discussion in AIHA, 2008, Chapter 18).

CONCLUSIONS

Both PCR and culture analysis of collected samples can result in similar building assessment conclusions (e.g., *Paecilomyces variotii* cross contamination in Building B). However, both methodologies are subject to limitations. For example, the utility of PCR data in a mold assessment is limited by missing target taxa as with *Cladosporium halotolerans* in the evaluation of Building A. Interpretation of PCR data is problematic, as in the ERMI approach, when a taxon is misclassified as to its water indicator or phyloplane status (Building A). The use of culture data is limited by the longevity of spore culturability among species in aged samples as with *Cladosporium cladosporioides* in Building B.

The most important aspect of a mold assessment is the physical inspection of the building and its HVAC system for evidence of dampness and biological growth (WHO, 2009; Morey, 2011; Miller, 2011). This study shows that the greatest limitation in mold assessments occur when there is an over reliance placed on data obtained from sampling.

REFERENCES

- AIHA, Field Guide for the Determination of Biological Contamination in Environmental Samples. American Industrial Hygiene Association, Fairfax, VA. 2005.
- AIHA, Recognition, Evaluation, and Control of Indoor Mold. American Industrial Hygiene Association, Fairfax, VA. 2008.
- Gravesen S, Frisvad J and Samson R. Microfungi, Munksgaard, Copenhagen, Denmark.. 1994.
- Miller JD. Mycological Investigations of Indoor Environments, In: Microorganisms in Home and Indoor Work Environments, 2nd Edn., CRC Press, Boca Raton, FL. 2011. p. 229-245.
- Morey PR. Studies on Fungi in Air-Conditioned Buildings in a Humid Climate. In: Proc. Intern Conference on Fungi and Bacteria in Indoor Air Environments, Eastern NY Occupational Health Program, Saratoga Springs, NY. 1995. p. 79-92.
- Morey PR. Remediation and Control of Microbial Growth in Problem Buildings, In: Microorganisms in Home and Indoor Work Environments, 2nd Edn., CRC Press, Boca Raton, FL. 2011. p. 125-144.
- Samson R, Houbraken J, Summerbell R, Flannigan B and Miller JD. Common and Important Species of Fungi and Actinomycetes in Indoor Environments. In: Microorganisms in Home and Indoor Work Environments, 2nd Edn., CRC Press, Boca Raton, FL. 2011. p. 321-513.
- Summerbell R, Green B, Corr D and Scott J. Molecular Methods for Bioaerosol Characterization. In: Microorganisms in Home and Indoor Work Environments, 2nd Edn., CRC Press, Boca Raton, FL. 2011. p. 247-263.

- Vesper S. Traditional Mould Analysis Compared to a DNA-Based Method of Mould Analysis. *Critical Reviews in Microbiology*, 2011:37, 15-24.
- WHO. Guidelines for Indoor Air Quality: Dampness and Mould, Regional Office for Europe, World Health Organization, Copenhagen, Denmark.. 2009.

FACTORS PROMOTING THE EXPOSURE TO BIOAEROSOLS AMONG SWISS CROP WORKERS

Hélène Niculita-Hirzel and Anne Oppliger

ABSTRACT

Agricultural workers are among the professional groups most at risk of developing acute or chronic respiratory problems. Despite this fact, the etiology of these occupational diseases is poorly known, even in important sectors of agriculture such as the crops sector. A chronic exposure to multiple microorganisms, such as different bacterial and fungal species, has been proposed to be the cause of these multiple respiratory pathologies. Nevertheless, these microbial communities are still partially known. The aim of this study is to characterize all fungal species inhaled by the crops workers during different grain related activities and identify the abiotic and biotic factors that reduce the growth of the toxigenic, irritative or allergenic microbial species. Here, we are presenting the factors promoting the exposure to bioaerosols during different wheat related activities: harvesting, grain unloading, baling straw, the cleaning of harvesters and silos. Total dust has been quantified following NIOSH 0500 method. Reactive endotoxin activity has been determined with Limulus Amebocyte Lysate Assay. All molds have been identified by the pyrosequencing of ITS2 amplicons generated from bioaerosol.

INTRODUCTION

Agriculture is considered one of the occupations most at risk of acute or chronic respiratory problems caused by biological agents. The statistics collected so far suggest that farmers have a higher morbidity and mortality due to respiratory diseases (rhinitis, asthma, hypersensitivity pneumonitis, chronic bronchitis, pulmonary toxicity) than the general population, despite a lower prevalence of smoking (Heller, Kelson, 1982; Greskevitch et al., 2007). Similarly, the prevalence of work-related respiratory symptoms such as wheezing, coughing and dyspnea, is abnormally high among farmers (23% to 50%) (Skórska et al., 1998). Despite of epidemiological data, the etiology of these occupational diseases is not well known, even in important sectors of agriculture as the cereal industry. Indeed, grain dust is a known respiratory health hazard in the grain industry. Health effects associated with organic dust exposure in grain handlers include mucous membrane and skin irritation, grain fever (organic dust toxic syndrome), and chronic bronchitis (do Pico, 1994), as well as grain dust asthma, allergic sensitization, and airflow obstruction (Chan-Yeung et al., 1992).

The respiratory diseases developed by crops workers were proposed to be induced by a complex combination of different microbial agents: endotoxins of Gram-negative bacteria, peptidoglycan of Gram-positive bacteria and fungi present in grain dust (Poole et al., 2010). The endotoxins and peptidoglycans have been recognized to be microbial agents that can induce inflammatory responses. Fungi have been proposed to be a source of allergenic agents, such as *Alternaria sp.* or *Cladosporium sp.*, or toxins (e.g. tricothecenes) such as *Fusarium sp.* complex. Nevertheless, although the exposure of grain workers to cultivable airborne microorganisms have been evaluated to be as high as over one million colony forming units per m^3 (Eduard, 1997; Swan, Crook, 1998). There is no data on the identification of these specific fungal species or of all fungal community members available. Thus, there is a need to determine the microbial communities' composition in these occupational aerosols, to evaluate the frequency of exposure to known allergenic, irritative or toxigenic fungi and to identify the grain related activities which are the most exposed to these microorganisms.

Few guidelines have been established to prevent the adverse effects of agricultural dust inhalation. The American Conference of Governmental Industrial Hygienists (ACGIH) has recommended threshold limit values (TLV) for total dust (10 mg/m^3) and grain dust (4mg/m^3, for oat, wheat, and barley) (ACGIH, 2007), although the total dust guideline does not account for the biological activity of airborne agricultural particulates. There is currently no workplace standard specifically for endotoxin exposure, although a limit of 90 endotoxin units (EU) /m^3 as an 8-h time-weighted average (TWA) was suggested (DECOS, 2010) on the basis of several studies, experimental as well as epidemiological, that have shown that endotoxins can cause respiratory effects at concentrations around this standard (50–100 EU/m^3) (Castellan et al., 1998; DECOS, 1998; Zack et al., 1998).These guidelines are based on 37 mm cassette total dust samplers and need to be adjusted for comparison to the Institute of Medicine (IOM) inhalable sampler (Reynolds et al., 2009). Similarly, there is no workplace standard for toxigenic fungi and their toxins, although International and European legislation on contaminants in food include measures to ensure protection of public health by setting down the maximum levels for these contaminants. Acquiring information on the levels of these toxigenic fungi and of their toxins in various environmental settings is critical in establishing future recommendation, guidelines and controls for reducing disease. As traditional culture methods to quantify microbial exposures have proven to be of limited use, non-culture methods and assessment for microbial constituents are recommended (Douwes et al., 2003; Rinsoz et al., 2008; Oppliger et al., 2011).

This study aimed to 1) investigate bioaerosols samples from different grain related activities (harvesting, grain unloading, making bales of straw, cleaning of combine harvesters and silos) for their fungal composition by a metagenomic approach and for their reactive endotoxin activity by an enzymatic approach, and 2) characterized the grain manipulations that exposed the most the workers. The goal of this manuscript is also to be used as a learning tool on occupational exposure to bioaerosols. Four indicators described here might help to improve grain worker safety and health:

1. The exposure level of inhalable dust and endotoxin as well as of the most frequent fungal species in the aerosols during different grain related activities
2. Data on the frequency of exposure to toxigenic, irritative or allergenic molds during wheat related activities
3. Identification of tasks that present more risk for worker health
4. Advices to improve grain worker protection

MATERIAL AND METHODS

Study design

As the proliferation of microorganisms is dependent on temperature, rainfall and human disturbance (e.g. usage of tillage, addition of fungicides), we sampled the bioaerosols during harvesting of summer 2010 on 100 wheat fields equally distributed in diverse weather conditions of the Vaud region of Switzerland in order to obtain a complete image of the fungal species present all over the geographical region followed in this study.

In order to evaluate personal exposure of grain workers for different wheat related activities: harvesting, grain unloading, making bales of straw, cleaning of combine harvesters and silos, we followed between 15 July to 15 August 2010 respectively: eight reapers working all over the Vaud region (five driving combine harvester with a cab and three without a cab), thirteen grain workers in eight geographically separated granaries (six collective granaries and two individual granaries), two workers making baling straw, two workers doing five independent combine harvester cleaning, one worker doing silo cleaning. Participation was based on free consent.

Dust and bioaerosols assessment

Temperature, barometric pressure and relative air humidity were measured at each sampling site with a thermohygrometer and barometer PCE-THB 40 (PCE Group Iberica).

Inhalable dust was sampled on 5μm pore size PVC filter at a flow rate of 2.0 l/min using pocket pumps (MSA Escort Elf, Mine Safety Appliance Company, Pittsburgh, USA or SKC pocket pump 210-1002, SKC Inc., USA) and IOM heads (SKC Inc.). The filters were pre- and post-weighed on an analytical balance (Mettler, 0.001 mg sensitivity). Bioaerosol measurement consisted of collecting bioaerosols in the work area in the same place as dust samples.

Endotoxins were sampled onto polycarbonate filters (37 mm diameter, 0.45 μm pore size) placed into a ready to use polystyrene cassette (endofree cassette, Aerotech Laboratories, Inc., Phoenix, USA). Personal exposure to endotoxins was monitored with a pocket pump (AirChek XR5000, SKC Inc.) calibrated at 2 l/min. Endotoxins were extracted by shaking the filters at room temperature for 1 h in 10 ml of non-pyrogene water in a 50 ml conical polypropylene tube. Filter extraction solutions were vortexed vigorously prior to drawing a sample for endotoxin analysis using the kinetic Limulus amoebocyte lysate assay (LAL) as described in Oppliger et al., (2005).

Fungal particles were sampled for DNA extraction onto gelatin filters placed into a clear styrene cassette 3 sections (25 mm diameter, SKC Inc.). Once back in the laboratory, they were stored for 5 days at 4°C until DNA extraction. Genomic DNA was extracted by shaking the filters at room temperature for 15 min in 6 ml of NaCl 0.9% in a 50 ml conical polypropylene tube. Filter extraction solutions were centrifuged for 15 min at 14,000 rpm. The fungal wall material present in the recovered pellet was mechanically disrupted with the Tissue Lyser (Qiagen) within the first buffer of the FastDNA® SPIN Kit for Soil (Bio101). The DNA extraction was done with the FastDNA® SPIN Kit for Soil (Bio101) in accordance with the manufacturer instructions. The amount of each extracted genomic DNA was measured with Quant-iT™ PicoGreen ® dsDNA (Invitrogen) in a Tecan Infinite M200.

Fungal species identification

Amplicon libraries were performed using a combination of tagged primers designed for the variable ITS-1 region, ITS1F and ITS2, as described for the tag-encoded 454 GS-FLX amplicon pyrosequencing method in Buée et al. (2009) but adapted for 454 GS-FLX Titanium. Pyrosequencing (from the ITS1F primer) on the GS-FLX Titanium 454 System (454 Life Sciences / Roche Applied Biosystems) at Microsynth (Switzerland) resulted in mean on 283 reads per sample that satisfied the sequence quality criteria employed (cf. Droege & Hill, 2008). Tags were extracted from the FLX-generated composite FASTA file into individual sample-specific files based on the tag sequence by Microsynth (Switzerland). Sequence editing and analysis of the reads by operational taxonomic unit (OTU) clustering have been done as described in Buée et al., (2009) with the exception that USEARCH 5.1 (Edgar, 2010) and UCHIME (Edgar et al., 2011) have been used in order to generate the consensus sequences. The identity of these OTUs is determined by BLASTn (Altschul et al., 1997) against a nonredundant GenBank database.

Statistical analysis

Standard statistical methods were employed in the analysis of data using R package.

RESULTS

Dust levels

Mean levels of total dust is higher than the legal Swiss limit value of 10 mg/m^3 for the inertial dust only for harvesting related activities when the workers are not protected by a cab (Fig. 1). The concentration of dust was of 24.7 $mg/m^3 \pm 24.8$

for the harvesting without cab (fixed sampling at the worker head level) although it was close to 0 for the harvesting within a cab (fixed sampling within the cab). Nevertheless, dust levels during all grain related activities when the workers are not protected within a cab or within a building are higher than the ACGIH recommended threshold of 4 mg/m^3 for grain dust.

Figure 1. Levels of inhalable dust monitored during different grain related activities. The dotted line shows the Threshold Limit Value of inert dust inhalable (10 mg/m^3).

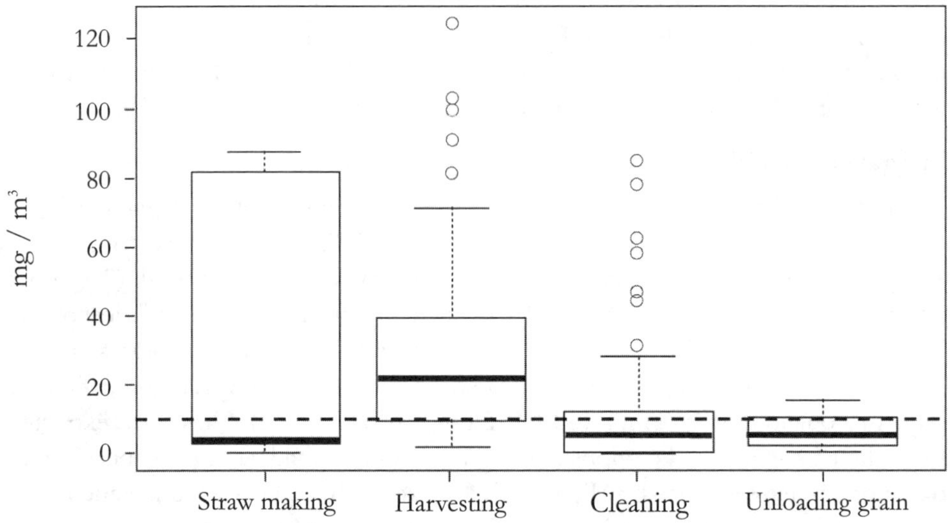

Endotoxins levels

Global results showed that exposure to endotoxin during all grain related activities when the workers are not protected within a cab or a building were above the 1,000 EU/m^3 recommended by the Swiss National Insurance (SUVA). But these levels were well below this recommended value for activities done within a cab or a building (Fig. 2). Moreover, exposure to endotoxins was correlated to the exposure to grain dust with the lowest endotoxin levels observed for the workers that were within a cab or within a room during all their activity (Geometric Mean = 134 EU/m^3). The endotoxin level increased when the workers were occasionally on direct contact with grain dust (Geometric Mean = 582 EU/m^3). Nevertheless this level is much lower than that of the workers that are in direct contact with grain dust all over their activity such as those that were continuously present on the unloading grain site (Geometric Mean = 4,150 EU/m^3) or those that were cleaning grain dust (Geometric Mean = 72,896 EU/m^3 for combine harvester cleaning, 15, 855 EU/m^3 for silo cleaning).

Figure 2. Active fraction of endotoxins in the air monitored during different grain related activities. The dotted line shows the recommended threshold value for the active fraction of endotoxin in the inhalable air, in Switzerland, that is much higher than that in Holland (90 EU/m^3).

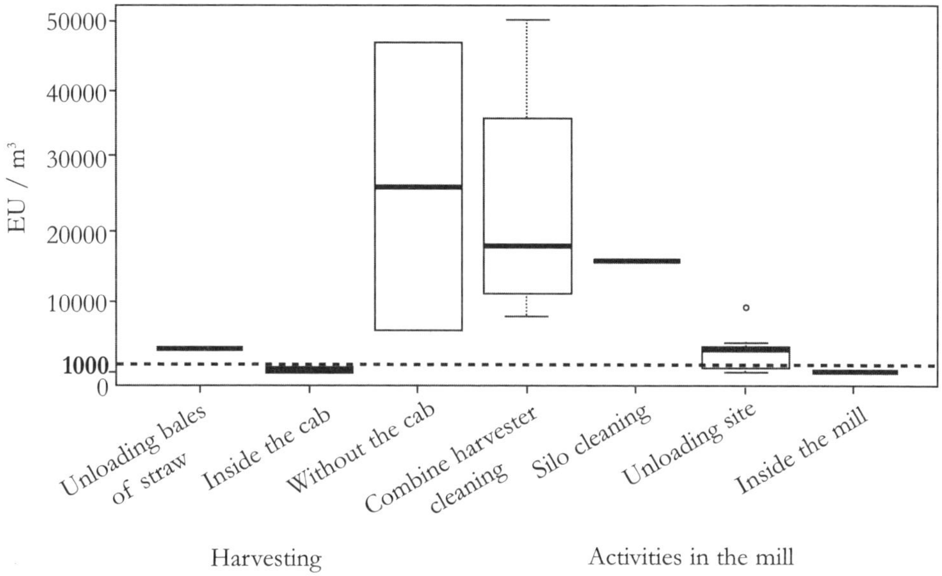

Dominant fungal species within grain aerosols

Fungi levels were previously described as very high: 10^6 - 10^7 CFU/m^3 for harvesting and 5. 10^5 – 5. 10^6 for grain handling (Eduard 1997). These values exceeded the SUVA recommended limit of 1,000 CFU/m^3 in every grain related activity. Our molecular analysis showed that two of the three dominant fungal species that have been systematically found in all bioaerosols samples are known to be respiratory sensitizers and allergenic: *Cladosporium spp.* and *Alternaria infectoria* (Fig. 3).

Figure 3. Average frequency of the dominant fungal species present systematically (A.) or sporadically (B.) in aerosol samples.

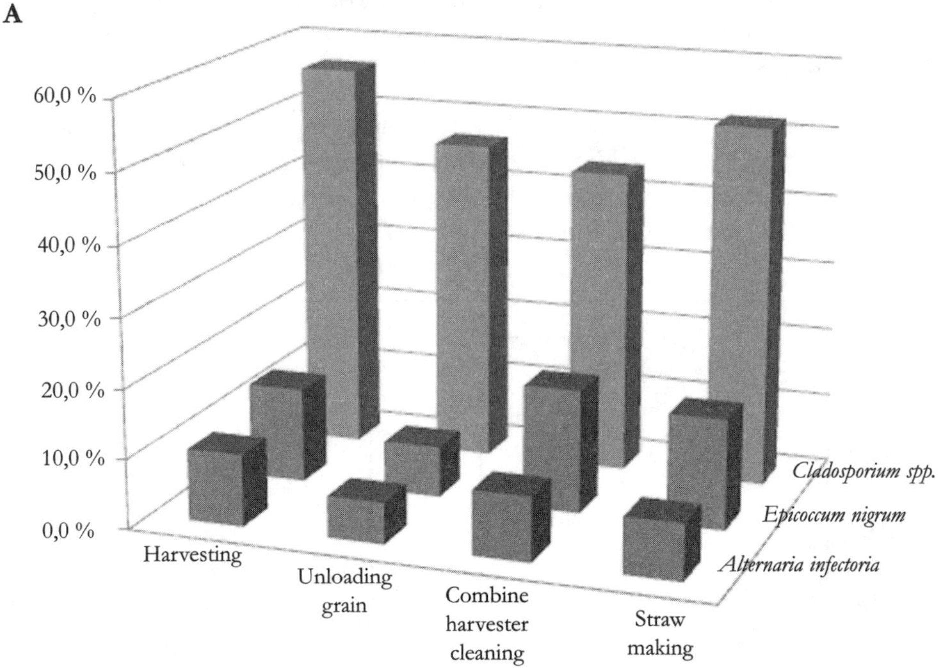

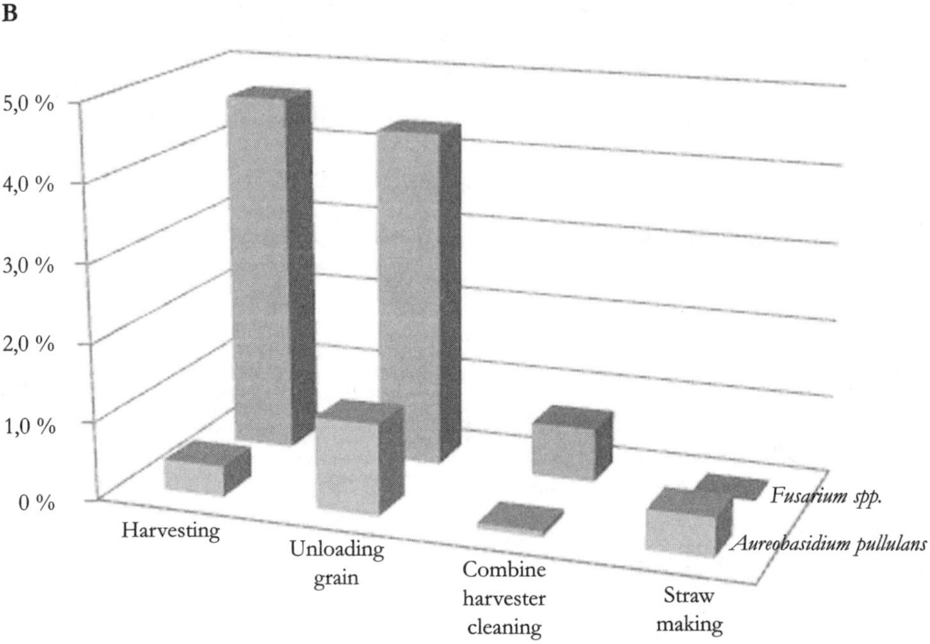

In addition, two of the dominant fungi that have been observed sporadically in the bioaerosols are a toxigenic fungus, *Fusarium spp.* that could deliver trichothecens, and *Aureobasidium pullulans* that is recognized to induce hypersensitivity pneumonitis. *Fusarium spp.* has been found in 62% of harvesting samples, 85% of unloading grain samples, 50% of cleaning samples and in none of straw making samples although *Aureobasidium pullulans* has been found in 51% of harvesting samples, 71% of unloading grain samples, 25% of cleaning samples and 100% of straw making samples. These results show that we can predict the risk of exposure to the toxigenic, allergenic or irritative fungal species when molecular tools were used.

CONCLUSIONS

This project's objective was to characterize crops worker exposure to pathogenic, irritative or allergenic microorganisms during different grain related activities. Our results confirm the previous quantitative data on the worker exposure to wheat dust and endotoxins. In addition, they show that crop workers are systematically exposed to complex mixtures of allergens, irritants or cytotoxic components. The novelty of our study is the systematic detection of dominant molds in the bioaerosols thanks to molecular tools. Indeed, these tools detect not only the cultivable fraction of fungi, but also the fungal fragments that are not cultivable which have been shown to be the major fraction of fungal particles in bioaerosols.

As a result of this research, recommendation might be done to crop workers in order to decrease their exposure to the microorganisms associated to wheat. Considering the particularly high level of endotoxins, but also of toxigenic, irritative and allergenic fungi, generated during grain or straw handling, direct contact with grain or straw should be limited as often as possible. Crop workers have to handle grain or straw at distance from a cab. If direct contact cannot be avoided, the wearing of a personal protective equipment, such as a P2 mask, is highly recommended.

ACKNOWLEDGMENTS

The authors greatly appreciate the support of all the grain workers who were part of this study. They also would like to thank Laure Bernardet, Briseis Castella, Baptiste Jimenez, Celine Pattaroni and Jeremie Projer for assistance on samples collection in the field and samples processing in the lab. This work was supported by a Swiss National Foundation grant PMPDP3-129027 to H.NH.

REFERENCES

- Altschul SF, Madden TL, Schäffer AA, Zhang J, Zhang Z, Miller W, Lipman DJ.Gapped BLAST and PSI-BLAST: a new generation of protein database search programs. *Nucleic Acids Research* 1997:25: 3389–3402.
- Buée M, Reich M, Murat C, Morin E, Nilsson RH, Uroz S, Martin F. 454 Pyrosequencing analyses of forest soils reveal an unexpectedly high fungal diversity. *New Phytologist*, 2009:184: 449–456.

- Castellan RM, Olenchock SA, Kinsley KB, Harkinson JL Inhaled endotoxin and decreased spirometric values. *New Engl. J. Med.* . 1987:317, 605–610.
- Chan-Yeung M, Ward H, Enarson D, Kennedy S. Five cross-sectional studies of grain elevator workers. *Am. J. Epidemiol.* 1992:136:1269–1279.
- DECOS, Health based recommended Occupational Exposure Limit for Endotoxins, Dutch expert committee on occupational standards, Health Council for the Netherlands, Rijswijk, 1998.
- DECOS, Health based recommended Occupational Exposure Limit for Endotoxins, Dutch expert committee on occupational standards, Health Council for the Netherlands, Rijswijk, 2010.
- do Pico GA. Grain dust. In: Rylander R, Jacobs RR, editors. Organic dusts: Exposure, effects, and prevention. 1994. Boca Raton, FL: CRC Press, Inc. p 193–218.
- Douwes J, Thorne P, Pearce N, Heederik D. Bioaerosol Health Effects and Exposure Assessment: Progress and Prospect. *Ann. Occup. Hyg.* 2003:47: 187–200.
- Droege M, Hill B. The Genome Sequencer FLX System--longer reads, more applications, straight forward bioinformatics and more complete data sets. *J Biotechnol.* 2008:136: 3-10.
- Edgar RC. Search and clustering orders of magnitude faster than BLAST. Bioinformatics. 2010.
- Edgar RC, Haas BJ, Clemente JC, Quince C, Knight R. UCHIME improves sensitivity and speed of chimera detection. Bioinformatics, 2011.
- Eduard W. Exposure to non-infectious microorganisms and endotoxins in agriculture. *Ann Agric. Environ. Med.* 1997:4:179–186.
- Greskevitch M, Kullman G, Bang KM, Mazurek JM. Respiratory disease in agricultural workers: mortality and morbidity statistics. *J. Agromedicine.* 2007:12: 5-10.
- Heller & Kelson 1982, Oppliger A, Rusca S, Charrière N, VuDuc T, Droz PO. Assessment of bioaerosols and inhalable dust exposure in swiss sawmills. *Ann. Occup. Hyg.* 2005:49: 385–391
- Oppliger A., Masclaux F., Niculita-Hirzel H. Assessment of airborne microorganisms by real-time PCR: optimistic findings and research. *Frontiers in Biosciences.* 2011:3: 445-453.
- Poole JA, Dooley GP, Saito R, Burrell AM, Bailey KL, Romberger DJ, Mehaffy J, Reynolds SJ. Muramic acid, endotoxin, 3-hydroxy fatty acids, and ergosterol content explain monocyte and epithelial cell inflammatory responses to agricultural dusts. *J. Toxicol. Environ. Health* A. 2010:73:684-700.
- Reynolds SJ, Nakatsu J, Tillery M, Keefe T, Mehaffy J, Thorne PS, Donham KJ, Nonneman M, Golla V, O'Shaughnessy P. Field and wind tunnel comparison of four aerosol samplers using agricultural dusts. *Ann. Occup. Hyg.* 2009:53: 585-594.
- Rinsoz T, Duquenne P, Oppliger A. Application of real-time PCR for total airborne bacterial assessment, comparison with epifluorescence microscopy and culture dependent methods. *Atmospheric Environment.* 2008:42: 6767-6774.
- Skórska C, Mackiewicz B, Dutkiewicz J, Krysińska-Traczyk E, Milanowski J, Feltovich H, Lange J, Thorne P. Effects of exposure to grain dust in Polish farmers: work-related

symptoms and immunologic response to microbial antigens associated with dust. *Ann Agric Environ Med.* 1998:5:147-53.

- Swan JRM, Crook B. Airborne microorganisms associated with grain handling. *Ann Agric Environ Med* 1998:5: 7–15.
- US Environmental Protection Agency (EPA). Compilation of Air Pollutant Emission Factors, 5th edition, Vol. I; Chapter 9: Food and Agricultural Industries. AP-42. 2003. The EPA website. Available at: http://www.epa.gov/ttn/chief/ap42/ch09/final/c9s0909-1.pdf.
- Zock JP, Hollander A, Heederik D, Douwes J. Acute lung function changes and low endotoxin exposures in the potato processing industry, *Am. J. Ind. Med.* 1998:33: 348–391.

THE INVESTIGATION OF MOLD OCCURRENCE INSIDE SELECTED DURBAN HOMES.

Nkala BA, Jafta N and Gqaleni N

ABSTRACT

Introduction

Dust collected from indoor environments has been reported to contain biological pollutants such as pollen, spores, molds, bacteria, viruses, allergens, dust mites and epithelial cells. This study was aimed at assessing the presence of molds in house dust from selected Durban residences.

Methods

One hundred and five (105) house dust samples were obtained from households. The samples were taken from three surface areas namely; living room couches, bed mattresses, and carpets. Well documented methods were used for the isolation, identification and quantification of mold.

Results and conclusion

Among the isolated genera in all three surface areas, *Rhizopus spp.* and *Penicillium spp.* were widely distributed throughout in greater proportion. The overall highest colony forming units per gram of dust (CFU/g) for *Penicillium spp.* range: 3400 – 62316 CFU/g, was obtained from living room couches, followed by *Rhizopus spp.* (5200 – 15990 CFU/g). The mold results were compared with the South African Occupational Health and Safety Act (OHSA) 85 of 1993 as amended suggested guidelines of 1,000, 000 CFU/g. The findings of this study suggest that the molds in the homes studied were below the suggested guideline. However, this does not imply that the indoor conditions are unsafe or hazardous. Instead, the findings act as an indicator of mold presence indoors. The type of airborne mold, its concentration and extent of exposure and the health status of the occupants of a building will determine the health effects.

KEYWORDS: Indoor air pollution; Molds; house dust; Guidelines

INTRODUCTION

Hallow, (2003) and Robins et al., (2002) demonstrated an association between the industrial emissions and poor health status of humans in the South Durban Basin. Jeena et al., (2002) suggested that informal settlements (shacks) in Durban have the ability to support mold growth. Thus, the residents could be exposed to pollutants in the dust particles within their homes, making it necessary to study house dust within residences. House dust may contain chemical and biological pollutants in indoor environments. Biological pollutants such as pollen, bacteria, viruses, cockroaches, dust mites and epithelial cells have been reported to contribute towards indoor pollution (Smith et al., 2000; Derek et al., 2004). These may enter a building in various ways: through open doors, windows, ventilation shafts or by becoming attached to pets, cloths, or other personal items that are taken into a building (Curits et al., 2004).

The indoor environment plays a crucial role in support of mold growth. A warm indoor environment with relative humidity greater than 50% is optimum for indoor mold growth (Deacon, 1997). Usual sources of moisture include kitchens and bathrooms which tend to have increased mold growth (Verheoff and Burge, 1997). The most commonly identified indoor molds species include *Cladosporium, Penicillium, Alternaria, Aspergillus* and *Mucor* (Ren et al., 2002; Beguin and Nolard, 2006).

Zureik et al., (2002) reported that adults living in different countries who were sensitized to *Alternaria* and *Cladosporium* developed severe asthma. The skin, eyes and lungs are the primary target organs through which humans will have contact with molds in house dust (Godish, 1985; European Communities, 1991; Bency et al., 2003). Direct mold contact with skin may result in contact dermatitis or mild eye irritation and may sensitise individuals to molds. Inhalation of mold dusts, depending on the size may lead to the development of a chronic respiratory disease (Stern et al., 1984). Exposure to chemicals such as mycotoxins on the molds may result in a condition known as mycotoxicoses.

The purpose of this study was to isolate, identify and quantify mold occurrence in selected house dust samples.

MATERIALS AND METHODS

Sample selection

The households for this study were drawn randomly from the South Durban Health Study (SDHS) which had 823 school children participating from 7 schools in Durban communities (Naidoo et al., 2007). In total 136 households, from the main SDHS were recruited to participate in this study. Of the 136 homes visited, dust samples were collected from 126 homes. We could not collect samples from 10 homes due to unavailability of electricity as required by the equipment we used at the time of visit. House dust samples were obtained between the period of May 2004 to October 2005.

Dust collection

A portable vacuum cleaner (Wap Combi Cleaner Vs. 300s, Germany) with a specialized unit that collects house dust into the filter paper (Macherey-Nagel, Germany) with particle retention of 8 μm was used. House dust was collected from both children's sleep areas and/or play rooms. Sleep area samples were collected from children's pillows and mattresses for 2 minutes and if there was a soft floor in the room a 1.0 m^2 area was vacuumed for 2 minutes as well. From the play area, samples were collected from the couches and carpets or rugs for 2 minutes over a 1.0 m^2 area. The collected dust sample was folded and covered with aluminium foil before being placed into a ziplock bag prior to transportation and storage at 4°C until analysis. For the purpose of this study 1 gram of dust sample was needed for the analysis, of which eleven (11) samples were insufficient for dust laboratory analysis. Consequently, one hundred and five dust samples (105) were considered valid for further analysis.

Media and saline preparations

Dichoran-glycerol (DG18) (Oxoid Ltd., Basingstoke, Hampshire, UK) agar was prepared as per manufacturer's instructions with slightly adjustment for the purpose of this study. In addition to the agar 0.10 g of chloramphenicol selective supplement SR78 was added to inhibit any bacteria growth that might have been present in the sample (Basalan et al., 2004). After sterilization medium was cooled aseptically poured out into sterile Petri dishes (90 mm diameter) and allowed to set. Malt extract agar (MEA) (Oxoid Ltd., Basingstoke, Hampshire, UK) was prepared according to manufacturer's instructions. The media were aseptically poured out into sterile Petri dishes (90 mm diameter) and allowed to set. Phosphate buffer solution (PBS) was prepared as per laboratory standard procedure with the addition of Tween 20. The mixture was aliquoted into 10 ml universal bottles and sterilized.

Mold isolation and analysis

Two hundred milligrams (200 mg) of each house dust sample was added to 10 ml of sterile saline solution and allowed to stand for 15 minutes. This was followed by shaking with rotor mixer for 2 hours. Samples were serial diluted with saline from (10^{-1}) into (10^{-4}). One milliliter (1 ml) aliquot was plated out from test tube (10^{-4} to 10^{-2}) onto a surface of DG18 medium and plates (90 mm diameter), swirled and labeled underside with the dilution factor. Petri dishes were individually wrapped with parafilm around the edges of the plate to prevent cross-contamination and incubated at 25°C for 7 days. All the work and analysis was carried out in triplicate. The aseptic technique was adhered to at all times as to minimize or prevent any contamination. Mold analysis was done under the laminar flow cabinet (Labotec, Midrand, SA). After the incubation period, mold colonies were visibly counted and expressed as colony forming units per gram (CFU/g) of house dust.

Cultivation of molds for classification purpose

Colonies identified from the samples were inoculated onto the surface of MEA and incubated at 25°C for 14 days. Fungal slides were prepared by placing a small sample of fungal mycelium into a clean slide. The hyphae were teased out with dissecting needles until completely separated, and 1-2 drops of lacto-phenol blue (LPB) was added into the stain. The cover slip was placed over and viewed under a light microscope as per the method previously described by Smith, (1946). The fungal colonies were classified on the basis of colony morphology and isolates were then identified to genus level under the light microscope. In addition mycology books were used for further classification (Smith, 1946; Onions et al., 1981; Deacon,1997). The isolated molds that dominated in this study were sent to the mycology laboratory at Allerton Provincial Laboratories, Pietermaritzburg for further confirmation.

Data analysis

The data was analyzed using software package of Microsoft Excel 2000 and SPSS for Windows (15.1). Non-parametric test (Mann-Whitney test and Kruskal-Wallis test) were computed to assess differences on mold levels in sampled surface areas. Box plot was used to assess mold distribution in sampled surface areas.

RESULTS

Most of the houses sampled were made of bricks (n=126), wood (n=7) and few were shacks (n=3) made with mixed materials. The roofing in these houses included asbestos, corrugated iron and roof tiles. None of the houses used air conditioner. Most of the houses (50%) were built 25 years or more, 39% built within the last 25 years. The age of the rest of houses was not known by the respondents.

House dust samples (n=105) from the Durban residential buildings were analyzed in triplicate for culturable mold. Of all types of different mold genera identified on different surfaces, living room couches samples had the most types of mold genera (56.2%) followed by >bed mattresses (37.1%) and then >carpets (6.7%). The least molds were recovered from carpets and possible explanations is that not all households that participated in the study were carpeted hence less samples were collected within this surface area.

In addition different type of genera that were isolated and identified are presented in Table 1. The predominant genera that were isolated and identified in this study from three surface areas are shown in Figure 1. The mold species as per surface areas namely: - living room couches: *Rhizopus spp.* (n=10), *Alternaria spp.* (n=5), *Penicillium spp.* (n=5), *Aspergillus spp.* (n=4); bed mattresses: *Rhizopus spp.* (n=4), *Aspergillus spp.* (n=3) and carpets: *Penicillium spp.* (n=2), *Spondonema spp.* (n=2). In all the analyzed dusts from households, there was not a single sample that was free of molds.

Table 1: Household dust samples (n=105) from surface areas (living room couches, bed mattresses, and carpets). The frequencies of genera isolated are in parenthesis.

SURFACE AREA	GENERA	NUMBER OF HOUSE-HOLDS	PERCENT-AGE
Living room couches	*Alternaria* (n=5) *Aspergillus* (n=10) *Byssochlamys* *Cladosporium* (n=2) *Coccidiodes* *Curvularia* (n=3) *Drechslera* *Fusarium* (n=2) *Geotrichum* (n=2) *Gliocladium* *Helminthosporium* *Modurella* *Moniliella* *Mortieralla* *Mucor* *Penicillium* (n=16) *Rhizopus* (n=10)	59	56.2 %
Bed mattress	*Absidia* *Acremonium* (n=2) *Alternaria* *Aspergillus* (n=6) *Chaetomium* *Cladosporium* *Curvularia* *Drechslera* *Fusarium* (n=3) *Geotrichum* (n=2) *Helminthosporium* *Mortierella* *Mucor* (n=2) *Myceliphthora* *Penicillium* (n=6) *Rhizopus* (n=4) *Spondenema* *Trichoderma* (n=2) *Ulocladium* (n=2)	39	37.1 %

Table 1: continued

SURFACE AREA	GENERA	NUMBER OF HOUSE-HOLDS	PERCENT-AGE
Carpets	*Myceliphthora* *Penicillium* (n=2) *Phialophora* *Rhizopus* *Spondonema* (n=2)	7	6.7 %

Figure 1: Predominant identified mold genera (n=10) that were isolated from Durban residential houses (south and north) in dust sample.

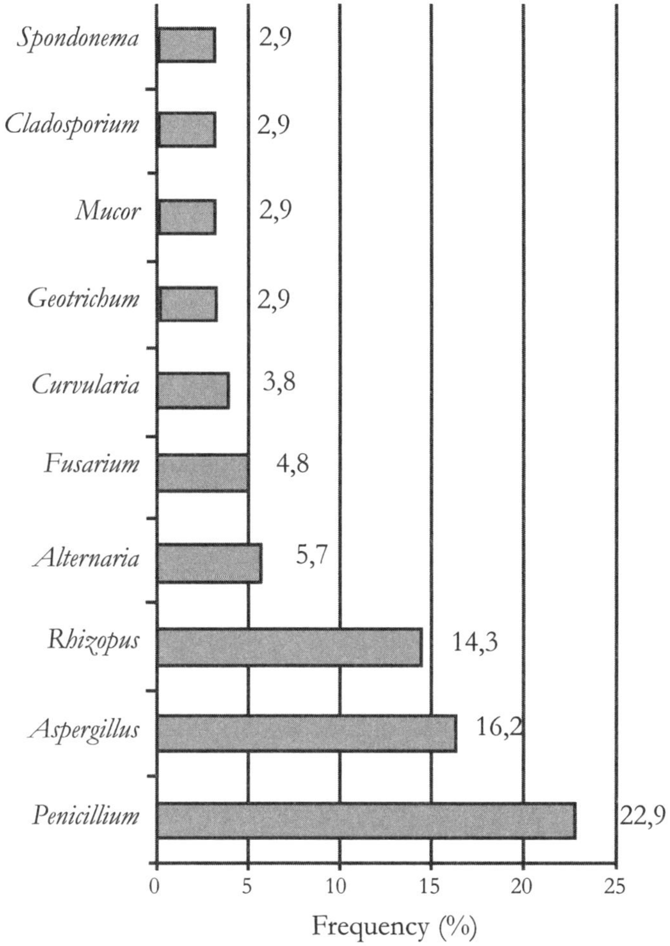

Mold incidences (105) were classified into mean versus genera (25) and the figure not shown. The isolated genera were as follows: *Absidia, Acremonium, Alternaria, Aspergillus, Byssochlamys, Chaetomium, Cladosporium, Coccidiodes, Curvularia, Drechslera, Fusarium, Geotrichum, Gliocladium, Helminthosporium, Modurella, Moniliella, Mortierella, Mucor, Myceliphthora, Penecillium, Phialophora, Rhizopus, Spondonema, Trichoderma,* and *Ulocladium*. The mold counts distribution ranged from 250 CFU/g to 640 000 CFU/g.

Further analysis using the Box and Whisker's plot (Figure 2) indicated that irrespective of the surface sampled, the CFU/g were marginally scattered from the median. Though there may be differences in the extent of distribution of CFU/g from the median, none of these were statistically significant. The same could be said for the differences in the range of CFU/g of specific mold recovered though there were three genera that seemed to predominate.

Figure 2: Box plot showing median concentration in (CFU/g of house dust) and distribution of total molds by location.

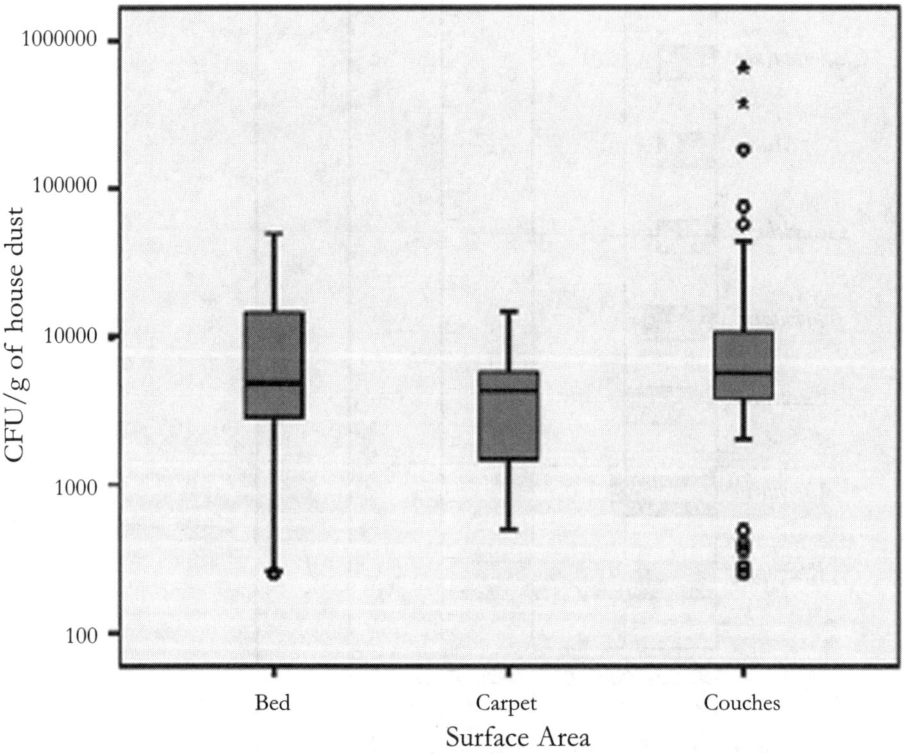

The horizontal line in each box is the median value thus the bottom of the box is 25th percentile and the top of the box is 75th percentile, respectively; the line extended above the box marks 1.5 median values. Circles, values, asterisk, values greater than 1.5 times the median value.

Analysis of the molds according to the risk category guidelines, by CFU range, compiled by Doe (2003) is shown in Table 2. According to these, mould CFU within a range of 10,000 or less are considered to constitute a low health risk, between 10,000 – 100,000 moderate, 100.000 – 1,000,000 high, and 1,000,000 or more extremely high. In this study, only two genera, namely *Aspergillus* and *Penicillium*, fell within the high risk category and were amongst the top three identified mold which includes *Rhizopus*. The results were below that reported by South African Occupational Health and Safety Act 85 of 1993 (OHSA) as amended, which states that molds above 1,000,000 CFU/g may be an indicator of the presence of unusual levels of molds. However, most isolated mold levels were below OHSA threshold limits despite that some were under classified on 'high risk' category when using the Doe (2003) classification.

Table 2: Mold genera that were within moderate and high risk category guidelines are represented in colony forming units of house dust per gram (CFU/g) inside the Durban houses sample.

SAMPLE AREA	GENERA	MOULD TOTAL COUNTS (CFU/g)	RISK CATEGORY
Couches	*Aspergillus*	27,000 – 370,000	High
	Penicillium	12,000 – 640, 000	High
	Rhizopus	44,000 – 75,000	Moderate
	Gliocladium	21,000	Moderate
	Modurella	18,000	Moderate
Bed mattresses	*Aspergillus*	17,000 – 39,000	Moderate
	Penicillium	12,000 – 50,000	Moderate
	Trichoderma	37,000	Moderate
	Spondenema	31,000	Moderate
	Cladosporium	26,000	Moderate
	Ulocladium	24,000	Moderate
	Fusarium	20,000	Moderate
Carpets	*Phialophora*	15, 000	Moderate

Low: 10,000 or less Moderate: 10,000 – 100,000 High: 100,000 – 1,000.000
Extremely High: 1,000.000 or more Doe (2003).

DISCUSSION

According to our knowledge, this is the first study in South Africa that describes mold occurrence in house dust. The reported top ten predominant genera that were isolated from indoor buildings were among the most commonly isolated in homes: - *Penicillium spp., Aspergillus spp.* and *Rhizopus spp.* These mold species have been previously reported in Canada, Germany and South Africa in problematic and non problematic households (Dales et al., 1997; Sekhotha, 2001; Jacob et al., 2002, Gansan, 2004; Jafta et al., 2012).

One South African study reported the occurrence of *Penicillium spp.* and *Aspergillus spp.* that dominated residential homes in south Durban from suspended dust (Sekhotha, 2001). Gansan, (2004) also reported mold infestation which is related with households dampness which were dominated by the following species *Cladosporium spp.* and *Penicillium spp.* The other study was conducted by Jafta et al. (2012) with no particular interest whether homes were problematic or not, and also reported on the *Cladosporium spp., Aspergillus spp.,* and *Fusarium spp.* being dominant in south Durban homes. In Canada, Dales et al., (1997) reported that *Aspergillus spp.* and *Penicillium spp.* were dominant in residence with water damaged buildings. Furthermore, similar findings were observed in a Germany study where *Cladosporium spp.* and *Aspergillus spp.* predominated in settled dust found in households (Jacob et al., 2002).

This study reported considerable high mold counts in CFU/g in settled house dust when compared to a study conducted by Hicks et al., (2005). It was noticeable that *Aspergillus spp.* and *Penicillium spp.* were classified into the high risk group according to Doe (2003). Despite these mold counts being reported as "high risk category" the colony counts were below the "contamination guidelines" of 1,000, 000 CFU/g (Brief and Bernath, 1988). The results in this study show that these genera are commonly detected indoors with elevated CFU/g (Jaffal, 1997; Jacob et al., 2002; Piecková and Wilkins, 2004; Haas et al., 2007). The reported CFU/g which are below the suggested guidelines may not pose any risk to an individual, but serves as an indicator of mold presence indoors.

Mold infestation in residential homes pose health risk such as allergic asthma and toxins production with known genera that were also identified in this study (predominant). Samson et al., (1994) and Horner et al., (2004) reported on *Absidia spp, Chaetomium spp., Curvularia spp., Helminthosporium spp., Rhizopus spp.*, and *Penicillium spp.* with the ability to cause allergic reaction in humans. Horak et al., (1996); Verhoeff et al., (1997); Pieckova and Jesenskà (1999) reported on *Alternaria spp.*, and *Cladosporium spp.* causing extrinsic asthma while *Aspergillus spp.* and *Fusarium spp.* have a potential to produce mycotoxins that cause mycotoxicosis. Some species of *Fusarium* reportedly cause disseminated infection which is aggressive in particular to the immune compromised individuals and penetrate the entire body and bloodstream (Nelson et al., 1993; Nelson et al., 1994; Placinta et al., 1999).

The limitation of this study was that outdoor mold was not analyzed. When studying indoor mold infestation it's crucial to understand the influence of outdoor mold. However, *Cladosporium spp.* and *Alternaria spp.* were isolated in this study which was reported by Jacob et al., (2002) to dominate outdoor environment. These molds have been previously reported to be isolated both in indoor and outdoor environment (Samson et al., 1994; Hameed et al., 2004; Horner et al., 2004). Therefore, this indicates mobility of spores in outdoor into the indoor environment. The presence of molds indoors is of health concern as the individuals exposed to them many develop negative health effects. The mold infestation in this study was widely distributed in all households sampled.

It can be concluded that objectives of this study were met by the identification and quantification of indoor molds. Outdoor investigation would have enabled a better understanding of the relationships between these two environments. However, indoor molds that were predominating in this study were identified and also reported on the ones that are usually found outdoors. This suggests that outdoor molds migrates indoors and may negatively impact this environment. The findings of this study show that the molds in the homes studied were below the suggested guideline. However, this does not imply that the indoor conditions are unsafe or hazardous. Instead, the findings act as an indicator of molds presence indoors. The type of airborne mold, its concentration and extent of exposure and the health status of the occupants of a building will determine the health effects on an individual. In future studies outdoor mold should be conducted for better understanding mold infestation. Furthermore, mold allergens in humans residing in sampled households should be tested.

CONFLICT OF INTEREST

We have no conflict of interest to declare.

ACKNOWLEDGMENTS

We also thank the National Research Foundation (NRF) for funding.

REFERENCES

- Basalan M, Hismiogullari SE, Hismiogullari AA, Filazi A. Fungi and aflotoxins B1 in horse and dog feeds in Western Turkey. *Review Medical Journal.* 2004:156(5):248-256.
- Beguin H, Nolard N. Mould biodiversity in homes I. Air and surface analysis of 130 dwellings. Earth and Environmental Science. 2006:10(2-3): 157-166.
- Bency KT, Jansy J, Thakappan B, Kumar B, Sreelekha TT, Hareendran NK, Nair PKK, Nair MK. A study on the air pollution related human disease in Thiruvanthpuram City, Kerala. In Martin Buch I., Suresh V.M. and Kumaran T.V., eds., Proceeding of the Third International Conference on Environmental and Health, Chennai, India, 15-17

December. 2003. Chennai: Department of Geography, University of Madra and Faculty of Environmental Studies, York University. Pages 15-22.

- Brief RS, Bernath T. Indoor pollution: guidelines for prevention and control of micro-biological respiratory hazards association with air conditioning and ventilation system. *Journal of Applied Industrial Hygiene.* 1988:3(1):5-10.
- Curitis L, Lieberman A, Stark M, Rea W, Vetier M. Adverse health effects of indoor. *Journal of Nutritional and Environmental Medicine.* 2004:14(3):1-14.
- Dales RE, Miller D, McMullen D. Indoor air quality and health: validity and determinants of report home dampness and moulds. *International Journal of Epidemiology.* 1997:26(1):120-125.
- Deacon JW. Modern mycology. 1997. 3rd Edition. Blackwell Science Ltd, UK Publishers.
- Derek GS, Claire B, Stephen B. Science-based recommendations to prevent or reduce potential exposure to biological, chemical, and physical agents in schools. Healthy Schools Network, Inc (HSN), 2004. Albany, NY.
- Doe J. Mould Check, Laboratory Report. 2003. http://www.testsymptomsathome.com/-PDF_Files/Mold-sample.pdf.Accessed April 18, 2008
- European Communities. Indoor air quality and impact on man: Report 10-effects of indoor air pollution on human health. Environmental and Quality of Life. http:www.inive.org/medias/ECA/ECA_Report10.pdf. Accessed 09 February 2008.
- Gansan J. Natural ventilation, dampness and mouldiness in dwellings in the Waterloo housing development (Durban Metropolitan Area): a case study of indoor air quality. Dissertation, Nelson R. Mandela School of Medicine, University of KwaZulu-Natal. 2004.
- Godish T. Air quality. Lewis Publishers, Inc. 1985.
- Haas D, Habib J, Galler H, Buzina W, Schlacher R, Marth E, Reinthaler FF. Assessment of indoor air in Austrian apartment with and without visible mould growth. *Atmospheric Environment.* 2007:41:5192-5201.
- Hallow D. Air Pollution in Selected Industrial Areas in South Africa. National Report on Community-based Air Pollution Monitoring in South Africa. 2003.
- Hameed AAA, Yasser IH, Khoder IM. Indoor air quality during renovation actions: a case study. *Journal of Environmental Monitoring*, 2004:6:740-744,
- Hicks JB, Lu ET, Guzman R, Weingart M. Fungal types and concentration from settled dust in normal residences. *Journal of Occupational and Environmental Hygiene.* 2005:2:481-492.
- Horak B, Dutkiewicz J, Solarz K. Microflora and acarofauna of bed dust from homes in Upper Silesia, Poland. *Annals of Allergy, Asthma and Immunology.* 1996:76(1):41-50.
- Horner WE, Worthan AG, Morey PR. Air and dust borne mycoflora in houses free of water damage and fungal growth. *Applied and Environmental Microbiology.* 2004:70(11):6394-6400.
- Jacob B, Ritz B, Gehring U, Koch A, Bischof W, Wichmann HE, Heinrich J. Indoor exposure to moulds and allergic sensitization. *Environnemental Heath Perspectives.* 2002:110(7):647-653.

- Jaffal AA. Residential indoor airborne microbial population in the United Arab Emirates. *Environmental International.* 1997:23(4): 529-533.
- Jafta N, Batterman SB, Gqaleni N, Naidoo RN, Robins. Characterisation of allergens and airborne fungi in low and middle-income homes of primary school children in Durban, South Africa. *American Journal of Industrial Medicine.* 2012.(in press)
- Jeena PM, Danaviah S, Gqaleni N. Airborne moulds prevalence in an informal settlement and its relationship to childhood respiratory health. Urban Health and Development Bulletin. 2002:4(2):27-32.
- Naidoo R, Gqaleni N, Batterman S, Robins T. Multipoint Plan: Project 4 South Durban Heath Study and Health Risk Assessment. Final Project Report. 2007.
- Nelson PE, Desjardins AE, Plattner RD. Fumonisins, mycotoxins produced by *Fusarium spp.*: biology, chemistry and significance. *Annual Review of Phytopathology.* 1993:31: 233-252.
- Nelson PE, Dignani MC, Anaissie EJ. Taxonomy, biology and clinical aspects of *Fusarium* species. *Clinical Microbiology Reviews.* 1994:7(4): 479-504.
- Onions AHS, Allsoo D, Eggins HOW. Smith introduction to mycology. 7th Edition. London: Edward Arnold.1981.
- Pieckovà E, Jesenskà Z. Microscopic fungi in dwellings and their health implications in humans. *Annals of Agricultural and Environmental Medicine.* 1999:6: 1-11.
- Pieckovà E, Wilkins K. Airway toxicity of house dust and its fungal composition. *Annals of Agricultural and Environmental Medicine.* 2004:11:67-73.
- Placinta CM, Mello JPF and Macdonald AMC. A review of world contamination of cere-al grains and animal feed with *Fusarium* mycotoxins. Animal Feed Science and Technology. 1999:78:21-37.
- Ren P, Janun TM, Belanger K, Bracken MB, Leadere BP. The relation between fungal propagules in indoor air and home characteristics. *Journal of Allergy.* 2002:56(5): 419-424.
- Respiratory Health Survey. *British Medical Journal.* 2002:325:1-7.
- Robins T, Batterman S, Lalloo U, Irusen E, Naidoo R, Kistnasamy B, Kistnasmy J, Baijanath N, Mentz G. Air contaminated Exposure, Acute Symptoms and Disease Aggravation among Students and Teachers at Settlers School in South Durban. Settle Primary School Health Study Draft Final Report Presented at Nelson R. Mandela School of Medicine on 22 November 2002.
- Samson RA, Flannigam B, Flannigan ME, Verhoeff AP, Adan OCG, Hoestra ES. Health implications of fungi in indoor environments. Air Quality Monographs 2. Elsevier Science BV, Amsterdam. (Skår 2). 1994.
- Sekhotha MM. An investigation into the conditions that lead to surface mould growth in urban residential homes and evaluation of methods used to combat this problem. Dissertation, Nelson R. Mandela School of Medicine, University of Natal. 2001.
- Smith G. An Introduction to industrial mycology. 3rd edition. Publishers Edward Arnold.1946.
- Smith KR, Samet JM, Romieu I, Bruce N. Indoor air pollution in developing countries and acute lower respiratory infections in children. Thorax. 2000:55: 518-532.

- Stern C, Boubel RW, Turnern DB, Fox DL. Fundamentals of air pollution. 2nd edition. Publishers Harcoupt Brake Jovanovich.1984.
- Verhoeff AP, Burge HA. Health risk assesment of fungi in home environments. Animals of Allergy, *Asthma & Immunology*. 1997:78: 544-556.
- Zureik M, Neukirk C, Leynaert B, Liard R, Bousquet J, Fraçoise F. Sensitisation to air-borne moulds and severity of asthma: cross sectional study from European Community respiratory health survey. *British Medical Journal*. 2002:325:1-7.

CONSERVING OUR CULTURAL HERITAGE: THE ROLE OF FUNGI IN BIODETERIORATION

Hanna Szczepanowska and A. Ralph Cavaliere

ABSTRACT

The objects of cultural heritage are composed of varied materials which can be affected by diverse microbial communities. The study of these complex and heterogeneous assemblies of materials and microorganisms require an inter- and multi-disciplinary approach. Development of a strategy towards prevention, mitigation of biodeterioration and removal of microorganisms, especially fungi begins with the understanding of the materials' fabric, assessment of causes behind the biodeterioration, and the context in which it occurs.

Three aspects of biodeterioration of cultural heritage are discussed: 1) the multitude of bio-agents' on cultural heritage materials, 2) fungal interaction with substrates, and 3) prevention and conservation of biodeteriorated artworks. The challenges of conservators' work in dealing with bio-degraded museum collections are discussed based on the case studies of biodeteriorated art on paper, exemplifying two types of fungal interaction with the substrate: 1) surface deposits of pigmented spores/conidia, and 2) pigmented fruiting structures embedded in the matrix of the substrate.

The microbial metabolites deteriorate the substrates on which they grow resulting in chemical and physical changes of the material bulk and surface, at times leading to structural weakening. We focused our studies on black stains which are prevalent on art rendered on paper, a subject that has received very little attention. Our techniques of analysis included three-dimensional topographic imaging and visualization, structural characterization and optical microscopy, scanning electron microscopy (SEM), and confocal laser scanning microscopy (CLSM).

INTRODUCTION

Conservation of art aims to prevent deterioration of objects which are in good condition and conserving those which have already been damaged. Biodeterioration which is one of many forms of artworks' deterioration is particularly complex because it involves living microorganisms interacting with a variety of highly complex heterogeneous materials. Biodeterioration in the context of cultural heritage can be defined as decay of culturally, historically or artistically significant objects induced by microorganisms, such as bacteria, fungi or lichens. It differs from biodegradation which refers to the breakdown of materials in nature.

Two main phases occur in the initial stage of biodeterioration- physical contact of spores with surface and attachment to surface. Biodeterioration starts as soon as microbial spores land on a surface where molecular water is present. Liquid water that is essential for microbial growth becomes available on the surface when it reaches dew point temperature, and water condenses on the surface (Dew point temperature, private communication with Dr. Marion Mecklenburg, Senior Scientist, Emeritus, MCI- SI). It is a simplified shortcut to a synergistic work of relative humidity, temperature and the type of material involved that leads to the formation of liquid water on the surface. In addition to the need for liquid water, numerous authors list physico-chemical involvements in the first stage of fungal attachment to surfaces. Furthermore, spores secrete adhesions, such as protenaceous and/or polysaccharide materials which aid in the attachment of spores to the surface.

Fungal pigments produced on paper

Paper is a network of plant fibers derived from cotton, flex, or wood combined with particles of mineral fillers such as kaolin or calcium carbonate added to produce a desired, specific surface. Paper is a structurally complex, heterogeneous material susceptible to biodeterioration caused by fungi, one of which is staining caused by fungal pigments. Many species of fungi produce pigments which serve different functions - varying from a protective action against lethal photo-oxidations (carotenoids- in bacteria) to protection against environmental stress (melanins- in fungi) and acting as cofactors in enzyme catalysis (flavins). In addition, bio-pigments often protect the organism against extremes of heat and cold and against natural antimicrobial compounds produced by other invading microbes (allelotic metabolites) (Liu and Nizet, 2009, Paolo et al., 2006, Medentsev and Akimenko, 1998). Phylogenetic diversity of pigmented species, and the chemical diversity of the pigments themselves, precludes a single unifying hypothesis for their evolution and persistence. In general, pigmentation produced by microorganisms results from complex bio-chemical processes occurring in their cell and are referred to as secondary metabolites.

It has been well established that dark pigments associated with melanins are produced in some fungi as well as in organisms across many kingdoms. Melanins are amorphous substances with a remarkable ability to absorb-infrared, visible and ultraviolet radiation which makes their study using optical, light-based instruments, particularity challenging (Buskirt et al., 2011). The biosynthetic pathways of pigment formation are known only to some extent and the molecular structure of the pigment and its organization still remains unclear (Bochenek and Gudowska-Nowak , 2003). Fungal, animal and synthetic melanins share a number of similarities in their chemical makeup. Many of the fungi producing melanin are filamentous forms which belong to the class Dothideomycetes, members of the large, well characterized Phylum (or Division) Ascomycetes (Sterflinger et al., 1999, Sterflinger, 2006).

Black pigmented fungi, which grow on art works, may also be pathogenic. Pigments in fungi are used in presumptive clinical diagnosis (Liu and Nizet, 2009) therefore have been researched extensively in the medical field (Jacobson 2000, Icenhour et al., 2003). Fungal pathogenesis in plant and animal hosts has been of main concern to agricultural studies. Some of these same pathogens of the Dothideomycetes have been identified on paper.

The extent of staining depends on where within their structure fungi produce pigments. It can occur in spores/conidia and mycelium as powdery deposits on surfaces, in their fruiting bodies, as inclusions in substrate, or as pigmented products of metabolism. Each scenario has impact on the conservation strategy (Fig.1).

Figure 1: 1920 etching, covered with black fungi induced stains along the top margin. Adhesive applied along the top to attach the artwork to a window mat provided nutrient for fungi.

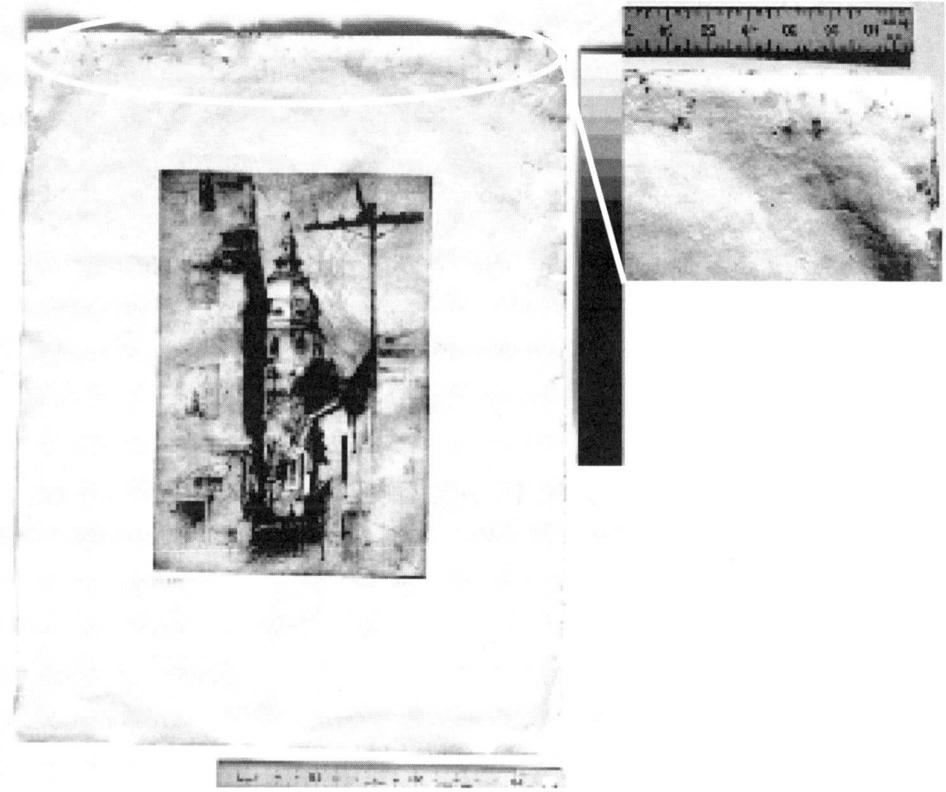

Two phenomena occur with respect to the interface of fungi and the matrix of paper. One is the attachment of pigmented spores and mycelium to paper fibers facilitated by secretion of adhesives. The other is intertwining of paper fibers with pigmented fungal elements, such as perithecial hairs and hyphae. We studied both types of fungus-paper interfaces on the original works of art on paper.

MATERIALS AND METHODS

Considering the limitations of examining the original artworks, and often strict rules applied to sampling the material, the examination protocol relied on the choices that are permitted in the real life situation. In addition to optical microscopy, a confocal laser scanning microscopy was utilized which is a novel application in the examination of fungal stains on paper.

Two, twentieth century, original artworks (1920 and 1958), and one ca 17th century study-paper, affected by fungi-induced black stains were selected for study. The papers' characteristics indicated compact and calendared paper (etching 1920), unsized, printing-paper (1958) and soft, blotter-like paper (17th c). Stains on these papers were characterized as follows:

1. 1920 etching: dark cells attached to the surface of paper fibers (*Taeniolella* (Torula) *sp.*)
2. 1958 etching: pigmented fruiting structures embedded into substrate (*Chaetomium sp.*)
3. 17th century paper: pigmented spores/conidia clustered among paper fibers (*Cladosporium sp.*) (*Cladosporium sp.* and *Taeniolella sp.* were identified by De-Wei Li, Ph.D. Research Mycologist, Connecticut Agricultural Experiment Station Valley Laboratory, Windsor, CT.)

The context in which biodeterioration occurred in each case was different and can be deduced from the analysis of the bio-stains location on the artwork, understanding the construction of the artwork and history of its handling.

1) The black stains on the 1920 etching were not coincidentally located along the top edge where adhesive is applied to attach the artwork to a window-mat. Once water became available, it dissolved the adhesive providing rich nutrients to the germinating spores (Fig. 1).
2) Fungal deterioration of the 1958 etching occurred as a result of hurricane Katrina (2005). Dark brown, nearly black pigmented fruiting bodies (perithecia) were entangled into the matrix of the paper. It is unknown how long the paper remained wet, however, long enough to permit formation of the fruiting bodies of the fungus *Chaetomium* (Szczepanowska and Cavaliere, 2000, 2003; von Arx et al., 1986). Consequently the invaded portion of the artwork on which *Chaetomium*, a common cellulolytic fungus grew, was severely structurally deteriorated (Fig. 2).

Figure 2: 1958 etching by Shigeru Kimura, damaged during hurricane Katrina 2005. Individual dark brown spots see inset, are perithecia of *Chaetomium sp.*(Artwork in private collection, used with permission.)

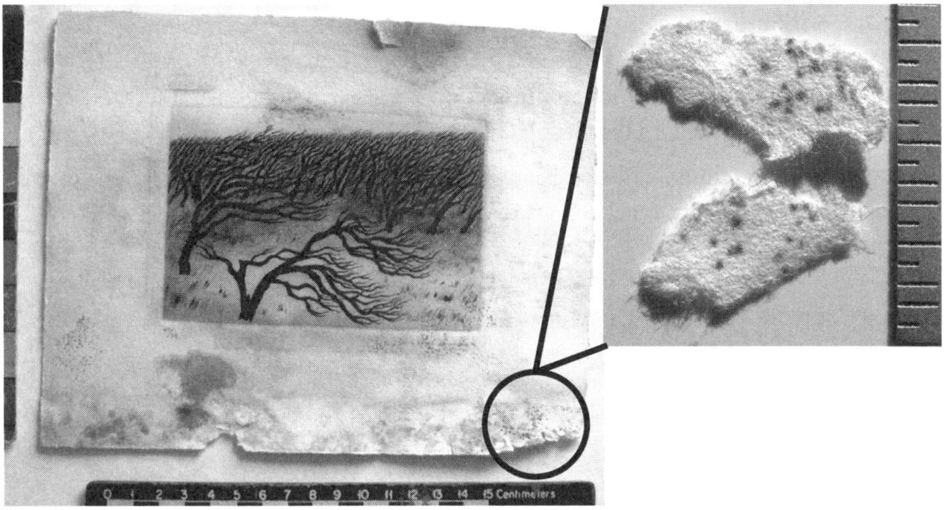

3) On the 17th century paper, the black stains are scattered throughout a fibrous matrix. Although its provenance is unknown, based on the calligraphy of a fragmentary inscription and remnants of pigment (minium- red lead pigment), it was deduced that the paper was manufactured between the 16th and 18th century (Fig. 3).

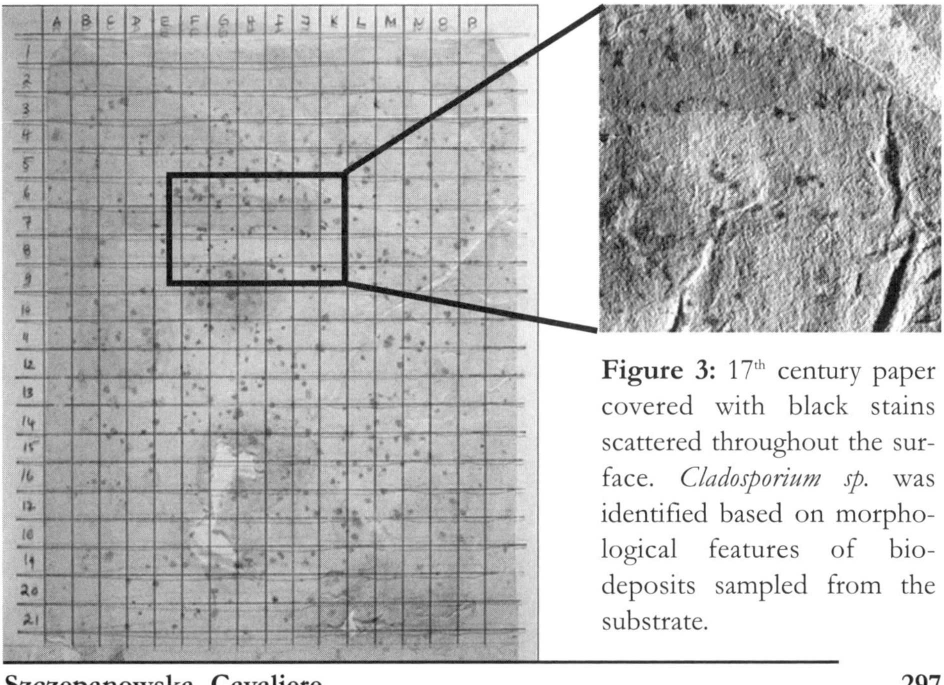

Figure 3: 17th century paper covered with black stains scattered throughout the surface. *Cladosporium sp.* was identified based on morphological features of biodeposits sampled from the substrate.

Surface characteristics of the bio-mass on the artworks were examined in respect to the fungal particles interaction with the paper fibers matrix. The evaluation regiment was as follows:

1. Bio-deposits distribution and surface topography were examined with the stereomicroscope, Wild M8 Heerbrugg, Zoom Stereomicroscope with LED light ring, Volpi, Swiss. Magnification ranged 6x to 50x. The digital camera, Leica EC 3, Microsoft Systems , Heerbrugg, captured the images (Figs. 4a, 5a, 6a).

Figure 4a: 1920 etching; bio-stain's surface topography examined with streomicroscope,50x magnification; black stain area ca. 05mm x 2mm.

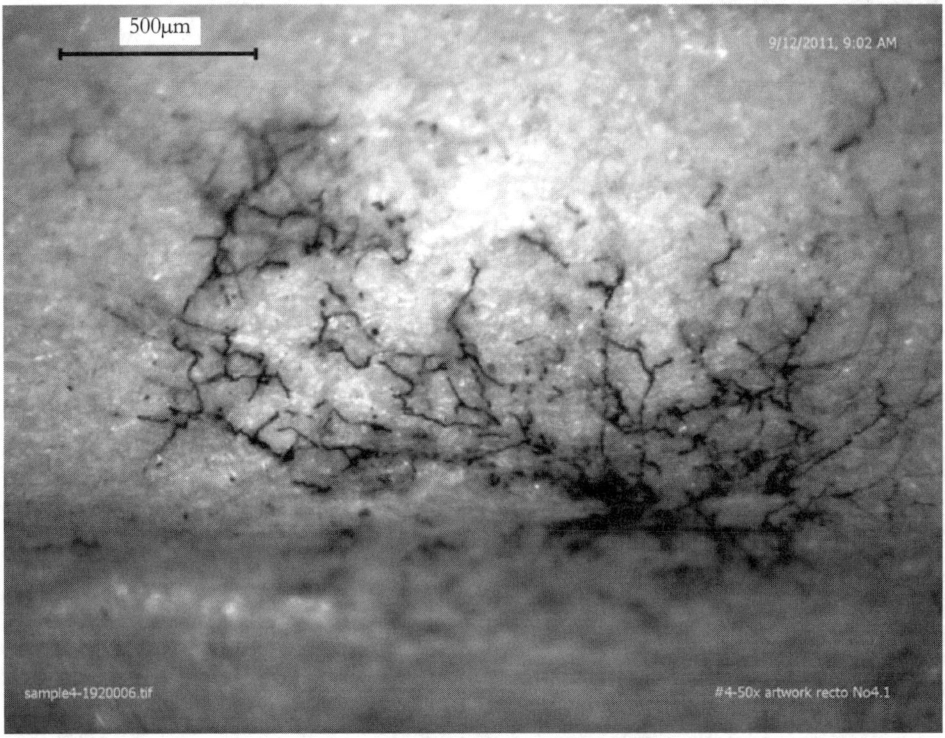

Figure 5: 1958 etching: pigmented fruiting structures embedded into substrate (*Chaetomium sp.*) **5a** Dark-brown perithecia visible as defined spots; stereo-micrograph, 50x. Size of perithecia: 0.3mm to-0.6mm.

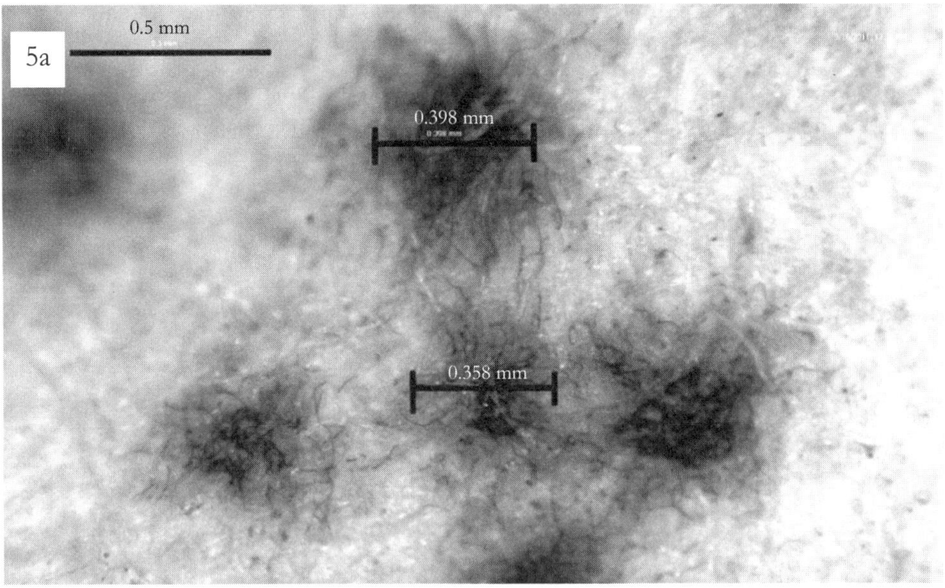

Figure 6: A 17th century paper: pigmented spores/conidia clustered among paper fibers (*Cladosporium sp.*). **6a** Stereo-micrograph, 50x of the black stains; stains size varied: 05mm x 2mm.

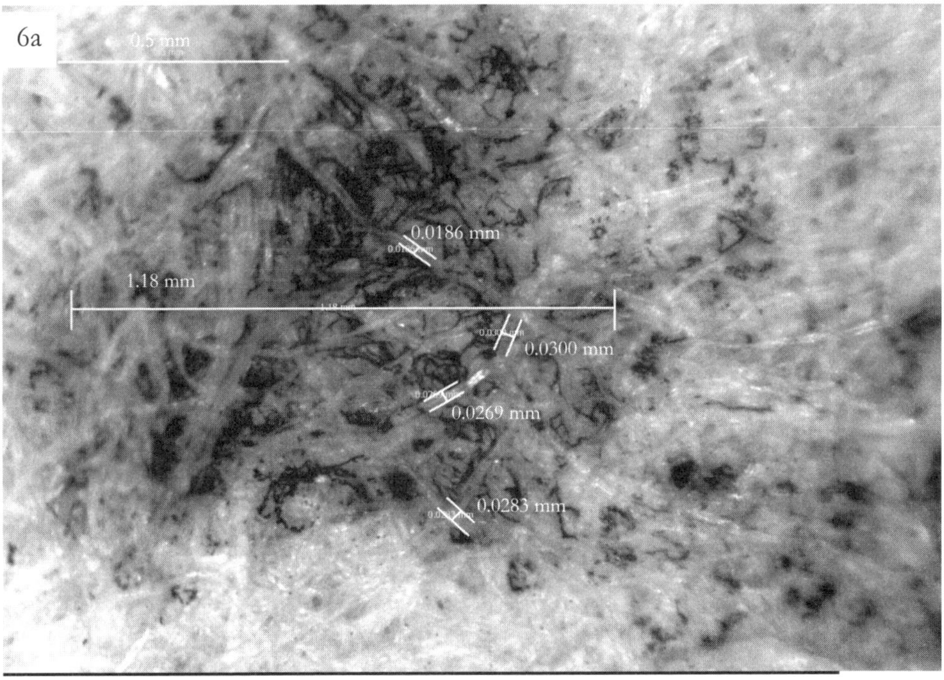

2. Bio-mass interaction with substrate and morphological features of bio-mass were examined with the transmitted light microscope, Leica DM LM, using phase contrast illumination, dark field and bright light illumination. Magnification ranged: 50x, 100x, 200x and 40ox, power source: Leica AC volts 0-15 (Figs. 4b, 5b, 6b).

Figure 4b: The same stain as in figure 4a examined in transmitted light microscope, 200x. Pigmented cells are trailing on the surface of paper; cells size: 3 µm-5 µm.

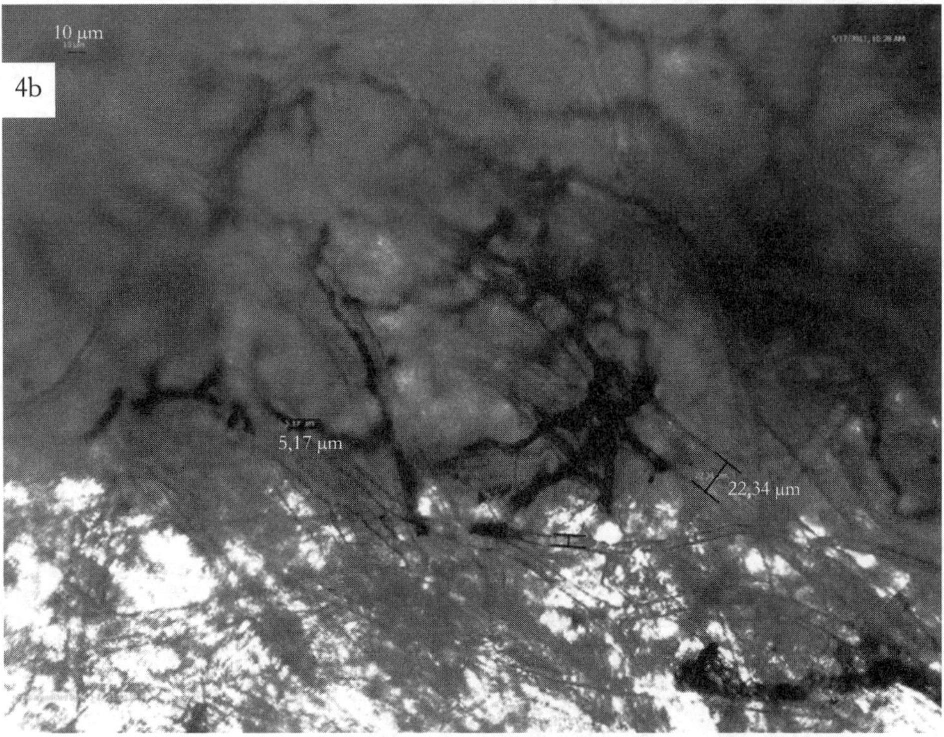

Figure 5b: Same specimen as in figure 5a transmitted light micrograph, 200x, an individual fruiting structure. Melanized, dark perithecial hairs are entangled with paper fibers, here visible as transparent light blue filaments.

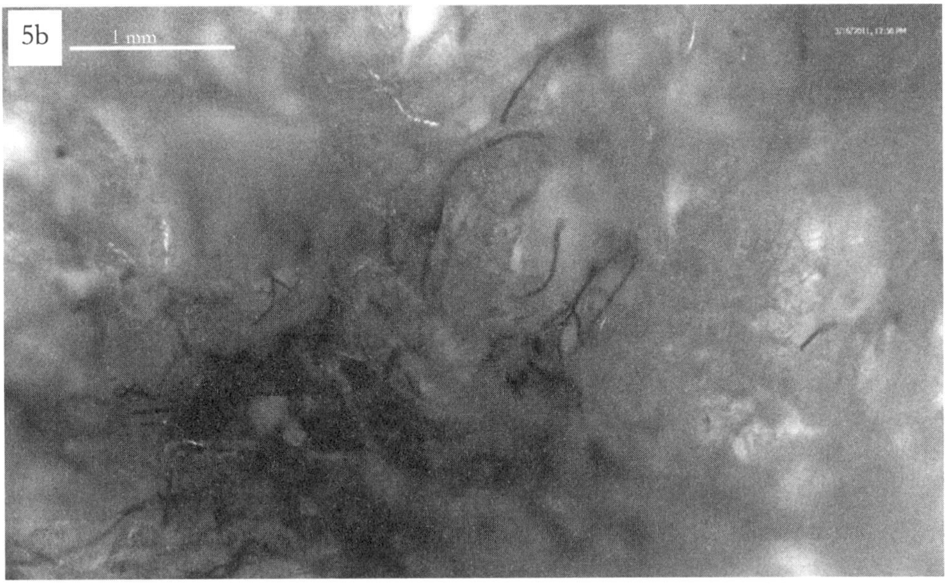

Figure 6b: The same stain as in figure 6a examined in transmitted light microscope, 400x; melanized dark brown cells are attached to paper fibers. Large pores in paper permit air supply necessary for the cells growth.

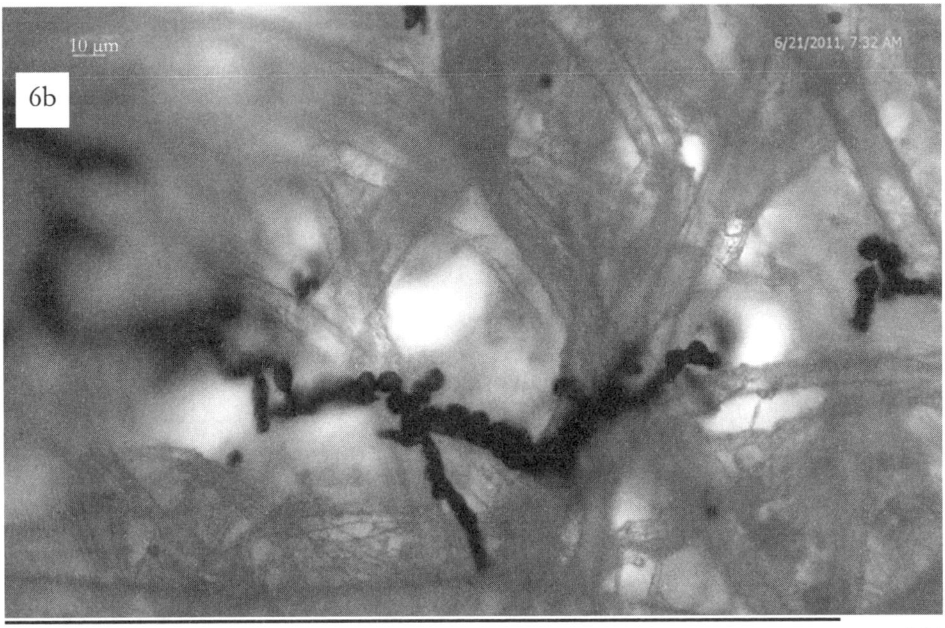

3. Bio-mass interaction with the substrate on a nano-scale was examined with the scanning electron microscopy in variable pressure. SEM-VP Hitachi S 3700N was used, magnification range: 500x to 3000x (Figs. 4c, 5c, 6c).

Figure 4c: Interaction of bio-mass with the substrate, the same stain as in figures 4a and 4b, SEM-VP, 500x magnification; bar 20 μm. Catenulate (chain-like) cell formations are underlined in white.

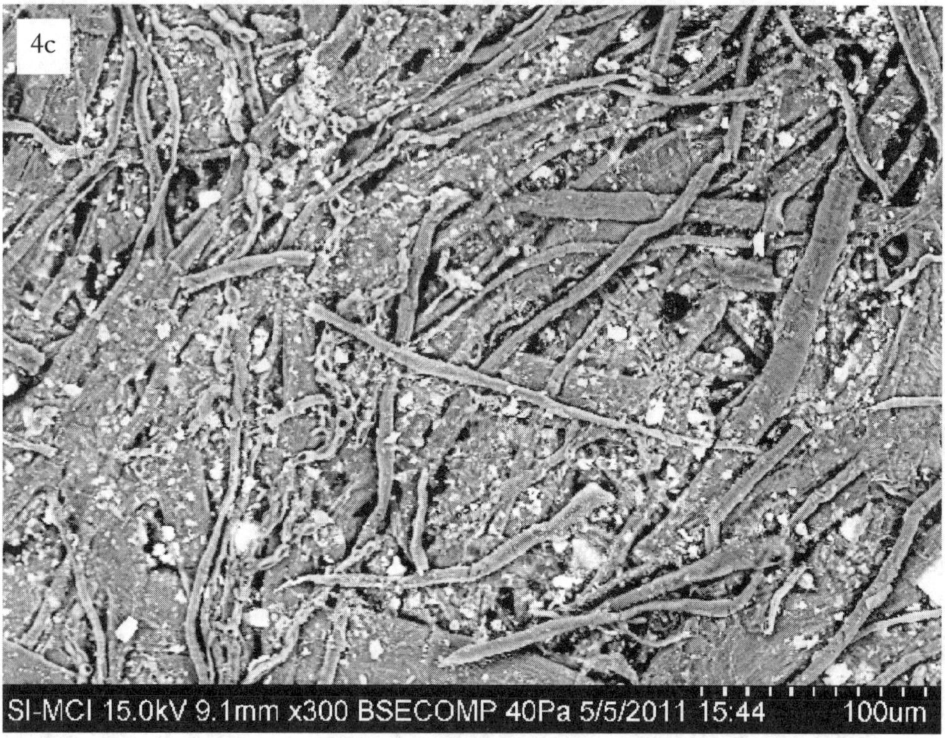

Figure 5c: The same specimen as in figures 5a and 5b SEM-VP 500x reveals characteristics of perithecial hairs, smooth walled, tubular, ca. 3μm in diameter.

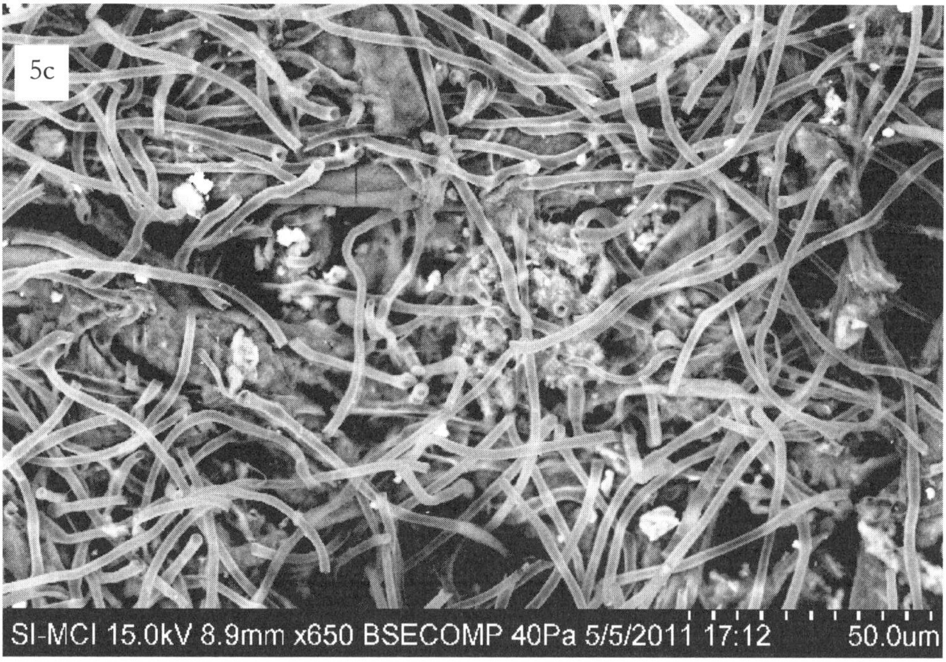

Figure 6c: The same specimen as in figures 6a and 6b SEM-VP, bar 4 μm; cells are clustered and attached to paper fibers; average cell dimensions: 5-9 μm.

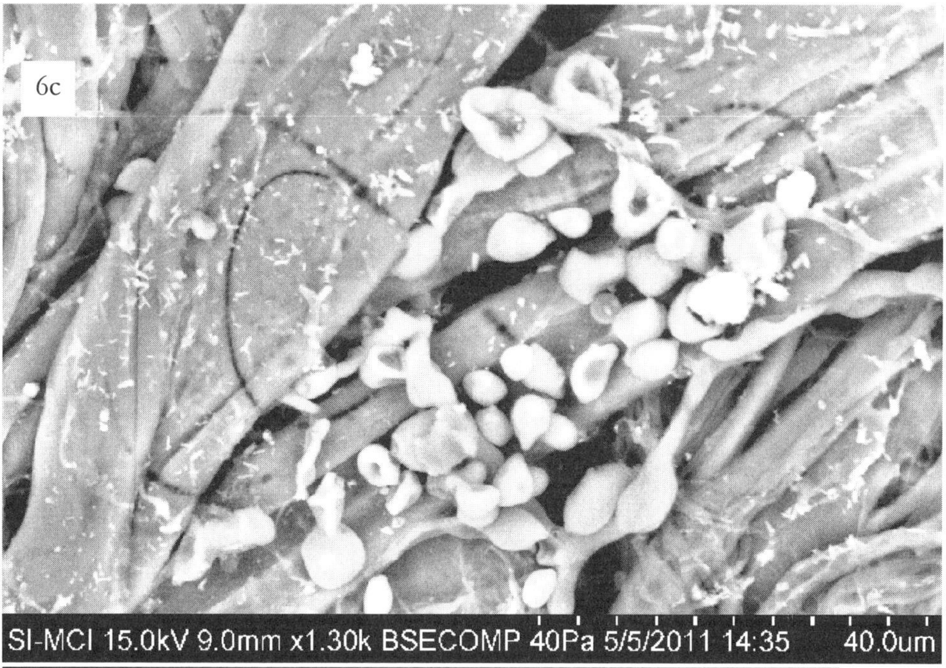

4. Surface morphology and topography with and without bio-mass was measured and profiled with confocal laser scanning microscope (CLSM) Keyence VKX at magnification 1000x (Figs.4d, 5d, 6d).

Figure 4d: The same specimen as in figures 4a - c, 3D mapping and surface topography, area measured: x:200 µmy: 100 µm z: 40 µm. Keyence VKX Laser Scanning Microscopy, 1000x. Catenulate cells and cells clusters are marked in black against light grey paper fibers.

Figure 5d: Same specimen as in figures 5a - c, CLSM, the marked area was measured to produce profile of bio-mass deposit; the highest concentration of the bio-mass of perithecial hairs is marked with white-circle.

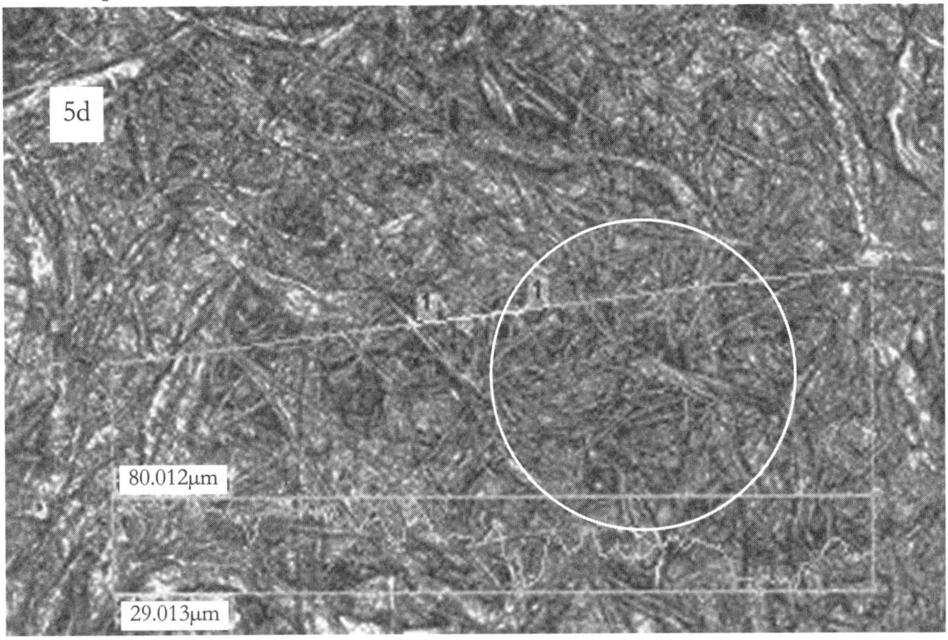

Figure 6d: Images of the stain seen in figures 6a - c, CLSM, 3D mapping of bio-mass distribution in relation to paper matrix. Cells are of light orange color against dark brown paper fibers. 1000x magnification, bar 10 μm.

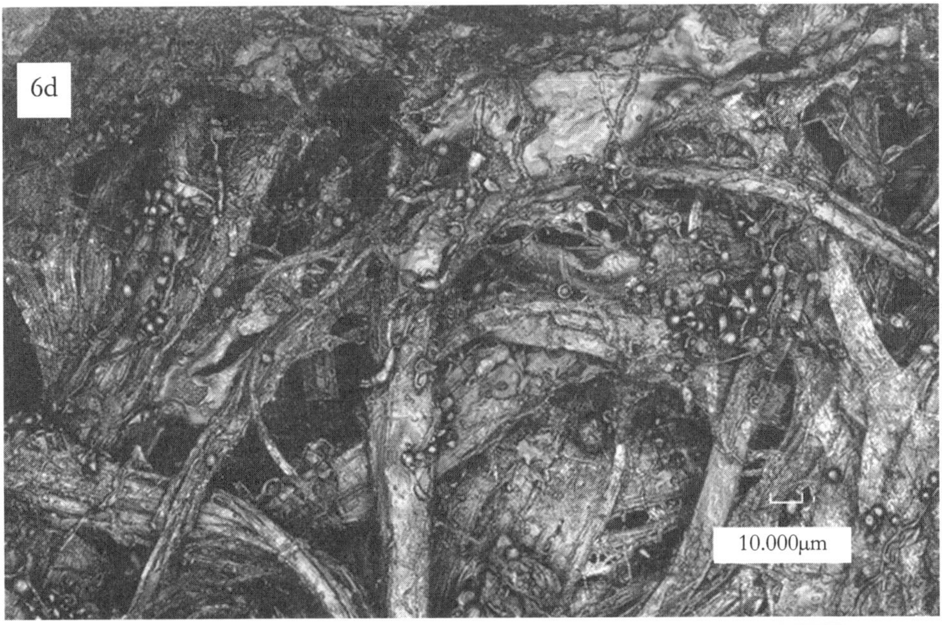

Transmitted light and incidental light microscopy has been traditionally employed on biological specimens. Confocal laser scanning microscopy (CLSM) one of the surface metrology tools was developed for applications in high precision industrial environment for the detection of surface imperfections among many others. By scanning the entire surface with a short-wavelength laser, full-focus observation is achieved at resolutions that cannot be attained with an optical microscope. The most important advantage of using CLSM is the reproducibility of the measurements. By measuring the height, width, and cross-section of the fungal elements we obtained a profile capturing quantitatively the differences in surface conditions of both papers with and without fungal deposits.

RESULTS

Each mode of investigation, evaluation of surface topography, morphology, and profiling fungal deposits complemented each other revealing new aspects of fungus stains and their interactions with the substrate. Two species of fungi in the class Dothideomycetes, *Cladosporium sp.* and *Taeniolella sp.* formed thick-walled, pigmented mycelia and conidia, all embedded into the paper matrix. The growth pattern of *Taeniolella* indicated attachment of the hyphae to the fiber surfaces of the paper. However, in 25 of the stained samples there was no evidence of mycelia penetrating into the interior of fibers. The latter disagrees with what was proposed by Nugari (2008) who suggested that mycelium grows in the internal structure of paper fibers.

Distribution of the dark brown spores of *Cladosporium sp.* followed the pattern of pores' location in paper. The spores nested in paper surface indentations which provide natural shielding against their mechanical removal. *Chaetomium,* which produces large perithecial heads, reaching over 0.3-0.4mm in diameter, was embedded in the paper matrix. However, there was no indication that the fruiting structures secreted pigments into the substrate. In all studied cases, the pigmentation was confined to the fungal elements, such as spores, hyphae and elements of the fruiting structures; no pigmented extracellular material was observed as being secreted into the substrate.

Preventive treatment and conservation of artworks

Knowing that moisture is essential for the germination and growth of fungi, all efforts should be made to control the environment and prevent occurrence of water condensation on surfaces. Constant low temperature in the environment in which artworks are exhibited or stored is one of the preventive methods against microbial growth.

Pigmentation induced by microorganisms on cultural material is resilient to removal, or to its reduction of color intensity by chemical means. That is understandable because pigmented bio-polymers are produced in many cases as fungal defense mechanism and cannot be dissolved in water or other solvents. Experiments with bleaching agents used in conservation for the reduction of discoloration and stains produce limited results. A mechanical removal of pigmented fungal structure, either fruiting bodies or spores is recommended only when no disturbance of the artwork occurs (Fig. 7) By mechanical removal it is understood that fungal elements are extracted with a surgical scalpel, tweezers or inoculating needle. Laser removal of pigmented fungal structures using 532 nm wavelength proposed in 1994 (Szczepanowska and Moomaw, 1994) was repeated in 2005 (Scholten, et al., 2005, Ochocinska-Komar et al., 2005; Pilch, et al., 2005) as another successful treatment option.

Figure 7: Removal of perithecia from the paper's surface. Weak ionic interfacial forces anchoring the bio-mass on the surface allow a mechanical separation of the bio-structures without disturbance of the original matrix.

Preliminary conclusions of work in progress

Detailed studies of fungus distribution in the various types of papers in our studies revealed a broad range of their interactions with the substrate. That diversity is directly related to the morphological characteristics of different fungal species, and the unique topography of each substrate on which they grew. Only partial understanding of the interfacial phenomena of the bio-pigments and substrates was gained using a broad range of analytical instruments. Maintaining the sequence of investigation, from gross (global) evaluation to more detailed (local) analysis was essential, as each technique provided complementary information. Combining optical light microscopy with surface metrological techniques and instruments, such as confocal laser scanning microscopy provided information about spatial inter-relations between the fungi and substrate, contributing information towards development of preventive and conservation strategy. However, insight into many unknown behavioral patterns of microorganisms, varied environmental conditions in which they grow and diverse composition of substrates is required to treat each case as a unique, multi-faceted problem. As the research continues more information will be available, elucidating this complex phenomenon of the interaction of fungi and their pigments with substrates.

REFERENCES

- Bochenek K,.Gudowska-Nowak E. Electronic properties of random polymers: Modeling optical spectra of melanins, In: *Acta Physica Polonica.* Warszawa. 2003:2775-2790.
- Buskirk AD, Hettick JM., Chipinda I,Law BF, Siegel PD, Slaven JE, Green BJ, Beezhold DH. Fungal pigments inhibit the matrix-assisted laser desorption/ionization time-of-flight mass spectrometry analysis of darkly pigmented fungi. Elsevier, *Analytical Biochemistry.* 2011:411:122-128. DOI: 10.1016/j.ab.2010.11.025.
- Icenhour C, Kottom TJ, Limper AH. Evidence for a melanin cell wall component in *Pneumocystic carinii.* In: Infection and Immunity. American Society for Microbiology. 2003:5360-3.
- Jacobson ES. Pathogenic roles for fungal melanins. In: *Clinical Microbiology Reviews.* 2000:708-17.
- Liu YG, and Nizet V. Color me bad: microbial pigments as virulence factors. In: Cell Press, Elsevier 2009. DOI: 10.1016/j.tim 2009.06.006.
- Medentsev AG, Akimenko VK. Naphthoquinon metabolites of the fungi. *Phytochemistry* . Elsevier Science: 1998:47: 6: 935-959.
- Nugari MP. Biodeterioration of paper, In: Plant Biology for Cultural Heritage, Biodeterioration and Conservation. Caneva G., M.O. Nugari and Ornella Salvadori (eds). The Getty Conservation Institute, Los Angeles. 2008.
- Ochocinska-Komar K, Kaminska A, Martin M, Sliwisni G. Observations of the post-processing effects due to laser cleaning of paper. In: Lasers in the Conservation of Artworks: LACONA V proceedings. Osnabrük, Germany, Springer. 2005:Sept.15-18. p. 29-34.
- Paola WF, Dadochova E, Manda P, Casadevall A., Szaniszlo PJ, and Nosanchuk JD. Effects of disrupting the polyketide synthase gene WdPKS1 in *Wangiella [Exophiala] dermatitidis* on melanin production and resistance to killing by antifungal compounds, enzy matic degradation, and extremes in temperature. *BMC Microbiology* 2006:6:55 DOI:10.1186/1471-2180-6-55
- Pilch E, Pentzien S, Madebach H, Kautek W. Anti-fungal laser treatment of paper; A model study with a laser wavelength of 532 nm. In: Lasers in the Conservation of Artworks: LACONA V proceedings, Osnabrük, Germany, Springer. 2005:Sept.15-18. p. 19-28.
- Scholten H, Schipper D, Ligterink FJ, Pedersoli JL, Rudoplh JrP, Kautek W, Havermans JBGA, Aziz HA , van Beek B , Kraan M, van Dalen P, Quillet V, Corr S., Hua-Strofer HY

Laser cleaning investigation of paper models and original objects with Nd: YAG and KrF laser systems. In: Lasers in the Conservation of Artworks: LACONA V proceedings, Osnabrük, Germany, Springer. 2005:15-18. p. 11-18.

- Sterflinger K. Black yeasts and meristematic fungi; ecology, diversity and identification. In: Biodiversity and Ecophysiology of Yeasts. The Yeast Handbook. 2006.
- Sterflinger K, de Hoog GS, and Haase G. Phylogeny and ecology of meristematic Ascomycetes. *Studies in Mycology.* 1999:43: 5-22.
- Szczepanowska H, Cavaliere AR. Fungal deterioration of 18th and 19th century documents: a case study of the Tilghman family collection, Wye House, Easton, Maryland; Elsevier, International Biodeterioration and Biodegradation. 2000:46:245-249.
- Szczepanowska H, Cavaliere AR. Drawings, Prints and Documents–Fungi Eat Them All!, in: Art, Biology, and Conservation of Works of Art: Biodeterioration of Works of Art. Koestler R, Koestler V, Charola AE, Nieto-Fernandez F (Eds.). Published by the Metropolitan Museum of Art. 2003:128-151.
- Szczepanowska H, Moomaw HW. Laser stain removal of fungus induced stains from paper. *Journal of the American Institute for Conservation* (JAIC) 1994:33: 25-32.
- Von Arx JA, Guarro J, and Figueras MJ. The Ascomycetete Genus Chaetomium. Beihefte zur *Nova Hedwigia.* Berlin, Stuttgart. 1986.

PART III:

PREVENTION / PUBLIC HEALTH

310

RISK AND HAZARD ASSESSMENT OF MOLDS GROWING INDOORS

Harriet M. Ammann

ABSTRACT

Learned Scientific Bodies have concluded that being exposed to mold indoors is unhealthful (IOM, 2004, Health Canada, 2007, WHO, 2009, AIHA, 2009). As early as 1996, AIHA recommended that immediate hazard assessment be performed when certain species of fungi were growing indoors in appreciable amounts. A few risk assessments have been attempted for indoor fungal growth, but have not provided credible results. This paper will delineate between hazard assessment and risk assessment and the basic kinds of knowledge required to achieve useful results.

Risk assessments, especially in methodology outlined by the U.S. E.P.A., are generally used to provide a guidance level of exposure that is tolerable or allowable in protecting the health of members the public, most often over long periods of exposure. Risk assessment is used in public health and regulatory efforts to prevent exposure. In contrast, hazard assessment speaks to a more immediate need for decisions about worker or public health protection. It is situation specific, assessing the nature and toxic potency of exposure, and the susceptibility of the particular population exposed, especially those who are vulnerable to the exposure in question. Hazard assessment is used to make immediate risk management decisions concerning removal of unhealthful agents from a site or removing vulnerable populations from exposure.

Since risk assessment is likely to be used for general public health protection or regulatory efforts, developing allowable or permissible exposure levels to human populations of variable susceptibility, its analyses are more extensive. According to EPA guidance, specific steps are involved: Hazard evaluation identifying a critical effect and critical study, Dose-Response Assessment using No-Observed Adverse Effect Levels (NOAELS) and Lowest-Observed Adverse Effect Levels (LOAELS), usually from chronic animal studies, Exposure Assessment, and Risk Characterization resulting in a Reference concentration that can be used in Risk Management.

Hazard Evaluation includes an extensive review of the literature to determine a critical effect that identifies that target organ or system that responds adversely at the lowest level of exposure in animal or epidemiological studies. It also identifies studies that satisfy criteria for adequacy to be used in quantitative dose-response assessments.

Dose-Response Assessment is performed using the critical effect and critical study in at least two relevant long-term animal or epidemiological studies that determine a NOAEL and a NOAEL.

Exposure Assessment requires that adequate measures of human exposure to the specific agents that have been identified in the previous analyses and which can be credibly quantified.

Finally, Risk Characterization using equations (mathematical models) developed to include both the measures of potency (NOAEL and LOAEL) as well as measures of exposure to healthy and susceptible populations, can be performed and may result in values that can be determined to be allowable or permissible exposure numbers. The quality of the risk assessment is determined by the quality of the input data, and always includes uncertainties that must be described.

INTRODUCTION

Learned Scientific Bodies conclude that living or working in damp and moldy environments is detrimental to health (National Academies of Science, Institute of Medicine [IOM, 2004], Canada Gazette, Health Canada, 2007, World Health Organization [WHO], 2009, American Industrial Hygiene Association [AIHA], 2008). Technical difficulties in determining and measuring putative agents of exposure have prevented the establishment of exposure levels deemed acceptable for susceptible populations. A few risk assessments have been attempted for some toxic products of mold, but the lack of appropriate long-term studies from which No- and Lowest-Observed Effect Levels could be derived, as well as the inability to identify and measure exposure to potential toxic agents, have made these attempts not credible. Despite this lack, industrial hygienists and occupational and public health officials have recognized that actions that protect occupants of damp buildings against the detrimental effects from exposure to organisms and their harmful products must be taken. Occupational health professionals also recognize that protective measures must be taken for workers who are engaged in remediating and cleaning environments contaminated with products from mold and bacterial growth, and other unhealthful organisms and products of damp indoor environments.

Hazard Assessment of the Individual Case

As early as 1996, AIHA recommended in its Field Guide for Determination of Biological Contaminants in Environmental Samples that immediate hazard assessment be performed when certain species of fungi known to cause occupational disease were found to be growing indoors in appreciable amounts. These fungi included toxigenic strains of *Stachybotrys chartarum*, and the infectious species *Aspergillus flavus* and *A. parasiticus*.

Hazard assessment is defined by AIHA (AIHA, 2008) as a prompt evaluation of the nature and extent of contamination, and the assessment of occupant susceptibility such as age and state of health (especially for pulmonary including asthma, chronic bronchitis, emphysema and other chronic lung disease, including but not limited to cystic fibrosis). If the occupants are deemed at risk either from excessive exposure or predisposing factors, immediate protective measures are to be implemented.

A number of agencies in the U.S. and elsewhere have issued guidance on what "appreciable amounts" of contamination means. The American Conference of Governmental Industrial Hygienists (ACGIH) offered guidance that involved numeric estimates of area of contamination deemed significant in 1989 (ACGIH, 1989), but withdrew these numbers in 1999 (ACGIH, 1999) because evidence of relationship of square footage of unspecified "visible" mold contamination to health outcomes seemed to be lacking. The New York City Department of Health provided guidance for the assessment and remediation of *Stachybotrys atra* (now referred to as *S. chartarum*) that included numerical area contamination estimates (NYCDOH, 1993). This guidance was revised in 2000 to fungi in indoor environments (NYCDOH, 2000), and was again revised in 2008 to include areas of hidden mold (not immediately visible during a walk-through) (NYCDOH, 2008). Guidance from this and other agencies including AIHA and ACGIH has evolved as understanding of microbial contamination in damp environments and risk to occupants, investigators and remediators has grown. Guidance is likely to change, especially as new tools for evaluating potential agents of detrimental effect and likelihood of exposure are developed, and as understanding of specific and aggregate agents of exposure and their health effects expands.

However, at present, risk managers must use reasonable criteria for establishing hazard when buildings are found to be moldy so as to be able to protect occupants and workers. Evaluation of the occupants' susceptibility may be the aspect of hazard assessment that is least emphasized in this process. Evaluation of the amount of exposure speaks to the concept articulated by Paracelsus (Phillipus Aureolus Theophrastus Bombastus von Hohenheim-Paracelsus, 1493-1541) that "the dose makes the poison," i.e., that the amount of exposure (as well as the potency of the poison) determines whether ill effect will occur. In the 21st century, we know that "the dose and the host" make the poison, because susceptibility among those exposed varies. Susceptible persons may suffer harm at levels of exposure far below those of healthy individuals.

The very young are both more susceptible and more exposed. Infants do not have the defenses against poisons that adults have. Depending on age (and stage of development), the lung, immune defenses, the blood-brain-barrier and a number of enzymatic detoxification pathways may not be functional, or fully developed. Infants, toddlers and young children in general also are more exposed to environmental contaminants than adults. They crawl on the floor and put their dirty hands in their mouths. They breathe more air per unit body weight than adults do, so that the same measured air concentration yields more exposure for the child than the adult.

Older occupants may have lung injury (i.e. from smoking or being exposed to smoke or other toxic air pollutants), chronic diseases such as circulatory or lung diseases, or diabetes which predispose them to toxic insult. They have may have lost detoxification abilities, immune capability, or lung defenses against particles and gases.

Those with pre-existing illness are not limited to the aged. Younger occupants may also suffer from significant lung disease such as cystic fibrosis, chronic bronchitis, or asthma, or suffer from diabetes or heart disease.

Because the factors of age, state of health and genetic predisposition contribute to the possibility that indoor biological contaminants may affect susceptible persons at lower exposure concentrations than healthy adults, hazard evaluation must include an assessment of building occupants, as well as determining extent of contamination. This process usually involves speaking with occupants or surveying them.

Hazard assessment, as meant in the AIHA guidance, involves individual case evaluation in order to make immediate risk management decisions. Each building and each situation is unique and requires that building and occupant evaluation be performed in order that appropriate protective measures can be taken. Such measures may include relocation of susceptible persons, while moisture assessment, remediation and clean-up proceeds. The area of contamination may need to be isolated and contained, and negative pressure air created in the contaminated area to prevent further contamination or exposure of occupants. If contamination in the building is severe or involves the air handling system, relocation of all occupants may be necessary. Certainly follow-up assessment of causes of moisture and contamination, and removal of contamination must occur to protect occupants.

Risk Assessment

Risk Assessment has a more general application. It involves a thorough evaluation of specific agents of exposure and a complete scientific evaluation of individual toxic agents (or mixtures), so as to develop allowable levels of exposure. These can then be compared with specific exposures to susceptible populations in buildings, to determine the potential or risk of harm to them. Risk assessment for non-cancer-causing contaminants is based on the assumption that there is a "threshold" of exposure, that is, a level at which host defenses cannot overcome toxic effect. (Risk Assessment for carcinogenic pollutants, on the other hand, is based on assessing the probability that certain levels of exposures may initiate or exacerbate cancer).

Because Risk Assessment is a process that will result in a number (with modifying text) that is meant to be protective of even very susceptible people during long-term exposure to a specific toxic agent or to a characterized mixture, health and regulatory agencies have developed rigorous processes for the development of such numerical levels. The U.S. Environmental Protection Agency has developed risk assessment guidance for non-cancer-causing toxic air pollutants that is considered rigorous and is used by other countries.

The ACGIH also uses a process for the protection of workers, principally in industry, to develop Threshold Limit Values (TLVs) and Permissible Exposure Limits (PELs) that are evaluated by the National Institute of Occupational Safety and Health (NIOSH) in the U.S., and recommended to the Occupational Safety and Health Office (OSHA), the regulatory agency for workplace exposures. The TLV and PEL numbers enforced by OSHA are not considered to apply to exposures to the general population because they are meant to protect healthy young workers, not infants, the ill or the aged, and because of the "healthy worker effect" which acknowledges that workers who become ill from exposures in the workplace often leave, and the remaining workforce is healthier as a result. Other agencies, including the California Office of Environmental Health Hazard Assessment (OEHHA) also have developed Risk Assessment guidance for non-carcinogenic air pollutants.

This paper is based on the U.S. EPA guidance *Methods for Derivation of Inhalation Reference Concentrations and Application of Inhalation Dosimetry* (1994). Reference Concentrations (RfCs) are used by states under EPA direction for regulatory purposes. Predicted emissions for facilities to be permitted are converted to air exposures that are compared to RfCs so as to characterize risk (Step 4 of the Risk Assessment process). Risk Characterization determines whether harm will be imposed on members of communities, and whether harmful emissions should be controlled or further restricted. How RfCs are developed is detailed in the section on Dose-Response Determination.

As RfCs are values and descriptions which will be compared to actual exposures in order to determine risk, the studies used for determining the critical effect of an individual toxicant, and the studies used to develop the RfC must be of high scientific value. EPA states that, at minimum, for a low confidence RfC, data used for this determination must include: "Well-conducted sub-chronic inhalation bioassay that evaluates a comprehensive array of endpoints, including portal of entry (respiratory tract effect), and an established and unequivocal NOAEL (no-observed adverse effect level) and LOAEL (lowest-observed adverse effect level)" for a critical effect. A critical effect is the endpoint in any organ or system in the most sensitive mammalian species that occurs at the lowest level of exposure. A high confidence RfC requires chronic inhalation bioassays that have clear and unequivocal NOAELS and LOAELs for the critical effect, and two-generation reproductive studies in two different mammalian species.

The Risk Assessment Process, according to EPA guidance consists of four steps:

1. Hazard Identification (Qualitative Evaluation of Data Base)

2. Dose-Response Assessment (RfC Development)
3. Exposure Assessment
4. Determination of Risk or Risk Characterization.

Step 1: Hazard Identification

The first step, Hazard Identification, requires a comprehensive review of the literature to locate the well-designed and executed chronic (at least 90-days exposure) study that identifies the critical effect. Both acute and chronic studies must be examined; acute primarily for mechanistic information, and chronic, because acute and chronic effects often differ in severity of effect and even target organ. Available inhalation studies, and instillation studies that approximate (but are not equivalent to) inhalation studies, such as intra-nasal, or intra-tracheal deposition of toxins, must be carefully reviewed to find the critical effect. This requires evaluation of studies of that include not only route of entry effects (respiratory system) but also reproductive and developmental, neurotoxic, immunotoxic, hepatotoxic, nephrotoxic, cardiotoxic effects. The available studies must have sufficient numbers of animals to be able to statistically establish dose-response for the most sensitive end-point, and to establish clear and unequivocal NOAELs and LOAELs. The most-human-like animal such as apes or monkeys would appear to be the subjects of choice. However these are generally no longer used because of cost of keeping them and ethical considerations, including that they are most human-like, and that many of such species are endangered.

Finding the most sensitive species is not intuitive. For instance in lethal exposure studies performed by various laboratories using T-2 toxin (a simple trichothecene produced by some *Fusarium* species, but not as yet identified in indoor air) rats, mice, and guinea pigs were used as subjects. Lethal concentration (LC_{50}) 24 hours after a 10 minute, one-time exposure in young adult mice was 0.08 ± 0.04 mg T-2 toxin/liter air, for mature mice 0.325 ± 0.1 mg T-2 toxin/liter air (Cresia et al., 1987), for rats 0.02 mg T-2 toxin/liter air, for guinea pigs 0.21 mg T-2 toxin/liter air (Cresia et al., 1990). The lower the LC_{50}, the smaller amount administered results in death, so the smaller the LC_{50}, the more potent the toxin is in that species. In these experiments, the order of susceptibility from high to low, judged by LC_{50}, is: rats, young adult mice, guinea pigs, and mature mice. It appears that (mature) rats are the most sensitive, but the animals cannot be directly compared.

What is required is that the Human Equivalent Concentration be determined for each animal so that direct comparisons can be made. Since all the animals in these experiments were exposed in a similar manner and metric units were used, conversion from ppm is not needed. Adjustments for exposure regimen would be difficult because of the lethal nature of the outcome. Usually adjusting 5 hours per day exposure to animals in chronic studies to 24-hour exposure in humans is done. The physical and chemical characteristics of the contaminant must be defined (i.e. particle or gas; size of particle, reactivity and solubility of particle or gas). For particles, size and shape determines where in the respiratory tract they are deposited, so area of deposition can be calculated, and residence time evaluated. For gases, (and for particles to some degree), reactivity can determine damage to the respiratory tissues, while solubility (primarily in the lipid portion of mucous membranes) determines absorption into the blood stream. Transport to target organs via the blood-stream, and uptake by the target organs, may also be predictable, depending on these parameters. Identification of the target organs is also necessary, which can be done by post-mortem analyses or through radioactive or luminescent tracers. Both the respiratory and remote targets must be identified. .

Once the HEC has been calculated, comparisons among the test species can be made to determine the most susceptible species and critical effect.

Step 2: Dose-Response Determination (Derivation of the RfC)

Step 2 involves the derivation of the RfC (Dose-Response Determination).

- "The EPA chooses the most appropriate NOAEL(HEC) of the critical effect from a well-conducted study on a species that is known to resemble the human in response to this particular chemical (e.g. by comparative pharmacokinetics).
- When the above condition is not met, the EPA generally chooses the most sensitive study, species, and NOAEL(HEC), as judged by an interspecies comparison of the NOAEL(HEC) and LOAEL(HEC)" (U.S.EPA, 1994).

A Reference Concentration (RfC) is defined by EPA as an estimate of a continuous inhalation exposure for a given duration to the human population (including susceptible subgroups) that is likely to be without an appreciable risk of adverse health effects over a lifetime. Great care must be taken to minimize uncertainty in deriving such an allowable exposure number that is "likely to be without appreciable risk."

Once a critical effect has been determined the RfC can be calculated by making exposure time and dosimetric adjustments between animal and human physiologic parameters such as minute volume, lung surface area, absorption etc., must be performed.

Great care must be taken to minimize uncertainty in deriving and allowable exposure number that "is likely to be without appreciable risk" to even sensitive subpopulations. The RFC is further modified from the NOAEL (HEC) for the critical effect by consistent application of Uncertainty Factors (UFs). These are applied to account for the uncertainties in extrapolating from experimental animal studies to an estimate that is appropriate to the human exposure scenario. The standard UFs are applied as necessary for the following extrapolations:

- Average healthy human to sensitive human
- Laboratory animals to humans

- Extrapolating from less than chronic NOAELs (sub-chronic) to chronic NOAELs
- Converting a $LOAEL_{(HEC)}$ to a $NOAEL_{(HEC)}$
- Incomplete data base (to account for the inability of any single laboratory study to adequately address all possible adverse outcomes in humans

The RfC is generally divided by UFs of an order of magnitude (factor of 10) each, although dosimetry adjustments or mechanistic data can reduce a UF to a lesser number. The UFs are added, and the composite is divided into the $NOAEL_{(HEC)}$. An RfC will not be derived if more than four areas of extrapolation exist. The composite UFs from four areas of extrapolation are usually reduced from 10,000 to 3,000 because the factors are not independent of each other.

Levels of confidence are also descriptively part of the RfC. These include discussion of:

"Adequacy of study design

- Is the route of exposure to relevant to humans?
- Were an appropriate number of animals of both sexes used for the determination of statistical significance?
- Was the duration of exposure sufficient to allow results to be extrapolated to humans under different exposure condition?
- Were appropriate statistical techniques applied?
- Were the analytical techniques sufficient to adequately measure the level of the test substance in the exposure protocol, including biological media?

Demonstration of dose-response relationships

- Were sufficient exposure levels use to demonstrate the highest NOAEL for the endpoint of concern?
- Is the shape of the dose-response curve consistent with the known pharmacokinetics of the test substance?
- Has the dose-response curve been replicated by or is it consistent with data from other laboratories and other laboratory animal species?

Species differences

- Are the metabolism and pharmacokinetics in the animal species similar to those for humans?
- Is the species response consistent with that of other species?
- Is the species from which the threshold value was derived the most sensitive species?

Other factors

- What number of biological endpoints were evaluated and associated with dose-response relationships?

- Was there sufficient description of exposure protocol, statistical tests, and results to make an adequate evaluation?
- What was the condition of animals used in the study?" (age, health, nutritional state, etc.)." (U.S.EPA, 1994)

Reference concentrations for Mycotoxins?

The scientific literature on the inhalation toxicity of mycotoxins is sparse. No chronic (more than 90 days of 4 to 5 hours, 5 days a week) exposure studies to either pure mycotoxins or mycotoxin-associated spores or fragments have been done. Sub-chronic studies (1 to 3 months exposure) have also not been done. One sub-acute (less than 1 month) study has been performed. Nikulin and colleagues (1997) instilled two strains of *Stachybotrys atra* (now called *S. chartarum*) spores intra-nasally, in concentrations of 1x 10^3, and 1x 10^5 spores into mice twice a week for 3 weeks, for a total of 6 exposures. One of the strains of *S. atra* was extremely toxic due to the presence of the trichothecene mycotoxins satratoxins G and H, stachybotrylactones and stachybotrylactams, while the other contained no satratoxins and only minor amounts of the other toxins. The exposure groups of 10 animals per dose group, plus controls exposed to the carrier aerosol only, were histologically examined and their blood chemistry evaluated. No histological changes were detected in the thymus, spleen, or intestines (the target organs of trichothecenes determined in other studies) in any of the groups of mice. The lungs of animals receiving 1 x 10^5 of either strain, and 1 x 10^3 the less toxic strain of *S. atra* showed inflammatory changes in the lung, while the lungs of mice receiving 1x 10^3 of the less toxic strain remained normal. The severity of changes in animals treated with the toxic and less toxic strains of spores was significant. Approximate concentrations of satratoxins G were 1μ/kg, body weight, for satratoxin H 2.6 μg/k, 2 mg/kg for spirolactone and 0.5 mg/kg for spirolactam body weight at the 1 x 10^5 dose. This study was important for determining differences in severity of effect, especially on the lung, between the more and less toxic strains of *S. atra*, especially as related to their toxins. It differed from acute studies of T-2 toxin, in that histologic changes found in those studies were not found in mice treated with the spores of *S. atra* carrying the highly toxic satratoxins, nor the less potent spirolactones and spirolactams. The study was not designed to determine LOAELS nor NOAELs.

Acute studies (defined as one-time, less than 24-hour exposure, but in the following being only 10 minutes or more) involved aerosolized T-2 toxin, a simple trichothecene shown to produce lesions in tissues other than the lung. Cresia and colleagues (1987, 1990) performed studies on mice, rats and guinea pigs, in which they forced the animals to breathe aerosolized T-2 toxin for 10 minutes to determine LC_{50} (Lethal Concentration$_{50}$) This is the concentration at which one-half of the experimental animals die at a set time after exposure. It can be used as a measure of relative potency from inhalation exposure compared to other routes. The study on mice examined susceptibility of young adult to mature mice as well (Creasia et al., 1987), and measured retention time for T-2 toxin. In addition to LC_{50}, these studies determined target organs for this toxin through post-mortem analyses. For mice, inhalation exposure was at least 10 times more potent than systemic administration and 20 times more potent than dermal exposure. The studies were not designed to determine NOAELs nor LOAELS.

In mice, no lesions visible through light microscopy could be seen in the lungs of any exposed animals. Lymphocytes within the cortex of the thymus were the most sensitive to cytolytic effects of the toxin, and other lesions were seen in the spleen intestines and in the adrenal gland. The study in rat and guinea pig determined that inhalation exposure was at least 20 times more potent than systemic administration for the rat, and at least twice as toxic for the guinea pig. While pulmonary edema was seen in rats after exposure, no evidence of gross hemorrhage or pulmonary disease was seen, but necrosis of the crypt cells in the large and small intestine and of the thymus and spleen occurred.

Pang and colleagues (1987) used inhalation exposures of 42- to 90 minute duration to T-2 toxin, by-passing the pig's upper respiratory system via tracheal intubation. The purpose of the study was to evaluate the effect of T-2 toxin on the function of alveolar macrophages and pulmonary and peripheral lymphocytes and to determine morphologic changes in the lungs and other organs. The researchers found that morphologic changes in the lung were mild, while the lymphoid tissues, gastro-intestinal tract, and gall bladder were severely affected. Vascular and alveolar epithelial injury occurred as well. They determined that 9 mg/kg nebulized T-2 toxin could result in death, and clinical signs and morphological changes resembled those reported in pigs given T-2 toxin intravenously at a dose of 1.2 mg/kg (LD_{50}) or greater. These and other studies (Pang et al., 1988, Mars et al., 1986) explored the effects of inhalation as compared to other routes of exposure. They are all acute studies at very high exposure levels that determined lethal concentrations for inhalation and\or target organ specificity (Thurman et al., 1988). None were designed to determine NOAELS, nor LOAELS, nor were more recent studies (Rao et al., 2000a, 2000b).

To date no chronic studies that meet the criteria in EPA guidance for determining an RfC for mycotoxins that have been measured in damp indoor spaces, or that are thought to occur there, have been performed. There are data from acute animal studies, and some human studies that indicate that the immune and nervous systems may be critical target organs for those mycotoxins such as aflatoxin and simple and complex trichothecenes that inhibit protein synthesis, but this has not been sufficiently established to be used in risk assessment.

Step 3: Exposure Analysis

Using the RfC to evaluate risk of exposed people involves evaluating their exposure in air and comparing it to the RfC to determine if their exposure subjects them to harm. Exposure Factor Handbook (U.S. EPA 2011) is often cited for default values of body weight, lung volume, breathing rate etc. for infants, children and adults. EPA has created models (equations) for calculating exposure, using measures of exposure agents in air and the default values from the Exposure Factor Handbook. The resulting exposure value can then be compared with the RfC for each specific toxin in order to characterize risk for the exposed population under consideration.

Step 4: Risk Characterization, or Determination of Population Risk

Comparison of the population exposure assessment with the RfC is done by dividing the appropriately measured population exposure by the RfC, which derives the Hazard Quotient. (HQ = Exposure Concentration / RfC). Since the RfC is an estimate of exposure at which no harm from long-term exposure is expected, if the quotient of this division is less than 1, no hazard to the population exposed is thought to occur. If the quotient is more than 1, harm may occur, but at what quantity above 1 harm begins, is not defined. For exposure to mixtures, hazard quotients for the individual toxicants are added if they affect the same target organ or organ system, to give an estimate of combined risk. Risk managers decide whether preventive action (i.e. via regulatory control, or removal of contamination or removal of exposed persons) is necessary.

The ability to characterize risk required both the ability to determine an RfC, and the ability to give a credible measure of exposure of the population in need of protection.

Can Risk from Mycotoxins (and other Bioaerosols) be currently Characterized?

Much depends on the ability to determine exposure of the populations in question. Determination of exposure to bioaerosols is highly problematical. The reasons remain much the same as those articulated by ACGIH in 1998 in the book *Bioaerosols Assessment and Control*, in the introductory discussion about why there are no TLVs for bioaerosols. These include considerations of sampling procedures, which have primarily been for intact fungal spores and bacteria (although bacterial endotoxins which are released when bacteria rupture, have also been sampled). Mostly sampling for these entities is for very short periods of time under artificial conditions. In reality, workers may be exposed from 8 to10 hours per working day, while home dwellers may be exposed for 24 hours per day. Measurements made under static conditions (so as to address repeatability of measurement) do not address the fact that human and other activity constantly occurs in commercial and domestic buildings, and that movement of occupants, as well as movement of the building itself, disturbs microbial growth and can aerosolize it, resulting in different air concentrations than may be measured under static conditions. Additionally, molds bloom at unpredictable times, and this biological activity can significantly change what is measured at any given short-time measurement.

Health effects studied have rarely correlated with measures of spores and bacteria counts in air. Researchers have measured spores and whole bacteria under the assumption that these were the agents of exposure. Short-term measures of mold spores have been used in attempts to determine exposure to molds, but health effects studied have rarely correlated with such measures. Other observations of dampness and microbial growth, and settled dust from contaminated surfaces have been associated more closely with health complaints.

We now know that the agents of exposure are very fine particles (Sorensen, 1989; Nielsen 2003; Górny et al., 1999; Górny et al, 2002; Toivola et al., 2002, Cho et al., 2005, Brasel et al, 2000a, b, Reponen et al., 2007; Huttunen et al., 2008; Pikäranta et al., 2008; Bloom et al., 2009 a, b). Up to 90% of exposure to fungi growing in indoors (as measured by β-D-glucan-associated particles in air) may in fact be to fine particles that include β-D-glucan itself, as well as enzymes and toxins secreted by toxigenic molds into dust from substrate on which the molds are growing (Foto et al., 2005; Miller and Rand, 2011)

Small particles present an entirely different inhalation exposure paradigm from that of intact spores and bacteria. Small particles in themselves have been associated with health effects ranging from mucous membrane irritants to death (Peters et al., 1997, Pope et al., 2002, 2004). Small particles in aggregate have a much larger surface area per unit mass than larger particles have, and thus can adsorb many toxicants and carry them deep into the lung and along the olfactory nerve to the brain (Kildesø et al., 2003), or into the blood stream for systemic distribution to remote target organs. Microbial particles include proteins such as enzymes and structural proteins, other structural elements including glucans, ergosterol and endotoxins and mycotoxins. Molds secrete enzymes and mycotoxins onto the substrate on which they grow (they digest before they ingest, in contrast to humans). These potentially allergenic and toxigenic molecules can adsorb to dust on the moldy surfaces and become aerosolized and inhaled when the substrate is disturbed.

Small particles (less than 1 micron to nanoparticle size) are more buoyant and have longer half-lives in air, and are readily transported on air currents. They are deposited in the terminal and respiratory airways and alveolar regions of the lung where physical defenses are fewer and where they have immediate access to the bloodstream. Intact spores and bacteria, in contrast, are more readily trapped and cleared from the upper airway. (Reponen et al., 2007).

As very few animal inhalation studies that examine toxic potency have been done for very few of the 400 or so mycotoxins known to be produced by molds, trying to determine toxic effects from mycotoxin exposure indoors is particularly problematical. Exposure to these toxins from air measurements has also proved very difficult. Recently, however, measurement of microbial toxins in air from damp buildings has been accomplished (Foto et al., 2005; Brasel et al, 2005, Panaccione and Coyle, 2004; etc.).

A number of experiments instilling a variety of toxins into animal respiratory tracts have shown specific inflammatory and cytotoxic effects at low exposures concentrations measured in air of damp interiors. (Rand et al. 2005; Viana et al., 2002; Gregory et al., 2003, 2004; Sorensen et al., 1981, 1987; Sorensen, 1989, 1999). However, for risk assessment purposes, the dilemma remains that toxin-producing molds grow variably in damp indoor spaces depending on which molds grow outside (climatic considerations), and that mycotoxins are produced variably depending, among other conditions such as the amount, composition and age and of the microbial community growing in the damp space. So the quandary remains: which molds and bacteria to measure by which procedures, which toxic products to determine, and then how to set protective numbers when the exposure agents vary over time and disturbance. At the current time, none of the exposure requirements set forth in EPA guidance can be met for microbial toxins produced by organisms growing indoors.

It is also not yet clear whether allergy or toxicity is the primary health problem, or whether both have a role in symptoms seen in mold-exposed persons.. Health studies have focused largely on route of entry (respiratory system) effects, yet the toxicological literature identifies a variety of target organs for different mycotoxins. Mechanisms of action, such as inhibition of protein synthesis, by aflatoxins and by trichothecene mycotoxins for instance, have led to findings of immune inhibition, decrease in growth, developmental detriments and behavioral changes in laboratory and field studies in animals. Immunotoxicity, neurotoxicity, and effects on growth and development in human health investigations, however, have been inadequately explored.

Until the appropriate chronic inhalation studies on the myriad mycotoxins potentially produced by molds can be performed, and sufficient information provided to determine critical effect and RfCs, and until credible exposure measures can be implemented, Risk Assessment for these substances is not feasible, or reasonable.

Yet risk managers have sufficient information to issue guidance about prevention of health effects in occupants of damp buildings and for workers investigating or remediating such spaces. The associations with dampness and the presence of microbial contamination assessed by visual, olfactory, and building history investigations are strong, and can be used to take protective measures when hazard assessment is adequately performed. Prevention of exposure is key. On a greater than an individual case basis, building and cleaning sciences can provide guidance on how to keep buildings dry and clean, and provide training and guidance for remediating moisture problems in buildings, and the resulting microbial growth.

REFERENCES

- ACGIH. Guidelines for Assessment of Bioaerosols in the Indoor Environment. American Conference of Governmental Industrial Hygienists, 1989. Cincinnati, OH.
- ACGIH. Bioaerosols Assessment and Control. Macher J (ed). American Conference of Governmental Industrial Hygienists, 1999. Cincinnati, OH.
- AIHA. Field Guide for the Determination of Biological Contaminants in Environmental Samples. American Industrial Hygiene Association. Dillon HK, Heinsohn PA, Miller JD (eds.). American Industrial Hygiene Association, 1996.Fairfax VA.
- AIHA. Field Guide for the Determination of Biological Contaminants in Environmental Samples, 2005. 2nd ed. American Industrial Hygiene Association. Hung L-L; Burton JD; Dillon HK (eds.)
- AIHA. Recognition, Evaluation, and Control of Indoor Mold. Prezant B, Weekes DM, Miller JD (eds.) American Industrial Hygiene Association. 2008. Fairfax VA. J.D. Burton, H.K. Dillon (eds.). Fairfax, VA: AIHA, 2005.
- Bloom E, Grimsley LF, Pehrson C, Lewis J, Larsson L. Molds and mycotoxins in dust from water-damages homes in New Orleans after hurricane Katrina. *Indoor Air,* 2009a. 19(2): 153-158.
- Bloom E, Nyman E, Must A, Pehrson C, Larsson L. Molds and mycotoxins in indoor environments – a survey of water-damaged buildings. *J Occup Environ Hyg.* 2009b.6: 671-678.
- Brasel TL, Douglas DR, Wilson SC, Strauss DC. Detection of airborne *Stachybotrys chartarum* macrocyclic trichothecene mycotoxins on particulates smaller than conidia. *Appl Environ Microbiol* 2005:71(1):114-122.
- Brasel TL, Martin JM, Carriker CG, Wilson SC, Straus DC. Detection of airborne *Stachybotrys chartarum* macrocyclic trichothecene mycotoxins in the indoor environment. *Appl Environ Microbiol* 2005:71(11):7376-388, (2005).
- Canada Gazette. Department of Health. Canadian Environmental Protection Act. Final Residential Indoor Air Quality Guidelines for Mould. 2007:141:13 March 31, 2007.
- Cho S.-H, Seo S.-C, Schmechel D, Grinshpun SA, Reponen T. Aerodynamic characteristics and respiratory deposition of fungal fragments. *Atmospheric Environment* 2005:39: 5454-5465.
- City of New York Department of Health. 1995. Guidelines on Assessment and Remediation of *Stachybotrys atra* in Indoor Environments. New York: City of New York Department of Health, 1993. [Also published in Fungi and Bacteria in Indoor Air Environments (edited by E. Johanning and C.S. Yang). Latham, N.Y.: Eastern New York Occupational Health Program, 1995. pp. 201-108.]
- Cresia DA, Thurman JD, Jones LJ III, Nealley ML, York CG, Wannemacher RW Jr, Bunner DL. Acute inhalation toxicity of T- mycotoxin in mice. *Fund Applied Toxicol* 1987:8 (2): 230-235.

- Cresia DA, Thurman JD, Wannemacher RW, Bunner DL. Acute inhalation Toxicity of T-2 mycotoxin in the rat and guinea pig. *Fund Applied Toxicol.* 1990:14: 54-59.
- Englehart S, Loock A, Skutlarek D, Sagunski H, Lommel A, Farber H, Exner M. Occurrence of toxigenic *Aspergillus versicolori* solates and sterigmatocystin in carpet dust from damp indoor environments. *Appl. Environ. Microbiol.* 2002:68: 3886-3890.
- Foto M, Vrijmoed LLP, Miller JD, Ruest K, Lawton M, Dales RE. A comparison of air-borne ergosterol, glucan and Air-O-Cell data in relation to physical assessments of mold damage and some other parameters. *Indoor Air* 2005:15: 257-266.
- Górny RL, Dutkiewicz J, Krysińska-Traczyk E. Size distribution of bacterial and fungal bioaerosols in indoor air. *Am Agric Environ Med* 1999:6: 105-113.
- Górny RL, Reponen T, Willeke K, Schmechel D, Robine E, Boissier, Grinshpun, SA. Fungal fragments as indoor air biocontaminants. *Appl. Environ. Microbiol.* 2002:68(7): 3522-3531 .
- Gregory L, Pestka JJ, Dearborn DG, Rand TG. Localization of satratoxin-G in *Stachybotrys chartarum* spores and spore-impacted mouse lung using immunochemistry. *Toxicol Pathol.* 2004:32: 26-34.
- Huttunen K, Rintala H, Hirvonen MR, Vepsäläinen A, Hyvärinen A, Meklin T, Toivola M, Nevalainen A. Indoor air particles and bioaerosols before and after renovation of moisture-damaged buildings: the effect on biological activity and microbial flora. *Environ. Res.* 2008:107(3): 291-298.
- IOM. Damp Indoor Spaces and Health. Institute of Medicine of the National Academies of Science. Committee on Damp Indoor Spaces and Health. 2004. National Academies Press.
- Kildesø J, Wűrtz H, Nielsen KF, Kruse P, Wilkins K, Thrane U, Gravesen S, Nielsen PA, Schneider T. Determination of fungal spore release from wet building materials. *Indoor Air.* 2003:13: 148-155.
- Larsson P, Tjalve H. Intranasal instillation of aflatoxin B (1) in rats: bioactivation in the nasal mucosa and neuronal transport to the olfactory bulb. *Toxicol Sci* 2000:55(2) 383-391.
- Marrs TC, Egginton JAG, Price PN, Upshall DG. Acute toxicity of T-2 toxin to the guinea pig by inhalation and subcutaneous routes. *Br J Exp Path* 1986:67:259-268.
- Miller JD, Sun M, Gilyan A, Roy J, Rand TG. Inflammation-associated gene transcription and expression in mouse lungs induced by low molecular weight compounds from fungi from the built environment. *Chem Biol Interact. Doi:* 2009:10.1016/j.cbi.2009.09.023.
- New York City Department of Health & Mental Hygiene - Bureau of Environmental & Occupational Disease Epidemiology. 2008. Guidelines on Assessment and Remediation of Fungi in Indoor Environments. (www.nyc.gov/html/doh/html/epi/moldrpt1.shtml)
- Nielsen KF. Mycotoxin production by indoor molds. *Fungal Genetics and Biology* 2003:39: 103-117.
- Nikulin M, Reijula K, Jarvis BB, Hintikka E-L. Experimental lung mycotoxicosis in mice induced by *Stachybotrys atra*. *Internatl J Experim Path* 1996:77: 213–218.
- Nikulin M, Rejula K, Jarvis BB, Veijalainen P, Hintikka E-L. Effects of intranasal exposure to spores of *Stachybotrys atra* in mice. *Fund Appl Tox.* 1997:35(2):182–188.

- Panaccione DG, Coyle CM. Abundant respirable ergot alkaloids from the common airborne fungus *Aspergillus fumigatus*. *Appl Environ Microbiol.* 2005:71(6): 3106-3111.
- Pang VF, Lambert RJ, Felsburg PJ, Beasley VR, Buck WB, Haschek WM. Experimental T-2 toxicosis in swine, following inhalation exposure: Effects on pulmonary and systemic immunity and morphological changes. *Toxicol Pathol* 1987:15: 308-319.
- Pang VF, Lambert RJ, Felsburg PJ, Beasley VR, Buck WB, Haschek WM. Experimental T-2 toxicosis in swine, following inhalation exposure: clinical signs and effects on hematology, serum biochemistry, and immune response. *Fund Applied Toxicol.* 1988:11: 100-109.
- Peters A, Wichmann HE, Tuch T, Heinrich J, Heyder J. Respiratory effects are associated with the number of ultrafine particles. *Am J Resp Crit Care Med.* 1997:155 (4): 1376-1383.
- Pitkäranta M, Meklin T, Hyvärinen A, Paulin L, Auvinen P, Nevalainen A, Rintala H. Analysis of fungal flora in indoor dust by ribosomal DNA sequence analysis, quantitative PCR, and culture. *Appl Environ Microbiol* 2008:74(1): 233-244.
- Pope CA 3rd. Epidemiology of fine particulate air pollution and human health: biologic mechanism, and who's at risk? *Environ Health Perspect* 2000:108: 713.723.
- Pope CA 3rd, Dockery DW. Health effects of fine particulate air pollution: lines that connect. *J Air Waste Manag Assoc.* 2006:56 (6): 709-742.
- Rand TG, Sun M, Gilyan A, Downet J, Miller JD. Dectin-1 and inflammation-associated gene transcription and expression in mouse lungs by a toxic (1,3)-β-D glucan. *Arch Toxicol* 2010:84: 205-220.
- Rao CY, Burge HA, Brain JD. The time course of responses to intratracheally instilled toxic Stachybotrys chartarum spores in rats. *Mycopath.* 2000a:149: 27–34.
- Rao CY, Brain JD, Burge HA. Reduction of pulmonary toxicity of *Stachybotrys chartarum* spores by methanol extraction of mycotoxins. *Appl Environ Microbiol.* 2000b:66(7): 2817-2821 (2000).
- Reponen T, Seo S-H, Grimsley F, Lee T, Crawford C, Grinshpun SA. Fungal fragments in moldy houses: a field study in homes in New Orleans and Southern Ohio. *Atmos Environ* 2007:41(37): 8140-8149.
- Sorensen WG, Frazer DG, Jarvis BB, Simpson J, Robinson VA. Trichothecene mycotoxins in aerosolized conidia of *Stachybotrys atra*. *Appl Environ Microbiol.* 1987:53: 1370-1375.
- Sorensen WG, Simpson JP, Peach MJ III, Thedeil TD, Olenchock SA. Aflatoxin in respirable corn dust particles. *J Toxicol Environ Health.* 1981:7: 669-672.
- Sorenson WG. Fungal spores: hazardous to health? *Environ Health Perspect* 1999:107 Suppl 3:469-72.
- Thurman JD, Cresia DA, Trotter RW. Mycotoxicosis caused by aerosolized T-2 Toxin administered to female mice. *Am J Vet Res* 1988:49 (11): 1928-1931.
- Toivola M, Alm S, Reponen T, Kolari S, Nevalainen A. Personal exposures and microenvironmental concentrations of particles and bioaerosols. *J Environ Monit.* 2002:4: 166-174.
- U.S. EPA. 1994. Methods for Derivation of Inhalation Reference Concentrations and Application of Inhalation Dosimetry. U.S. EPA Washington D.C. EPA/600/8-90/066F.

- U.S. EPA. 2011. Exposure Factors Handbook 2011 edition (final). U.S. EPA Washington D.C. EPA/600/R-09/052F.
- Viana ME, Coates NH, Gavett SH, Selgrade MJK, Vesper SJ, Ward MDW. An extract of *Stachybotrys chartarum* causes allergic asthma-like responses in a BALB/c mouse model. *Toxicol Sci* 2002:70:98-109.
- WHO. 2004. Health and Environmental Briefing # 42 for local health officials. World Health Organization, Bonn, Germany. [available in printed form only]
- WHO. 2010. WHO guidelines for indoor air quality: dampness and mold. World Health Organization. Geneva CH

RECOMMENDATIONS FOR DETECTION AND REMEDIATION OF MOLD GROWTH IN INDOOR ENVIRONMENTS IN GERMANY

Christiane Baschien, Heinz-Jörn Moriske, Kerstin Becker, Marike Kolossa-Gehring and Regine Szewzyk

German Environmental Survey for Children

Moldy buildings are a relevant health problem in Germany. This became especially clear from the results of the German Environmental Survey for Children (GerES IV). This survey was the German Federal Environment Agency's fourth Environmental Survey and the environmental module of the German Health Interview and Examination Survey (German acronym: KiGGS) of the Robert Koch Institute (RKI). Conducted from 2003 to 2006, the objective of this cross-sectional nationwide study was to produce an extensive and representative data base to characterize the exposure of children in Germany to environmental factors. This large scale population study included a representative sample of 1790 children aged 3-16 years. The questionnaire-based interviews conducted in GerES IV for all children also asked about mold in the home and characteristics of the building. Visible mold growth was found in 15 % of the homes. Influencing factors for the occurrence of visible mold growth were age, type and location of the building. Mold growth occurred significantly more often ($p \leq 0.001$) in apartment blocks, multi-family buildings, old buildings and in urban areas.

In a case-control study, a sub-sample of GerES IV was used to investigate the correlation between exposure to mold spores in a home and sensitization of children to certain mold fungus species. Homes of all participants were thoroughly checked for visible mold and participants were interviewed about indicators of potential exposure to mold. In addition, samples were taken in the children's rooms or the living rooms to determine concentrations of fungi in indoor air and floor dust. From this study it became clear that 14 % of the dwellings had visible mold growth. Additionally, likely problems with hidden mold sources were discovered by air sampling methods in 17 % to 27 % (depending on the parameters analyzed) of the dwellings according to the criteria of the German Federal Environment Agency's guidelines (see below).

Indoor mold and health

Epidemiological studies have clearly established a link between dampness / mold and health risks for the residents but no studies could define any dose-effect relationships between concentrations of indoor fungi or bacteria and health problems. Subsequently, limit values for acceptable concentrations of indoor fungi and bacteria based on health risk assessment are absent. Nevertheless, actions have to be taken to remediate damp/moldy buildings since growth of mold in dwellings has to be considered a hygienic problem.

The German FEA Statement

The commission on indoor air hygiene at the German Federal Environment Agency has clearly stated that growth of molds in indoor environments represents a possible health risk and should not be tolerated according to the precautionary principle. Based on this assessment, the Agency has started several research, standardization and guideline projects to deal with the detection and remediation of mold growth in indoor environments in Germany.

Recommendations for detection and remediation of mold in indoor environments

Two "mold guides" have been published with recommendations on how to prevent, investigate, eliminate and evaluate indoor fungi. The "mold guides" are addressed to experts such as renovators, microbiologists and public health specialists but provide also information for tenants, landlords and the interested public. The first guide (2002, http://www.umweltbundesamt.de/uba-info-medien/2199.html) explains the correlation between dampness and fungal growth, describes methods for detection and enumeration of indoor fungi as well as an assessment of hygienic risks by mold growth with the subsequent need for remedial actions. The first step when investigating a building for mold growth is a thorough walk-through by a skilled person with knowledge in building physics and engineering. In case of visible mold growth and clear cause further measurements are not necessary but remedial measures have to be taken directly. Depending on the question and the local situation different methods (e.g. culturable fungi, total fungi, air samples, dust samples, material samples, mold-dog) may have to be used for the detection of hidden fungal sources e.g. in case of moldy odor without visible mold growth. Determination of the fungal genera/species present is recommended because it gives more information on the presence of dampness and possible health effects than the concentration of fungi alone. A table is provided in the guide (extended in the second guide) to classify the measured concentrations of fungi into three categories: mold source in indoor environment (i) probable, (ii) cannot be excluded, (iii) not probable (see Tables 1 and 2). If a fungal source is being discovered in the building, remediation actions have to be taken to deal with the problem.

The second "mold guide" (2005, http://www.umweltbundesamt.de/uba-info-medien/2951.html) is divided in two parts: The first part outlines the causes of indoor fungal growth with emphasis on constructional influencing variables and ventilation practice. The second part provides recommendations for remedial actions and gives advice on how to avoid and minimize risks for residents and workers during remediation actions.

Technical standards

In addition, research and standardization projects have been conducted in Germany to harmonize and validate methods for the detection and enumeration of fungi. Inter-laboratory trials have been organized by different institutions to validate sampling and detection methods. These activities have provided an important basis to obtain comparable results and also for the establishment of national and international standards for detection and enumeration of fungi in indoor environments. The overall measurement strategy is outlined in the German standard VDI 4300 Part 10 (2008): "Measurement strategies for determination of mold fungi in indoor environments". The following international standards have been published or are under development:

- ISO 16000 Part 16 (2008): Detection and enumeration of molds - Sampling by filtration
- ISO 16000 Part 17 (2008): Detection and enumeration of molds - Culture based method
- ISO 16000 Part 18 (2011): Detection and enumeration of molds - Sampling by impaction
- ISO 16000 Part 19 (2012): Sampling strategy for molds
- ISO 16000 Part 20 (in preparation): Detection and enumeration of molds - Total spore count
- ISO CD 16000 Part 21: Detection and enumeration of molds - Sampling from materials

Summary and Outlook

The policy behind all these activities in Germany is to harmonize (i) the methods for detection and enumeration of fungi, (ii) the strategy for detection of fungal sources in indoor environments, (iii) the methods for assessment of hygienic risks, and (iv) the methods of remediation and clearance. Future activities include improving on-site help for people with mold problems. In particular, the access to an expert´s opinion and the assertion of remediation need to be improved.

Table 1: Assessment scheme for culturable fungi in indoor air samples

Parameter	**Indoor source unlikely Background level**	**Indoor source possible Further investigations required**	**Indoor source probable Immediate further investigations required**
Cladosporium and other genera which may reach high concentrations in the outdoor environment (sterile mycelia, yeasts, *Alternaria*, *Botrytis*).	Concentration (cfu/m^3) of one genus in the indoor air is lower than 0,7 to 1,0 times the concentration in the outdoor air $I_{type\ A} \leq A_{type\ A}$ x 0,7 (+0,3)	Concentration (cfu/m^3) of one genus in the indoor air is lower than 1,5 $\pm$ 0,5 times the concentration in the outdoor air $I_{type\ A} \leq A_{type\ A}$ x 1,5 ($\pm$ 0,5)	Concentration (cfu/m^3) of one genus in the indoor air is more than 2 times the concentration in the outdoor air $I_{type\ A} \leq A_{type\ e}$x 2
Sum of the concentration for those species that are unlikely to occur in the outdoor environment	Concentration in the indoor air is not more than 150 cfu/ m^3 above the concentration in the outdoor air $I_{\Sigma untype\ A} \leq A_{\Sigma untype\ A} + 150$	Concentration in the indoor air is not more than 500 cfu/ m^3 above the concentration in the outdoor air $I_{\Sigma untype\ A} \leq A_{\Sigma untype\ A} + 500$	Concentration in the indoor air is more than 500 cfu/ m^3 above the concentration in the outdoor air $I_{\Sigma untype\ A} > A_{\Sigma untype\ A} + 500$

Table 1. continued

Concentration of one genus (sum of all species) unlikely to occur in the outdoor environment	Concentration in the indoor air is not more than 50 cfu/ m^3 above the concentration in the outdoor air $I_{Euntyp\ A} \leq A_{Euntyp\ A} + 100$	Concentration in the indoor air is not more than 100 cfu/ m^3 above the concentration in the outdoor air $I_{Euntype\ A} \leq A_{Euntype\ A} + 300$	Concentration in the indoor air is more than 100 cfu/ m^3 above the concentration in the outdoor air $I_{Euntype\ A} > A_{Euntype\ A} + 300$
Concentration of one species that is unlikely to occur in the outdoor environment	Concentration in the indoor air is not more than 50 cfu/ m^3 above the concentration in the outdoor air $I_{Euntyp\ A} \leq A_{Euntyp\ A} + 50$	Concentration in the indoor air is not more than 50 cfu/ m^3 above the concentration in the outdoor air $I_{Euntype\ A} \leq A_{Euntype\ A} + 50$	Concentration in the indoor air is more than 100 cfu/ m^3 above the concentration in the outdoor air $I_{Euntype\ A} > A_{Euntype\ A} + 100$
Concentration of one species that is unlikely to occur in the outdoor environment and that possesses spores that do not become airborne easily e.g. *Phialophora, Stachybotrys chartarum*	Concentration in the indoor air is not more than 50 cfu/ m^3 above the concentration in the outdoor air $I_{Euntype\ A} \leq A_{Euntype\ A} + 30$	Concentration in the indoor air is not more than 100 cfu/ m^3 above the concentration in the outdoor air $I_{Euntype\ A} \leq A_{Euntype\ A} + 50$	Concentration in the indoor air is more than 100 cfu/ m^3 above the concentration in the outdoor air $I_{Euntype\ A} > A_{Euntype\ A} + 50$

Table 2: Assessment scheme for total fungi (microscopy) in indoor air samples

Parameter	Indoor source unlikely Background level	Indoor source possible Further investigations required	Indoor source probable Immediate further investigations required
Type of spores which may reach high concentrations in the outdoor enviroment e.g. Type ascospores Type *Alternaria/Ulocladium* Type basidiospores *Cladosporium* spp.	Concentration of one type in indoor air is lower than 1 to 1,2 times the concentration in outdoor air $I_{type\ A} \leq A_{type\ A}$ x 1 (+0,2)	Concentration of one type in indoor air is lower than 1,6 ± 0,4 times the concentration in outdoor air $I_{type\ A} \leq A_{type\ A}$ x 1,6 (± 0,4)	Concentration of one type in indoor air is more than 2 times the concentration in outdoor air $I_{type\ A} > A_{type\ A}$ x 2
Type *Penicillium/Aspergillus*	Concentration in indoor air is not more than 300 spores/m^3 above concentration in outdoor air $I_{\Sigma P+A} \leq A_{\Sigma P+A} + 300$	Concentration in indoor air is not more than 800 spores/m^3 above concentration in outdoor air $I_{\Sigma P+A} \leq A_{\Sigma P+A} + 800$	Concentration in indoor air is more than 800 spores/m^3 above concentration in outdoor air $I_{\Sigma P+A} > A_{\Sigma P+A} + 800$

Table 2. continued

Type *Chaetomium* spp.	Equal concentration in indoor air and outdoor air $I_{Chaetom} \leq A_{Chaetom}$	Concentration in the indoor air is not more than 20 spores/m^3 above the concentration in outdoor air $I_{Chaetom} \leq A_{Chaetom} + 20$	Concentration in the indoor air is more than 20 spores/ m^3 above the concentration in outdoor air $I_{Chaetom} > A_{Chaetom} + 20$
Stachybotrys chartarum	Equal concentration in indoor air and outdoor air $I_{Stachy} \leq A_{Stachy}$	Concentration in the indoor air is not more than 10 spores/m^3 above the concentration in outdoor air $I_{Stachy} \leq A_{Stachy} + 10$	Concentration in the indoor air is more than 2 spores/ m^3 above the concentration in outdoor air $I_{Stachy} > A_{Stachy} + 10$
several non characteristic spore types not belonging to the types basidiospores, Alternaria/Ulocladium, Cladosporium spp., or ascospores	Concentration in the indoor air is not more than 400 spores/m^3 above concentration in outdoor air $I_{divers} \leq A_{divers} + 400$	Concentration in the indoor air is not more than 800 spores/m^3 above concentration in outdoor air $I_{divers} \leq A_{divers} + 800$	Concentration in the indoor air is more than 800 spores/ m^3 above concentration in outdoor air $I_{divers} > A_{divers} + 800$
Pieces of mycelia	Concentration in the indoor air is not more than 150 pieces/m^3 above the concentration in outdoor air $I_{Myzel} \leq A_{Myzel} + 150$	Concentration in the indoor air is not more than 300 pieces/m^3 above the concentration in outdoor air $I_{Myzel} \leq A_{Myzel} + 300$	Concentration in the indoor air is more than 300 spores/ m^3 above the concentration in outdoor air $I_{Myzel} > A_{Myzel} + 300$

Legend to Table 1

All lines of the table have to be included for a comprehensive assessment!

cfu=colony forming units; I=concentration in indoor air in cfu/m^3

A=concentration in outdoor air in cfu/m^3

type A= species that are likely to occur in the outdoor environment; untyp A=species that are unlikely to occur in the outdoor environment (e.g. indicator species for dampness like *Acremonium sp.*, *Aspergillus versicolor*, *A. penicillioides*, *A. restrictus, Chaetomium sp., Phialophora sp., Scopulariopsis brevicaulis, S. fusca, Stachybotrys chartarum, Tritirachium (Engyodontium) album, Trichoderma sp.)*

ΣuntypeA=sum of species that are unlikely to occur in the outdoor environment;

EuntypeA=one species that is unlikely to occur in the outdoor environment

Legend to Table 2:

All lines of the table have to be included for a comprehensive assessment!

I = concentration in indoor air in number of spores/m^3

A = concentration in outdoor air in number of spores/m^3,

type A = type of spores which may reach high concentrations in outdoor air e.g. ascospores, *Alternaria/Ulocladium*, basidiospores, *Cladosporium sp.*

ΣP+A = Sum of the spores type *Penicillium* and type *Aspergillus*

Chaetom = Sum of the spores type *Chaetomium sp.*

Stachy = Sum of the spores type *Stachybotrys chartarum*

divers = Sum of several non-characteristic spores not belonging to the spore types ascospores, *Alternaria/Ulocladium*, basidiospores or *Cladosporium sp.*

REFERENCES:

- Szewzyk R, Becker K, Hünken A, Pick-Fuß H, Kolossa-Gehring M. Kinder-Umwelt-Survey (KUS) 2003/06. Schriftenreihe Umwelt & Gesundheit, 2011;1-91.

INDOOR WATER AND MOLD DAMAGE - INVESTIGATION AND DECONTAMINATION PRACTICES IN GERMANY

Wolfgang Lorenz

INTRODUCTION

It is well known, that different events can lead to high humidity in buildings and on materials such as furniture and other contents. Mold growth can occur with a minimum of 80% relative humidity over days on materials or surfaces. Water may enter a building from the outside such as the roof, outside walls or basement. Water pipes or showers may also leak and brake. Dampness from high water content in the indoor air can condensate on cold surfaces or humid outdoor air can cause condensation in the HVAC Systems.

In Germany, the occurrence of water and mold building damage appeared to have increased significantly in the last 15 years. The reasons for this are not known. One key factor may be the tightness of modern buildings as mandated by the German government to conserve energy. Often older buildings are in a bad condition, because owners do not invest sufficient money in necessary building repairs. New buildings are constructed as fast as possible to save money. For example, when the occupants move in, the foundation concrete may not even be sufficiently dry. Of note, one hundred years ago the Municipality of Berlin ruled that building occupants may not move into a newly constructed house but they have to wait 9 month after the construction was completed! Another factor may be that more hidden or invisible mold damages are identified by improved investigation methods compared to 15 years ago. Furthermore, the importance of mold prevention for improved health received more attention in public over the last years.

When large quantities of water are released into a building, mold damage may be controlled by speedy technical drying. Different specialized equipment is nowadays available and a variety of methods exist to dry up construction materials rapidly. However, it is not uncommon that damages resulting from the water entry are hidden, or overlooked and ignored.

It is necessary to define "water damage", before discussing analytical methods and remediation technologies. It should not only be defined as indoor growth of fungi. An earlier study conducted by this author showed that in most cases of water damage inside buildings, in addition to fungi also bacteria, in particular actinomycetes could be identified in the field samples, see table 1.

TABLE 1: Fungi and bacteria detected in material samples taken from moist building materials

Number of Samples	Fungi or Bacteria not detected	Only Fungi detected	Fungi and Bacteria detected	Only Bacteria detected
612	51	94	453	14
100%	8,3 %	15,4 %	74 %	2,3 %

Moreover, as the results of a study by Sanders et al. (2010) below show, also mites were present on water-damaged buildings material and in dust samples. Of the 50 samples that were analyzed for mite-antigens (ELISA-test) all (100%) contained mites and in 36 samples mites were identified via microscope.

TABLE 2: Mites and mite-antigens in 50 samples taken from mold damages in homes (Sander et al., 2010).

	Dermatophagoides farinae and pteronyssinus	Tyrophagus spp., Acarus siro, Glycyphagus and Lepidoglyphus	Dermatophagoides and/or Tyrophagus, Acarus siro, Glycyphagus, Lepidoglyphus
Detection of mites via microscope	7	36	36
Detection of mites via ELISA-Tests	5	49	50

Some indoor air experts are of the opinion that mold damage must be visible. However, a statistical review of 100 investigations in which we were involved indicated that in most of our cases the damages were not readily visible, see table 3.

TABLE 3: Mold damages must not be visible. Most of our cases are hidden.

Results of indoor inspection of 100 cases	**Cases in %**
Visible microbial growth	16 %
Moldy or musty smell without visible microbial growth	11 %
Moist building material, no visible mold, no musty smell	43 %
Special construction (creep cellar, flat roof, covered walls), no visible mold, no musty smell	16 %
Only typical health complaints, no other indication	14 %

For example in some investigated cases the mold was hidden and not readily visible, see figure 1. In other cases mold contamination could only be identified with the help of the microbial laboratory with appropriate samples, see figure 2.

FIGURE 1: Visible but hidden mold damage on a wall behind the wall paper.

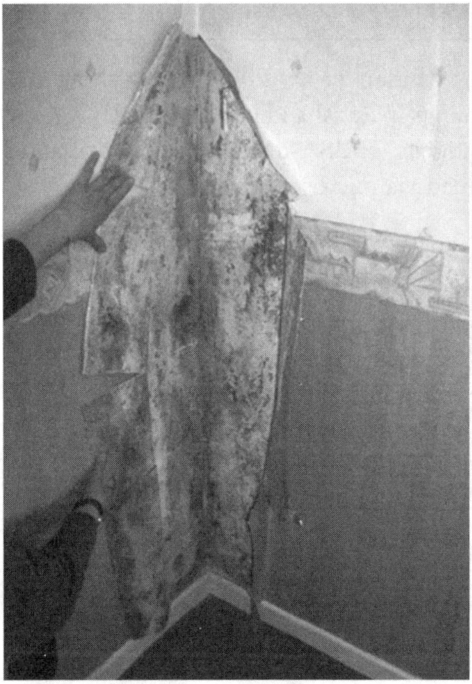

FIGURE 2a: Black line: Wall area with high moisture - but no visible mold.

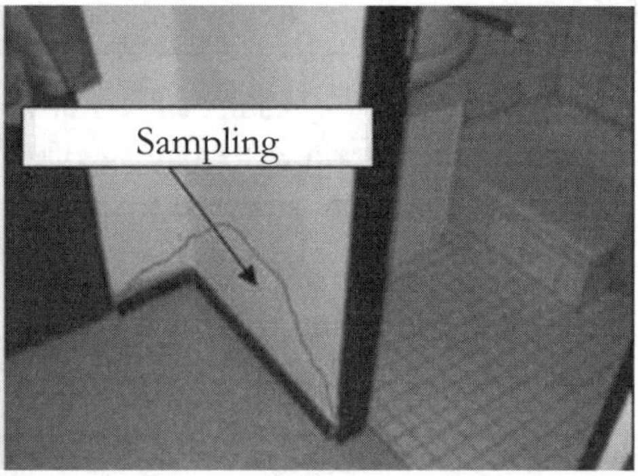

TABLE 4: Results of microbial analysis of the wall plaster.

Fungi / Bacteria	CFU/g
Aspergillus ochraceus	200
Aspergillus sydowii	120.000
Aspergillus versicolor	440.000
Penicillium spp.	2.000
Engyodontium album	80.000
Actinomycetes	400.000
Bacillus spp.	40
Not identified bacteria	2.000.000

Methods of Investigations

Useful information can be obtained by careful inspection of the building and interviews of inhabitants and building managers. One should routinely measure the humidity of walls, floor, and ceiling during inspection with non-destructive methods, using a high frequency or conductivity meter. We were able to detect 90% of the moisture damage by this approach. It appears not beneficial to measure routinely airborne fungi or microbial volatile organic compounds (MVOC), when the water damage is apparent and mold growth is clearly visible or the unique mold odor can be readily detected.

FIGURE 3: Wide spread mold damage visible

TABLE 5: Measurements of airborne fungi did not indicate water damage in 28% of cases, even though severe water damage had occurred.

	Results of Airborne Fungi Measurements					
	No Indication		Soft Indication		Indication	
Damage	n=	%	n=	%	n=	%
No Damage	2	4	0	0	0	0
Small Damage	3	6	1	4	1	2
Large Damage	13	28	5	11	21	45

It may be faster and more economical to search for visible hidden water and mold damage first and not start with the air sampling analyses first and followed by the search for sources.

TABLE 6: Methods to identify and localize mold damages.

Situation	Objective	Method
Visible mold damage	Cause of humidity?	Quantitative measurement of humidity in construction materials, thermo graphic analysis, long-term climate measurements.
	Determine extent of damage?	Moisture measurements. Material analysis and analysis of tape samples by microscopy.
Localized moisture damage	First proof of microbial growth by microscopy or culture of microorganisms. If microbial growth is confirmed same steps as above in case of visible mold.	
Neither mold nor moisture damage is evident – but there are other clear indicators of water damage.	Localization of moisture damages	Detailed building investigation with comprehensive physical measurements to localize moisture damages or use of a dog trained in mold detection.
	Localization of older moisture damages	Dog trained in mold detection

TABLE 7:: Laboratory analysis considerations

Method	Pros	Cons	Application
Microbial sample analysis by dilution method and culture	Quantitative results of genera and species	Result after minimum 10 days; costly, only a few laboratories in Germany are able to calculate results based on own scientific works	Standard method to analyse materials in case of non-visible mold
Microbial sample analysis by microscope	Fast, inexpensive, detection of cultivable and non cultivable fungi.	Only semi-quantitative and species cannot be identified. Most bacteria will not be detected. Need of an excellent trained and meticulous biologist	Standard method to analyze dimension of damage and subsidiary method
Tape sample analysis with microscope			Standard method to control decontamination
Contact – Sample collection with Rodac-plates	Inexpensive	It is not possible to differentiate between mold growth and contamination by dust	Not practical to analyses damages. Useful to analyses infectious bacteria or to control disinfection work
Measurement of airborne fungi (viable spores analysis by culture method)	Results appear convincing to lay people. In cases of „positive“ results sources are often obvious.	Uncertain method with unreliable results. More than 20% of damages / sources are overlooked. Expensive. In case of mold identification – source must be found.	To analyse for infectious microorganisms and secondary method to control disinfection work. Established for controlling air at working places with high concentration of microorganisms.

TABLE 7: continued

Method	Pros	Cons	Application
Air measurement of total microbial particles	Also non-viable particles are detected.	Method not validated. Damages will be overlooked.	Inferior method.
MVOC-air measurement	Sensitive method with proper sampling and correct assessment of results; detectable microbial sources are not overlooked.	Interpretation of results is difficult. Current labs produce different results.	No routine & simple method. Must be in done only by trained specialists and must be analyzed by special labs.
Dust analysis by culture	Easy sampling method. More reliable than analysis of airborne microorganisms.	No validated procedures for sampling and quantification are available.	Limited experience

Decontamination

In general the goal of remediation and decontamination efforts ought to be:

- Repair of all defects / water source control (= cause/s of damage)
- Complete removal of any moldy material and waste
- Drying of the moist/wet and non-moldy materials
- Detailed clean-up of all rooms by removing contaminated dust and removal of microbial substances

In Germany special regulations must be observed to assure health and safety requirements for workers and third persons at remediation site.

Based on a case specific risk analysis and assessment a variety different precautions and actions are necessary.

Key points of the German health and safety regulations are in summary:

- Enclosure of working area
- Ventilation of working area (low pressure system)
- Tools with integrated or adapted exhaust systems
- Protective clothing and equipment

FIGURE 4: Enclosure technologies

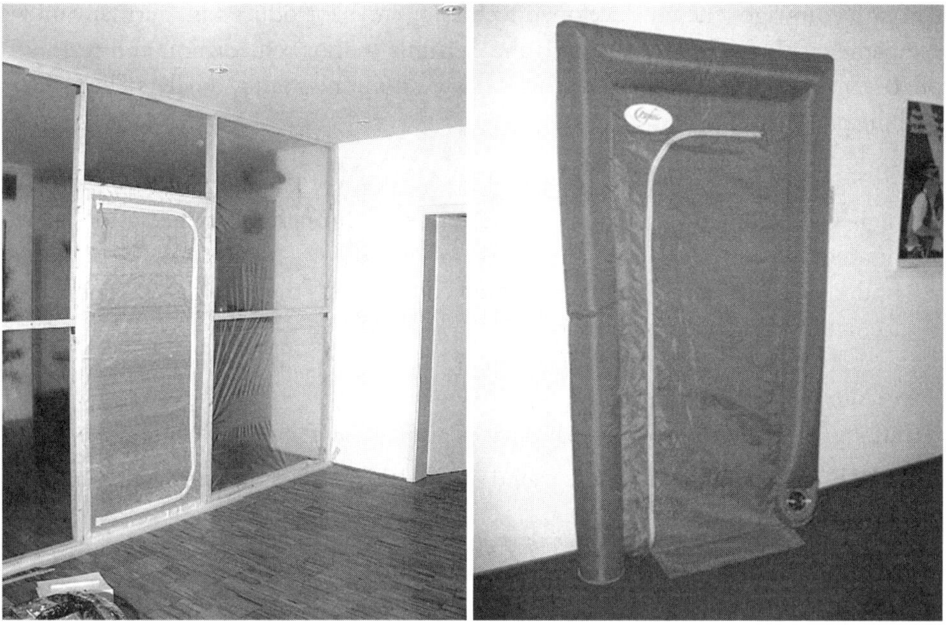

FIGURE 5: Standard system for exhaust ventilation (negative pressure system).

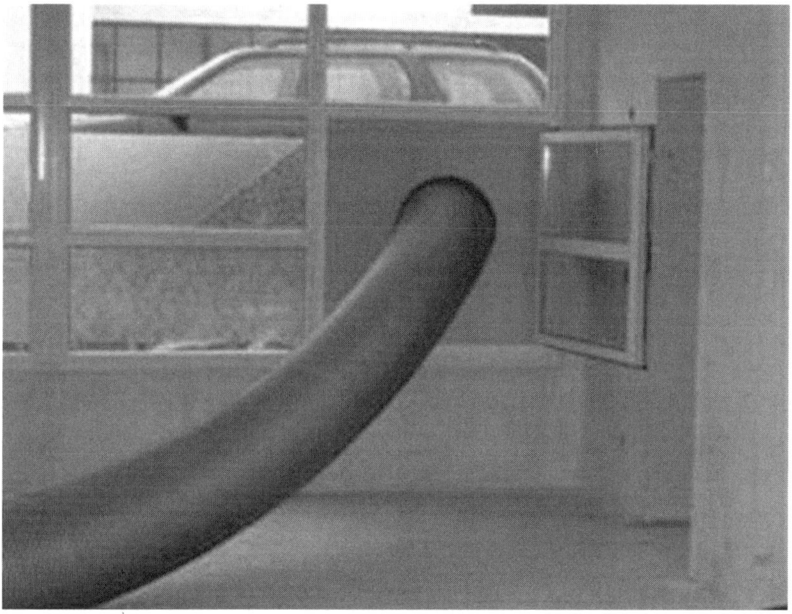

State of the art technologies

About 3 years ago a newly developed technology was introduced in Germany under the name of 'D-MIR' that refers to the dynamic indoor remediation and restoration of microbial contamination (http://www.d-mir.net/dmir/de/darstellung/darstellung.php).

The distinguishing factors of 'D-MIR' are:

- high volume air flow from clean indoor areas through the remediation site to outdoors
- optimized exhaust systems mounted directly at the tools and connected to high power vacuum cleaner
- vacuum cleaner fitted with HEPA-filter
- vacuum cleaner are placed outdoor
- additional high volume exhaust devices at place of dust release.

FIGURE 6: Driller with integrated exhaust system

FIGURE 7: High power vacuum cleaner – located outside

FIGURE 8: Ventgate-Technology with optimized exhaused ventilation at working site

The 'Berufsgenossenschaft der Bauwirtschaft' (http://www.bgbau.de), the statutory accident insurance for construction work, tested this technology using air measurements and gave it a positive rating.

In summary, the advantages of the 'D-MIR' remediation technology appear to be:

- no personal respirators are necessary for worker protection
- no special enclosures of the remediation site appears necessary
- in the building non-contaminated office space/rooms can be occupied while remediation work takes place in contaminated areas.

Currently, no disadvantages of the 'D-MIR' methods are known compared with conventional remediation practices. However, remediation worker have to be trained in the use of the 'D-MIR' system but usually the training cost are typically lower than for conventional systems.

Methods not recommended:

Based on our experience we do not recommend the use of chemicals, biocides or ozone gas in remediation and restoration projects:

- using chemicals, particularly disinfectants to kill microorganisms in remediation projects is not recommended. This may lead to unforeseen health problems in building occupants or workers (i.e., allergic or toxic reactions; pathogen-associated molecular patterns (PAMP) related immune response).
- Using ozone generators is not recommended, because ozone can destroy active cells but not spores, toxins or PAMPS. Low ozone concentration will not work and high ozone concentrations are too toxic for human and animals and can lead to respiratory hazards. Ozone may reduces moldy odors but will not lower microbial air pollution.

REFERENCES:

- Berufsgenossenschaft der Bauwirtschaft (BG Bau), Handlungsanleitung – Gesundheitsgefährdungen durch biologische Arbeitsstoffe bei der Gebäudesanierung (BGI 858), 2006.
- Landesgesundheitsamt (LGA) Baden-Württemberg (Hrsg.), Schimmelpilze in Innenräumen – Nachweis, Bewertung, Qualitätsmanagement, Stuttgart. 2001. / 2004 (http://www.gesundheitsamt-bw.de/SiteCollectionDocuments/40_Service_Publikationen/Schimmelpilze_in_Innenraeumen.pdf 8/12)
- Lorenz W, Trautmann C, Dill I: Analysis and impact of actinomycetes and other bacteria in indoors (published in German: Nachweis und Bedeutung von Actinomyceten und sonstigen Bakterien in Innenräumen. in: Handbuch für Bioklima (eds. Moriske, Turowski), ecomed Verlag, Landsberg am Lech, 2003:10:12:3-4.4.14
- Lorenz W, Sigrist G, Shakibaei M, Mobasheri A, Trautmann C. A hypothesis for the ori gin and pathogenesis of rheumatoid diseases. *Rheumatol. Int.* 2006:26;641-654.

- Mehrer A, Lorenz W, Gareis, Trautmann, Kroppenstedt, Stackebrandt. Cytotoxicity of different actinomycetes isolated from building materials. 5th International Conference on Bioaerosols, Fungi, Bacteria, Mycotoxins and Human Health, Saratoga Spring, N.Y. USA. 2003.
- Robert-Koch-Institut, Mitteilung des RKI: Schimmelpilzbelastungen in Innenräumen - Befunderhebung, gesundheitliche Bewertung und Maßnahmen. 2007. (http://www.apug.de/archiv/pdf/Schimmelpilze-BGBL-10-2007.pdf 8/12)
- Sander I, Franz J-T, Schies U, Zahradnik E, Kolk A, Schneider G, Wattrodt P, Kespohl S, Lorenz W, Bach C, Fleischer C, Flagge A, Brüning T, Raulf-Heimsoth M. Quantification of mites and mite allergens in residences with mold damages (published in German: Quantifizierung von Milben und Milbenallergenen in Wohnungen mit Schimmelpilzbefall; Dokumentationsband der 50. Jahrestagung der DGAUM, Aachen, 2010:6; 588 -589
- Umweltbundesamt (UBA) (Hrsg.), Leitfaden zur Ursachensuche und Sanierung bei Schimmelpilzwachstum in Innenräumen („Schimmelpilz-Sanierungsleitfaden"), Dessau 2005. (http://www.umweltdaten.de/publikationen/fpdf-l/2951.pdf 8/12)
- Umweltbundesamt (UBA) (Hrsg.), Leitfaden zur Vor-beugung, Untersuchung, Bewertung und Sanierung von Schimmelpilzwachstum in Innenräumen, Dessau. 2002. (http://www.apug.de/archiv/pdf/Schimmelpilze_Leitfaden.pdf 8/12)

UPDATE OF CANADIAN AND INTERNATIONAL MOLD GUIDELINES AND STANDARDS

Donald M. Weekes

KEYWORD: Mold; guidelines; standards; Canada; international

Background:

Since 2008 when the AIHA Green Book (Recognition, Evaluation and Control of Indoor Mold) was published, there have been numerous international (including Canadian) mold and moisture guidelines, reports and standards that have been promulgated by a variety of governmental entities, professional organizations and international agencies. Many of the standards and guidelines since the publication of the AIHA Green Book have utilized this AIHA book as a reference. During the September, 2011 Saratoga Bioaerosols conference presentation, these recent publications were summarized and compared to each other, and to the previous mold documents, including the AIHA 'Green Book'. Available information about forthcoming documents to be published was also discussed.

Materials:

The following guidelines and other documents were described in brief during the September, 2011 Saratoga Bioaerosols conference presentation:

- New York City Department of Health (NYCDOH) Microbial Guidelines (updated 2008)
- General Accounting Office (GAO) – Indoor Mold Report (2008)
- World Health Organization (WHO) Report – Dampness and Mold (2009)
- Environmental Abatement Council of Ontario (EACO) Mold Abatement Guidelines (updated 2010)
- US Occupational Safety and Health Administration (OSHA) – A Brief Guide to Mold in the Workplace (March, 2010)
- NY State Toxic Mold Task Force report (2010)
- American Society of Testing and Materials (ASTM) - Standard Guide for Assessment of Fungal Growth in Buildings (January, 2011)
- American College of Occupational and Environmental Medicine (ACOEM) Mold Position Statement (March, 2011)
- Unified Facility Guide Specifications (UFGS) – Mold Remediation Specification (May, 2011)
- Center of Disease Control (CDC) – Dampness and Mold in Buildings (May, 2011)

- American Society of Heating, Refrigeration, and Air Conditioning Engineers (ASHRAE) – Mold Position Statement (2011-2)
- American Industrial Hygiene Association (AIHA) – Revised Facts of Mold for Consumers and Professionals (January, 2012)
- Canadian Municipal Marijuana Grow Op (MGO) Guidelines (2008-2011)
- National Institute of Occupational Safety and Health (NIOSH) – Preventing Occupational Respiratory Diseases from Exposures caused by Dampness in Office Buildings, Schools and other Non-Industrial Buildings (March, 2011)
- State of California - Statement on Building Dampness, Mold, and Health (September, 2011)

Each of these documents will be discussed in brief, particularly with regards to occupant health information and statements. Also, pertinent details about microbial sampling in each document will be discussed, particularly with regards to post-remediation verification (PRV).

1) NYC DOH Mold Guideline (2008)

http://www.nyc.gov/html/doh/html/epi/moldrpt1.shtml

The NYC DOH Mold Guideline indicates the following: 'Environmental sampling is not usually necessary to proceed with remediation of visually identified mold growth or water-damaged materials.' 'In all situations, the underlying moisture problem must be corrected to prevent recurring mold growth.' Indoor mold growth can be prevented or minimized, however, by actively maintaining, inspecting, and correcting buildings for moisture problems and immediately drying and managing water-damaged materials.

The Guideline lists the Quality Assurance Indicators from the AIHA publication – "Assessment, Remediation, and Post-Remediation Verification of Mold in Buildings".

2) GAO – Indoor Mold Report (2008)

http://www.gao.gov/new.items/d08980.pdf

The GAO Indoor Mold Report 'concluded that certain adverse health effects are more clearly associated with exposure to indoor mold than others.' The GAO also concluded that 'enough is known that federal agencies have issued guidance to the general public about health risks associated with exposure to indoor mold, how to minimize mold growth, and how to mitigate exposure.' Therefore, the GAO called for a coordinated federal research effort into indoor mold and mold mitigation procedures.

3) WHO Report – Dampness and Mold (2009)

http://www.euro.who.int/__data/assets/pdf_file/0017/43325/E92645.pdf

The WHO Report states that: 'Although it is plausible that the exposures listed (in the report) are the main causal factors of the health effects associated with damp buildings, this has not been proven.' Also, the Report indicates that 'While microbial growth and health outcomes are consequences, their common denominator is undesired moisture behavior (i.e. excess moisture in building assemblies or on surfaces).' 'Moisture control, including ventilation, is the main method for containing mold.'

4) EACO Mold Abatement Guidelines (2010)

http://www.eacoontario.com/pdf/2010/eaco_mould-abatement-guidelines_book.pdf

The Environmental Abatement Council of Ontario (EACO) Mold Abatement Guidelines denoted three (3) levels of mold remediation – small, medium and large. There are also training requirements in the Guideline for remediation personnel, and it recommends medical pre-screening for all workers prior to mold remediation work.

For medium and large projects, the Guideline recommends that a Health & Safety Professional be consulted to provide quality assurance of the mold abatement. The Guideline also recommends that a Competent Supervisor be utilized for all medium and large projects.

For the clearance of the project, the Guideline recommended a post-remediation visual inspection and sampling (air & surface).

The Guideline also contains guidance on bird and bat droppings cleanup.

5) OSHA – A Brief Guide to Mold in Workplace (March, 2010)

http://www.osha.gov/dts/shib/shib101003.html

OSHA makes it clear that this Guide is NOT a Standard or Regulation. In the Guide, it states that 'Concern about indoor exposure to mold has increased along with public awareness that exposure to mold can cause a variety of health effects and symptoms, including allergic reactions.'

There are also Sections in the Guide on health effects, remediation methods, prevention, remediation plan, and personal protective equipment (PPE). It is noteworthy that the Guide does not recommend microbial sampling if visible mold is present.

6) NY State Toxic Mold Task Force (2010)

http://www.health.ny.gov/environmental/indoors/air/mold/task_force/docs/final_toxic_mold_task_force_report.pdf

The NYS Toxic Mold Task Force report states the following: 'Exposure to building dampness and dampness-related agents including mold has been recognized nationally and at the state and local level as a potential public health problem.'

- 'Evidence does not exist supporting clear distinctions between a category of "toxic mold" species versus other "non-toxic" mold species or between "toxic mold" health effects and health effects associated with other molds.'
- 'The development of reliable, health-based quantitative mold exposure limits is not currently feasible due to a number of technical challenges.'

7) ACOEM Mold Position Statement (February, 2011)

http://www.acoem.org/AdverseHumanHealthEffects_Molds.aspx

The American College of Occupational and Environmental Medicine (ACOEM) published an update of their 2002 Mold Position Statement in February, 2011. The 2002 Position Statement had been controversial with regards to the effect on human health of airborne mold spores and mycotoxins, and this update continues in the same vein. The 2011 Mold Position Statement states the following:

- 'Molds and other fungi...effect human health in three processes: 1) allergy, 2) infection, 3) toxicity'
- 'Current scientific evidence does not support the existence of a causal relationship between inhaled mycotoxins in home, school, or office environments and adverse human health effects.'
- 'To reduce the risk of developing or exacerbating allergies, mold should not be allowed to grow unchecked indoors.'
- 'There is scientific evidence that in certain cases, molds and other fungi may adversely affect human health, and mold has been associated with health issues ranging from coughs to asthma to allergic rhinitis.'

8) ASTM - Standard Guide for Assessment of Fungal Growth in Buildings (2011)

http://www.astm.org/Standards/D7338.htm

The American Society of Testing and Materials (ASTM) published a peer-reviewed Standard Guide in 2011. This Guide provides the IAQ practitioner with practical guidance regarding the inspection, evaluation and assessment of mold in a variety of buildings. For ASTM members, the Guide costs $40.00 USD.

The Guide states the following:

- 'Minimum steps and procedures for collecting background information on a building in question, procedures for evaluating the potential for moisture infiltration or collection, procedures for inspection for suspect fungal growth, and procedures beyond the scope of a basic survey that may be useful for specific problems.'
- 'Assessments for fungal growth may be useful wherever fungal growth is suspected, excess moisture has been present or when there are concerns regarding potential fungal growth.'
- 'Applicable to buildings including residential (for example, single or multi-family), institutional (for example, schools, hospitals), government, public assembly, commercial (for example, office, retail), and industrial facilities.'

9) UFGS – Mold Remediation Specification (May, 2011)

http://www.wbdg.org/ccb/DOD/UFGS/UFGS%2002%2085%2000.00% 2020.pdf

The Unified Facilities Guide Specifications (UFGS) are published periodically by a consortium of US governmental agencies, including the Naval Facilities Engineering Command (NAVFAC), the US Army Corp of Engineers (USACE), the Air Force Civil Engineering Support Agency (AFCESA), and the National Aeronautics and Space Agency (NASA). The UFGS Mold Remediation Specification is periodically updated by NAVFAC with the approval and participation of the other governmental agencies. The May, 2011 version is the most recent edition of the specification.

The specification is divided into various sections and appendices, including:

- Section 3.8.3.1 Clearance – clearance criteria for the majority of mold remediation projects. Clearance will be based on visual assessment (all visible mold removed, all visible dust removed, based on a "white glove" test) by Contracting Officer. "White glove" test shall consist of wiping the surface with a clean cloth of color suitable to reveal expected type of dust. For most surfaces, a white cloth is suitable. For gypsum wall board (GWB) dust, a dark cloth may be more appropriate.
- b. Failed remediation areas will be recleaned and the air filtration units (AFU's) kept in operation another 12 hours, followed by another visual assessment. Subsequent failures will follow the same routine until a pass condition is secured.

In the definitions section, the criteria for a Microbial Assessor are listed. It is noted that American Council for Accredited Certification (ACAC), American Board of Industrial Hygiene (ABIH) and the Board of Certified Safety Professionals (BCSP) accreditation are acceptable to be listed as a Microbial Assessor. Also, an individual can list two (2) years of microbial work experience with a Bachelors' degree engineering, architecture, building construction, occupational health, microbiology, occupational safety, or a related natural or physical science. An individual with four (4) years of work experience and an Associate's degree with a concentration in environmental, natural or physical sciences can also be qualified as a Microbial Assessor.

Appendix B: Mold Remediation Clearance Criteria for Buildings Housing Sensitive Populations

- Only for or mold remediation projects in buildings that will be occupied by sensitive and/or high risk populations, such as hospitals, child care centers, certain treatment centers, or when specified by the local medical support staff.
- Visual Assessment: 'White Glove' test. No visible mold or dust.
- Surface Sampling: A minimum of 5 samples per 93 square meters of (1000 square feet) of gypsum wallboard removed shall be collected.
- Air Sampling: Andersen N-6 single stage samplers. Compare results of remediation area samples with the results of samples collected in outdoor air [and adjacent areas providing makeup air].

10) CDC – Dampness and Mold in Buildings (May, 2011)

http://www.cdc.gov/niosh/topics/indoorenv/mold.html

The US Center for Disease Control (CDC) published this webpage on dampness and mold under its 'Indoor Environmental Quality' website. The following is stated on the site:

- 'Certain molds are toxigenic, meaning they can produce toxins (mycotoxins), but the molds themselves are not toxic, or poisonous. Contradicting research results exist regarding whether toxigenic mold found indoors causes unique or rare health conditions such as bleeding in the lungs. Research is ongoing in this area.'
- Health problems associated with excessive damp conditions and mold include: allergies, hypersensitivity pneumonitis and asthma.
- The document refers to NYC Mold Guidelines regarding mold remediation.

11) ASHRAE Mold Position Statement (2011-2)

This Position Statement is under development at this time. It is expected to be published in the second quarter of 2012. It will focus on the mechanical engineering concerns in heating, ventilation and air conditioning (HVAC) systems that can result in mold growth. Although health effects will be mentioned, it will not be the focus of the Position Statement.

12) AIHA – Revised Facts of Mold for Consumers and Professionals (2011)

http://www.aiha.org/news-pubs/newsroom/Documents/Facts%20About%20Mold%20December%202011.pdf

This AIHA webpage is an update of the 'Facts about Mold' document first published by AIHA in 2003. The document is split between 'Facts' for consumers and for professionals.

The document includes:

- Consumers: Focus on mold in homes and the methods to remediate the mold and fix the water intrusion issue.
- It states: 'For health outcomes, there are no available exposure assessment methods that provide useful information on individual health assessments. This is primarily because the response of each person to a chronic mold exposure is unique'.
- 'Some of these fungi produce toxic metabolites (mycotoxins), and almost all molds that grow in the built environment can produce triple helical glucan, both of which are toxic to lung cells. Many studies in appropriate laboratory animals have demonstrated that very low exposures of these compounds can result in inflammation. The health effects of breathing mycotoxins indoors are not well understood and they continue to be under study. This research is done to better understand why epidemiological studies consistently show increased asthma among occupants of damp buildings not associated with atopy.'
- Professionals: Description of the steps to be taken to inspect a building for possible mold contamination.
- 'Air, surface or bulk sampling may not be necessary, depending on the goal of the investigation. If visible mold is present, then it should be remediated, regardless of what species are present or whether or not samples have been collected. In situations where visible mold is present and there is a need to have the mold identified, surface or bulk sampling may be warranted. In specific instances, such as cases where health concerns are an issue, litigation is involved, or the source(s) of contamination is unclear, sampling may be considered as part of a building evaluation.'

13) Canadian Municipal MGO Guidelines (2008-2011)

The Ottawa MGO Guidelines were issued in April, 2010. This Guideline will be used as an example of the type of Guidelines being issued by municipalities in Canada regarding the investigation, the evaluation and the remediation of marijuana grow operations (MGO's) found in primarily in residences.

All environmental assessment reports require that both a P. Eng. (Professional Engineer) and an Air Quality Consultant (usually a CIH) stamp and sign all reports.

Pre-remediation investigation and report must include:

- Structural systems.
- Electrical systems – entire system.

- Mechanical systems – Heating, Ventilation and Air-conditioning, plumbing and sewage.
- Geotechnical assessment where foundations have been exposed to unheated conditions, or as directed, in relation to residue disposal onto lands or into septic systems.
- Air Quality - environmental molds and moisture, in accordance with the City's Guidelines.
- Use & Occupancy – Occupancy Requirements of the Building Code.

Post-Remediation Assessment - Written Guidelines for 'Clearance' -

- Up to 150 CFU/m^3 is acceptable if there is a mixture of species reflective of the outdoor air spores.
- Up to 500 CFU/m^3 is acceptable in summer if the species present are primarily Cladosporium or other tree and leaf fungi.
- The visible presence of fungi in humidifiers and on ducts, moldy ceiling tiles and other surfaces requires investigation and remedial action regardless of the airborne spore load.

Unwritten Guidelines for Clearance -

- 'Significant Numbers' – Greater than 1300 CFU/m^3 for total spores, greater than 750 CFU/m^3 for Asp/Pen-like spores.

14) NIOSH – Preventing Occupational Respiratory Diseases from Exposures caused by Dampness in Office Buildings, Schools and other Non-Industrial Buildings (March, 2011)

http://www.cdc.gov/niosh/docket/archive/pdfs/NIOSH-238/0238-033011-draft.pdf

This document is currently in review. It is very similar to CDC document, 'Dampness and Mold in Buildings', reviewed previously in this document. This document includes case studies involving extensive mold growth. It also mentions sarcoidosis as possibly related to dampness and mold. Finally, it discusses Preventative Building Design, Construction and Commissioning as a method to prevent mold growth in buildings.

The draft document states: 'Exposures from building dampness and mold have been associated with respiratory conditions, asthma, hypersensitivity pneumonitis (HP), and sarciodosis in research studies.'

It is expected that this document will be finalized in 2012.

15) State Of California Department of Public Health - Statement on Building Dampness, Mold, and Health (September, 2011)

http://www.cal-iaq.org/

This recent document from California updates the 2005 Report to the Legislature provided by CDPH on mold. In that report, the following was stated: 'sound, science based PEL's for indoor molds cannot be established at this time'.

It states: 'CDPH has concluded that the presence of water damage, dampness, visible mold, or mold odor in schools, workplaces, residences, and other indoor environments is unhealthy. We recommend against measuring indoor microorganisms or using the presence of specific microorganisms to determine the level of health hazard or the need for urgent remediation. Rather, we strongly recommend addressing water damage, dampness, visible mold, and mold odor by (a) identification and correction of the source of water that may allow microbial growth or contribute to other problems, (b) the rapid drying or removal of damp materials, and (c) the cleaning or removal of mold and moldy materials, as rapidly and safely as possible, to protect the health and well-being of building occupants, especially children.'

'Human health studies have led to a consensus among scientists and medical experts that the presence in buildings of (a) visible water damage, (b) damp materials, (c) visible mold, or (d) mold odor indicates an increased risk of respiratory disease for occupants. Known health risks include: the development of asthma, allergies, and respiratory infections; the triggering of asthma attacks; and increased wheeze, cough, difficulty breathing, and other symptoms.'

16) LEED 2012

This revision to the LEED program is currently in its 'Public Comment' period. It is not expect that there will be any specific requirements for mold in this revision. However, LEED 2012 will require that an IAQ Management Plan for construction and pre-occupancy phases of the project be developed. Also, all Health Care projects will require a moisture control plan. For the LEED EQ 3.2 (Pre-Occupancy IAQ Measurements), it is anticipated that the new requirements will be that all airborne contaminants will be below 2.5% of the applicable PEL's, TLV's or REL's. This credit will be now worth two (2) points instead of the current one (1) point. (See: http://www.usgbc.org/DisplayPage.aspx?CMSPageID=1988; 08.12)

CONCLUSIONS

In review of the various guidelines, standards and position statements that have been published since the AIHA Green Book was published in 2008, it is clear that there is no consensus in these documents as to the 'Best Practices' for the inspection, assessment and remediation of dampness and mold in buildings. On the contrary, it is clear that there is a variety of approaches to mold contamination issues, and that these approaches are quite different, depending on the entities that are issuing the document.

Various governmental agencies and professional associations have issued position statements, guidance documents, standards and specifications regarding mold in the past three (3) years. However, these documents, in some cases, contradict each other, and they do not agree on the best approach to dampness and mold in buildings.

The following are examples of mold-related issues where different approaches and methods are promulgated:

1) Microbial Air Sampling –

- It is not generally recommended for initial building surveys, except for municipal MGO guidelines.
- Microbial air sampling is not recommended for post-remediation verification by AIHA, NYC DOH, and OSHA. However, microbial air sampling is required for some projects (UFGS, MGO's, and EACO) for 'clearance' as part of a post-remediation verification.

2) Health Effects –

- It is agreed upon by most of the cognizant authorities that mold exposure can cause adverse health effects, including allergies, hypersensitivity pneumonitis and asthma.
- It is not agreed upon that any 'toxic effects' can be associated with exposure to airborne mold spores in buildings. In particular, the ACOEM Position Statement specifically states that 'Current scientific evidence does not support the existence of a causal relationship between inhaled mycotoxins in home, school, or office environments and adverse human health effects.'
- Most of the documents do agree that further research is needed to determine what 'toxic effects', if any, are associated with exposure to mold spores.

3) Sampling Results Guidelines-

- Some have specific 'numbers' (CFU/m^3) for air samples (UFGS, MGO municipal guidelines) and surface samples.
- Other Guidelines and Standards discourage sampling but recognize that sampling will be completed for 'clearance'.
- Bottom line: it is necessary that an Air Quality Professional must know the requirements for each project in each jurisdiction.
- Also, the Air Quality Professional must decide in advance what criteria will be used for each project, and inform all involved, including contractor, prior to the start of the project.

4) Canadian and US Approaches – Differences

- One critical difference between Canadian and US approaches to microbial contamination is that, in Canada, there is an acceptance and understanding that mold and damp buildings are associated with a variety of adverse health symptoms.
- Canadian 'guidelines' issued by governmental entities are considered to be enforceable as law. For example, the MGO 'guidelines' issued by Canadian municipalities are enforced by the municipal building officials, or the health ministry's officials. In the US, 'guidelines' (even those issued by governmental entities such as OSHA) are not generally 'enforceable'.
- Legal cases in Canada involving mold, especially those with claims of adverse health effects, are very few. Canadian law generally does not allow toxic torts, particularly those of a class action. In the US, there are a considerably more lawsuits, including class actions, for both bodily injuries and property damage due to the presence of mold in a building.
- Practitioners in Canada generally follow the 'guidelines' issued regarding mold evaluation and remediation as issued by cognizant authorities such as Health Canada. In the US, there is no consensus on what are the best guidelines or standards to follow regarding mold inspections, evaluations and remediation. Hence there are a wide variety of 'guidelines' and standards issued by federal, state and municipal governments, as seen by this cursory review of the current documents issued since 2008.

Other Guidelines Include:

- American Conference of Governmental Industrial Hygienists "Bioaerosols"
- US EPA: "Mold Remediation in Schools and Commercial Buildings"
- AIHA: "Report of the Microbial Growth Task Force" and "Assessment, Remediation and Post-Remediation Verification of Mold in Buildings"
- US Department of Housing and Urban Development: "Guide of Housing in Indian Countries"
- US Occupational Safety and Health Administration: "Brief Guide in Mold in the Workplace"
- Canadian Construction Association: "Mold Guidelines for the Canadian Construction Industry"
- Institute of Inspection, Cleaning and Restoration: "Standard and Reference Guide for Professional mold Remediation"

REFERENCES

- New York City Department of Health (NYCDOH) Microbial Guidelines. updated 2008. http://www.nyc.gov/html/doh/html/epi/moldrpt1.shtml

- General Accounting Office (GAO) – Indoor Mold Report. 2008. http://www.gao.gov/new.items/d08980.pdf
- World Health Organization (WHO) Report – Dampness and Mold. 2009. http://www.euro.who.int/__data/assets/pdf_file/0017/43325/E92645.pdf
- Environmental Abatement Council of Ontario (EACO) Mold Abatement Guidelines. updated 2010. http://www.eacoontario.com/pdf/2010/eaco_mould-abatement-guidelines_book.pdf
- US Occupational Safety and Health Administration (OSHA) – A Brief Guide to Mold in the Workplace. March, 2010. http://www.osha.gov/dts/shib/shib101003.html
- NY State Toxic Mold Task Force report. 2010. http://www.health.ny.gov/environmental/indoors/air/mold/task_force/docs/final_toxic_mold_task_force_report.pdf
- American Society of Testing and Materials (ASTM) - Standard Guide for Assessment of Fungal Growth in Buildings. January, 2011. http://www.astm.org/Standards/D7338.htm
- American College of Occupational and Environmental Medicine (ACOEM) Mold Position Statement. March, 2011. http://www.acoem.org/AdverseHumanHealthEffects_Molds.aspx
- Unified Facility Guide Specifications (UFGS) – Mold Remediation Specification. May, 2011. http://www.wbdg.org/ccb/DOD/UFGS/UFGS%2002%2085%2000.00%2020.pdf
- Center of Disease Control (CDC) – Dampness and Mold in Buildings. May, 2011. http://www.cdc.gov/niosh/topics/indoorenv/mold.html
- American Society of Heating, Refrigeration, and Air Conditioning Engineers (ASHRAE) – Mold Position Statement. 2011;2.
- American Industrial Hygiene Association (AIHA) – Revised Facts of Mold for Consumers and Professionals. January, 2012. http://www.aiha.org/news-pubs/newsroom/Documents/Facts%20About%20Mold%20December%202011.pdf
- Canadian Municipal Marijuana Grow Op (MGO) Guidelines. 2008-2011.
- National Institute of Occupational Safety and Health (NIOSH) – Preventing Occupational Respiratory Diseases from Exposures caused by Dampness in Office Buildings, Schools and other Non-Industrial Buildings. March, 2011.
- http://www.cdc.gov/niosh/docket/archive/pdfs/NIOSH-238/0238-033011-draft.pdf
- State of California - Statement on Building Dampness, Mold, and Health. September, 2011. http://www.cal-iaq.org/

DEFINING "CLEAN" IN TERMS OF THE UNSEEN FRACTION: A REPRESENTATIVE MARKER IN SCHOOLS

Richard Shaughnessy, Eugene C. Cole, Ulla Haverinen-Shaughnessy

INTRODUCTION

Cleanliness requirements for public buildings, or specific operations within such buildings, often require facilities to be kept in a "clean" and "sanitary" condition, as typically determined by visual inspection. Such an assessment, however, remains inadequate concerning the removal of many unseen and unwanted pollutants (i.e. biological, chemical, particulate residues), thereby failing to reduce the burden of exposure and health risk to the building's occupants.

Dusts that accumulate in school buildings from track-in on shoes and clothing, fallout from HVAC systems, and from self-generating sources, such as insects and classroom animals, can trigger a variety of allergy and asthma responses in children. Asthma is now recognized as the leading cause of school absences in the US; and studies have shown a variety of potential triggers present in dusts collected from carpeted and hard surface floors in schools, to include pollen, as well as mold, cat, dog, mite, and cockroach allergens, among others (Smedje and Norback, 2001; Abramson et al., 2006). Increased efforts at improved cleaning of floors and desks in schools have been shown to reduce upper respiratory symptoms (Walinder et al., 1999).

Research has indicated that the rapid spread of viral disease in crowded classrooms is associated with the cleanliness of high contact inanimate objects. Inadequate cleaning and maintenance practices in schools, compounded by the effects of emerging infectious disease agents and climate change factors, can severely alter the school building ecosystem and put students' health at increased risk. CDC Guidance on Influenza in schools (2009-2010) states that "school staff should routinely 'clean' areas that students and staff touch often…" The challenge in setting practitioner-based cleaning protocols today is more related to how we define "clean" as it applies to health. The routine cleaning protocol based on visual assessment remains inadequate concerning the removal of the unseen fraction, thereby failing to reduce the burden of exposure and health risk to the building's occupants.

Preliminary research has identified the measurement of adenosine triphosphate (ATP) as a rapid, real-time, quantitative, and economical approach to the characterization and ultimate reduction of potentially harmful contamination on a variety of surfaces and materials. ATP is the energy driver for biological systems and can be measured through an enzymatic luciferin/luciferase reaction detected and quantified as bioluminescence. The method converts ATP into a light signal which is measured by an instrument that provides a quantitative measurement of ATP from biomass in Relative Light Units (RLU). Whereas ATP does not directly monitor viruses in an environment, it does measure a mixture of biological forms that indicate human cellular material, along with bacteria and fungi. Such material includes epithelium from upper respiratory mucus membranes (mouth, throat, nasal passages) from saliva and exudates and associated material from coughs and sneezes from persons with viral, as well as bacterial infections. Viruses are associated with living cells as viruses need them to replicate. Again, ATP is an overall generic marker of biological contamination, and it allows us to monitor potential viral contamination (from viral infections) "indirectly". ATP has more recently been used as a marker for contaminant loads in both hospital settings and the food industry. This paper addresses preliminary measurements of ATP in occupied school conditions.

MATERIAL AND METHODS

Thirty-five (35) elementary schools from a 70-school district in the Southwestern United States were randomly selected to be available for the surface sampling effort. Determination of the final number of actual schools necessary for the district surface sampling study was premised upon initial measurement in an 8-school subset of the 35 schools. Based on resultant data from the eight schools, power analyses revealed the adequate number of schools necessary to assess surface contamination pre- and post- cleaning to be 25-30. The final number included in this study was 27 schools.

Surfaces were measured before and after cleaning by three commercially available and widely used ATP systems. Sampling included measurements from four types of surfaces, including classroom desks, cafeteria tables, bathroom stall doors, and sink areas. In order to observe levels of total culturable bacterial contamination as compared to ATP measurements, samples from each surface/site were also taken using replicate organism detection and counting agar plates (RODAC). A total of 8601 measurements (6480 ATP, 2121 RODAC) were taken throughout the 27 schools. Standardized pre-cleaning sampling (both ATP and RODAC) was done on each surface before cleaning, followed by the use of an EPA-registered one-step cleaner/disinfectant used in conjunction with microfiber cleaning cloths. For the post-cleaning measurements, surfaces were first wiped with a Microfiber towel, and then sprayed with the cleaner/disinfectant routinely used in the school district, covering the desired area. The surface was then wiped until dry with another Microfiber cloth, after which post-cleaning ATP and RODAC samples were collected.

RESULTS AND DISCUSSION

Analysis of the 6480 ATP samples has shown the potential for ATP sampling to serve as a means of defining and quantifying cleaning effectiveness based upon a percent reduction approach, and similar reduction data based on culturable bacteria showed the ATP results indicative of confirming sanitization. Compared to pre-clean measurements, post-cleaning data showed, on logarithmic scale, approximately one order of magnitude (90%) reduction, as shown in Figure 1.

FIGURE 1. Collective all-surfaces pre- and post-cleaning results for ATP and RODAC data.

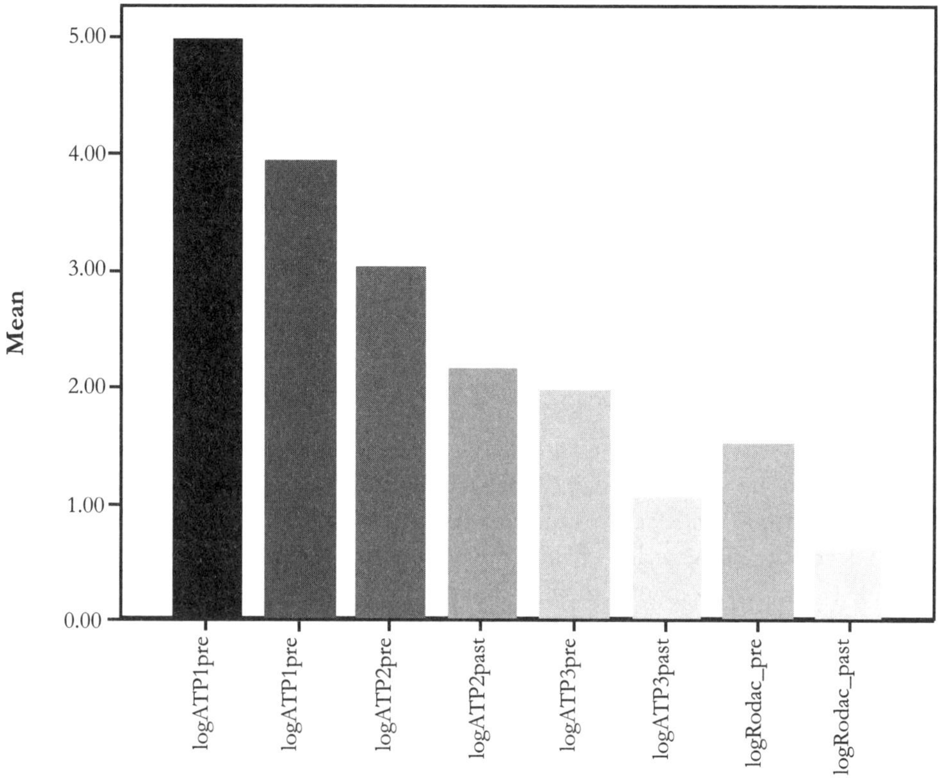

The data also revealed a reduction of culturable bacteria concomitant with ATP reduction after cleaning, indicating that viral load (associated with human cellular material) has also been reduced. Specific ATP values/ranges appear to be a function of both surface material type and location within a school building, and further study is expected to confirm unique values for such critical compartments of the school building ecosystem. The log-normal distributions of pre and post cleaning values for the three ATP systems and RODAC outcomes are to be reported in a subsequent paper by the investigators.

In addition, while the cleanliness of surfaces in school buildings and the standardization of cleaning effectiveness are critical to the building ecosystem, it is recognized that they are inextricably linked to other fundamental indicators of indoor environmental quality (IEQ). In this regard, ventilation has been shown as important for student exposures to airborne indoor pollutants and perhaps poor academic performance (Shaughnessy et al., 2006; Haverinen-Shaughnessy, 2011). It remains prudent that when researching surface cleanliness in school buildings, that cleanliness and the overall "healthiness" of the school environment be interpreted relative to the various factors; this includes parameters such as temperature and humidity, ventilation, settled dust, dampness and mold, among others.

CONCLUSIONS

The collected data provide a basis for understanding the distribution of ATP RLU values that can be expected pre- and post- cleaning in a school. It is anticipated that "normal" ranges can be established for each surface type, and that these may be used in the routine monitoring of cleaning effectiveness in school buildings. The data also revealed a reduction of culturable bacteria coincident with ATP reduction after cleaning, which indicates that viral load (associated with human cellular material) has also been reduced in the process. Specific ATP values/ranges appear to be a function of both surface material type and location within a school building, and further study is expected to confirm unique values for such critical compartments of the school building ecosystem.

ACKNOWLEDGEMENTS

This study was supported by a collaborative grant from the Cleaning Industry Research Institute International (Albany NY, USA) and ISSA-The Worldwide Cleaning Industry Association (Lincolnwood IL, USA) as part of their initiative toward development of a clean standard for schools K-12. We express our immense gratitude to the facility management personnel of the participating school district in the southwestern United States. We also thank Alesia Bailey and Randy Smith (University of Tulsa, Tulsa, OK, USA), and Keith Leese and Cathy Richmond (LRC Indoor Research & Testing, Cary, NC, USA), for their dedicated efforts, and without whom this project would not have been successful.

REFERENCES

- Abramson SL, Turner-Henson A, Anderson L, Hemstreet MP, Tang S, Bartholomew LK, Joseph CLM, Tyrrell S, Clark NM, Ownby D. Allergens in school settings: Results of environmental assessments in 3 city school systems. *J Sch Health.* 2006;76(6):246-249.
- Haverinen-Shaughnessy U, Moschandreas D, Shaughnessy R. "Association between Sub-Standard Classroom Ventilation Rates and Students' Academic Achievement". *Journal of Indoor Air,* Feb. Issue. 2011.
- Shaughnessy RJ, Haverinen-Shaughnessy U, Nevalainen A, Moschandreas D. A preliminary study on the association between ventilation rates in classrooms and student performance. *Indoor Air.* 2006;16(6):465-468.
- Smedje G, Norback D. Irritants and allergens at school in relation to furnishings and cleaning. *Indoor Air.* 2001;11(2):127-133.
- Walinder R., Norback D, Wieslander G, Smedje G, Erwall C, Venge P. Nasal patency and lavage biomarkers in relation to settled dust and cleaning routines in schools. *Scand J Work Environ Health.* 1999;25(2):137-143.

EFFECTIVE RISK COMMUNICATION CONCERNING ENVIRONMENTAL CHANGE WITH COMMUNITIES AND PATIENTS EXPOSED TO EXCESSIVE INDOOR MOISTURE OR MOLD

Paula Schenck and Robert DeBernardo

ABSTRACT:

Risk communication with many different stakeholders is required to facilitate environmental change when indoor moisture is a concern. Moisture indoors is associated with respiratory disease, especially asthma and corresponding economic burden (Fisk et al., 2007; Mudarri, Fisk, 2007). Intervening on mold and moisture indoors provides an opportunity to improve symptoms, possibly prevent asthma initiation, and address a contributing factor to the disparity in respiratory disease (US asthma prevalence increases with decreasing economic status from 6.4% (richest Americans) to 10.3% (those at or below poverty levels) (Moorman et al., 2007)). When there is moisture damage, there are a number of stakeholders, i.e. the owner of the building, the occupants, and the community. Each has vested interests which may conflict. It is the goal of the risk communicator to inform all the stakeholders of the health and building risks involved, the options and costs, and to propose and negotiate a mutually agreeable solution. Because of the complexity of the exposure and effect on health (Heseltine, 2009), the cost associated with intervention, differences and controversy around published guidelines and policies, and the emotional nature that characterizes this topic, mold and moisture health risks are challenging to explain and manage. Risk communication requires presentation of complex information and consequences in understandable and relatable language. To explain the action of environmental response as an appropriate and critical intervention to prevent illness, the communicator must embody an understanding of risk communication principles as well as knowledge of the subject. It is especially challenging because treatments for individuals and pharmacological remedies are efficacious for symptom relief but may not address probable contributing factors to the cause; and the underlying condition may worsen even while undergoing symptomatic treatment.

Risk communication strategies are discussed with consideration of problem-solving experience with school and office communities and clinical case studies. As our understanding of the health risks associated with chronic moisture in buildings deepens and concerns for patients and the public increase, physicians and health officials will be increasingly tasked to determine the scope of the public health problem related to mold and moisture in indoor environments and to communicate information about the risk to individuals and groups. Two case studies, one involving a school community and the other reflecting the iterative communication of a practitioner with a patient are discussed where risk communication supported environmental change. Risk communication skills are important in efforts to include environmental change in homes, schools and offices as part of patient treatment and community response.

INTRODUCTION:

Health and exposure to mold and moisture in indoor environments is a growing subject of interest and concern to public health professionals, medical providers and the public. State and local health officials are often the deciding voice on the significance of hazards to the public, especially in school communities. They are tasked to determine the scope of the public health problem and to communicate information about the health risk. Increasingly patients with symptoms and health concerns that require understanding about mold and moisture and consideration of environmental factors, turn to medical professionals for direction. This is especially so after Katrina flooding and with the increasingly frequent weather events that result in building moisture intrusion. Treatments for individuals and pharmacological remedies provide symptom relief. However while providing immediate comfort, they often do not address probable contributing factors to the cause and the underlying condition may worsen. Reducing exposure to environmental agents is a critical part of medical treatment and today's health providers need to be skilled risk communicators and knowledgeable about environmental contributors to illness. This is especially important given the limited time allocated for patient interactions in the modern clinical setting. There are many factors that add challenge to communications when indoor mold is a concern. The assignment of environment-relatedness to illness is complicated and relies on experience and judgment. Some quantitative information may be used, but the best practice for identification of environmental factors largely follows qualitative methods (Prezant et al., 2008). The background topics are complex (air pathways, exposure dose, health effects) and some health effects of concern attributed to mold exposure, reported by patients are under debate in the medical literature (Bush et al., 2006; Craner 2008), in the general media and on the web. Published guidelines and policies on mold/moisture and buildings abound from: medical societies i.e. American Academy of Pediatrics, American College of Preventive Medicine, the American College of Occupational and Environmental Medicine, the American Academy of Allergy, Asthma and Immunology; public health agencies i.e.-National Institute of Medicine (IOM 2004; IOM 2011), World Health Organization (Heseltine, 2009), and Centers for Disease Control and Prevention (Brandt et al., 2006; Brown et al., 2004); environmental agencies i.e. US Environmental Protection Agency (EPA); and trade and professional organizations i.e. American Industrial Hygiene Association (Prezant et al., 2008) and American Society of Heating, Refrigeration and Air Conditioning Engineers (Mudarri, 2009). Each seeks to provide information from the perspective of the organization to further public health, environmental quality and/or address clinical needs. These guidance documents differ as to the urgency to address the environment and the complexity of this debate is significant. The cost of remediation and assignment of responsibility has fueled interest beyond those directly affected, to the insurance, consulting (medical and environmental), and legal communities. In some litigious situations where communication is controlled, the opportunity for effective discussion of risk may be limited. An approach that utilizes sound risk communication principles is needed to effectively discuss the risk and appropriate actions.

METHODS:

The Center for Indoor Environments and Health at the University of Connecticut Health Center utilizes risk communication strategies in problem solving with school and office communities and in facilitating environmental change as part of patient treatment when mold and moisture intrusion in indoor environments is indicated as a contributor to illness. Methods used have been adapted from literature on risk communication and reflect feedback from practitioners and the authors' experience.

The industrial hygiene field manual for the Navy and Marine Corps begins with this definition of "Risk Communication is an interactive process or exchange of information and opinions among interested parties, or stakeholders, concerning a risk, potential risk, or perceived risk to human health, safety, or the environment." Their stated objective goes beyond "educating personnel and their families about mold and the potential health effects" to "engaging workers and residents in efforts to prevent mold growth and providing timely and appropriate information in the event of a mold "communication crisis" during which workers or residents are experiencing health effects or are concerned" (Navy and Marine Corps Public Health Center, 2011). Table 1, adapted from Health Services Administration, 2002 Communicating in a Crisis: Risk Communication Guidelines for Public Officials page 20 demonstrates some factors that act to determine the acceptability of the risk (USHSA 2002).

Table 1.

Risks perceived to...	...Are more accepted than risks perceived to...
be voluntary	be imposed
be under an individual's control	be controlled by others
have clear benefits	have little or no benefit
be distributed fairly	be unfairly distributed
be natural	be man-made
be statistical	be catastrophic
be from a trusted source	be from an untrusted source
be familiar	be exotic
affect adults	affect children

Adapted from Health Services Administration, 2002

The effective communicator will empathize while demonstrating competence, expertise, honesty, openness and commitment (USHSA 2002). There are a number of important factors to consider in engaging the stakeholders and in crafting the message. The risk communicator must strike a delicate balance between informing the stakeholders of the health risks without alarming them to incite a point of panic. Once all interested parties, the information that they need and their perceptions about risk are identified, a plan will discuss the modes of communication. Are there committee and/or town meetings? Bulletin boards-newsletters, list serves and web sites? How will the plan allow for all their voices (no matter how broad the concern or whether "real" or perceived)? How will timeliness be balanced with accuracy and completeness? How will stakeholders be involved with an eye to build support and share ownership of solutions? How will communication be sustained and evaluated, corrected and followed through? How will the media be included? (Lin, Petersen 2007) Table 2 provides background resources used in planning the risk communication encounters.

Table 2

- New York City Guidelines on Assessment and Remediation of Fungi in Indoor Environments
  http://www.nyc.gov/html/doh/html/epi/moldrpt1.shtml#communication
- American Industrial Hygiene Association, 2008, IMOM08-679 Recognition, Evaluation, and Control of Indoor Mold.
  http://iweb.aiha.org/iweb/Purchase/ProductDetail.aspx?Product_code=IMOM08-679 page 186-9
- U.S. Environmental Protection Agency
  IAQ reference guide- Effective Communications
  http://www.epa.gov/iaq/schools/tfs/guide3.html
  Risk Communication in Action: The Tools of Message Mapping
  http://www.epa.gov/nrmrl/pubs/625r06012/625r06012.pdf
- Agency for Toxic Substances and Disease Registry
  A Primer on Health Risk Communication Principles and Practices
  http://www.atsdr.cdc.gov/HEC/primer.html
- Canadian Construction Association Standard CCA 82-2004, Mould Guidelines for the Canadian Construction Industry
  http://www.cca-acc.com/documents/cca82/cca82.pdf page 30-1
- Health Services Administration, 2002
  Communicating in a Crisis: Risk Communication Guidelines for Public Officials
  http://www.riskcommunication.samhsa.gov/RiskComm.pdf
- Navy and Marine Corps Public Health Center
  Industrial Hygiene Field Operations Manual Chapter 13 Risk Communication and Mold
  http://www-nehc.med.navy.mil/downloads/IH/IHFOM/IHFOM_CH13-5.pdf

RESULTS:

Risk communication played a crucial role in these two case studies: one involved a school building community where a local health director needed assistance to advise a school district in which a group of parents had called for the closing of a public school, and the second describes a patient with significant building-related respiratory disease, who worked in an office building with a history of water incursion.

Risk communication in a school setting

A school community in a poor neighborhood is in crisis by mid October. The elementary school serves about 400 children, a quarter of them poor enough to qualify for reduced fee lunch in stark contrast with other sections of the affluent New

England community. The situation is emotional; health symptoms and illnesses among staff and children have been significant. Asthma episodes have sent children and the librarian to the hospital in ambulances a number of times since the beginning of the fall school year. One child had been hospitalized. A few of the teachers are under care of an allergist who with concern over mold exposure has removed some of his patients from work including a new teacher who is pregnant and another who has taught in the school for many years.

Over the previous summer a new roof was constructed over part of the building but the gutters were improperly reconnected. Water streamed into the building after a storm in mid-September which resulted in water intrusion in the partitions and ceilings, and into three classrooms and the cafeteria. After the carpet in the media center was shampooed in August, the room was closed up without air exchange for one week and the media center carpet became visibly moldy. The school facility staff had completed a number of actions they thought would remediate the mold and moisture damage: windows, casings, carpet and flooring were removed from two classrooms; cafeteria columns were washed with bleach solution; media center carpet was removed; and the air handling system was cleaned. These construction activities occurred during school hours and although the specific rooms were closed to school activities, there were no containments or separation of the areas from occupants.

Environmental information is available in state agency and consultant reports that detail air sampling results that include *Stachybotrys* sp. at low levels in the hallway outside the water damaged classroom. *Cladosporium* sp., *Penicillium* sp., *Aspergillus* sp., *Alternaria* sp., and *Phoma* sp. are present with *Cladosporium* levels similarly measured inside and out, and with *Penicillium* levels significantly higher inside by the classrooms. This information, without adequate explanation of the significance, contributes to the community's overall anxiety. The superintendent has temporarily closed the building, arranged for other District schools to accommodate the children, and has enlisted the local health director to advise her on communications with the community and on next actions. The partially remediated class rooms have water damaged areas, but mold is not visible on surfaces. The media has arrived and is waiting for an announcement. Will the school reopen? Is it safe? The town government representative is calling for the school to permanently close. The risk communication challenges are many. -What do you advise the Health Director? How do you discuss the technical information? What are the meaningful messages for the parents? The community? The District?

The risk communication strategy employed multiple tiers of planned encounters. A team was formed to plan next steps and included members from all recognized stakeholders: school administration, parents, teachers, students, town government, consultants and health providers. Discussions were facilitated among the key stakeholders to establish mutual understanding of the breadth of concerns and agendas. The team organized a public meeting with technical and medical experts who answered questions forthrightly, admitted the difficulty in interpreting the environmental information and acknowledged the mistakes made in the remediation activities. Individuals with specific medical concerns were given the opportunity for private meetings with knowledgeable health experts. A building science consultant was engaged to assess the building and progress was reported on the District's web site. Although the consultant recommended an action plan which would have resulted in reopening the school, the school was permanently closed. The action initially thought of as a troubling option to some of the community had become generally acceptable in part because of the effective communication program.

The clinical setting:

This case study is presented to highlight an example of provider and patient risk communication concerning environmental change. Some descriptive clinical information is presented to describe the significance of the patient's health status. Clinical care (other than environmental change as part of the clinical treatment) is not addressed.

A 42 year old woman presented with: a 2 year history of sneezing and nasal congestion primarily while at work; dizziness, fatigue, and headaches occurring at work resolve in 1-2 hours after leaving for the day; and intermittent skin rashes on her forearms and upper chest. Non-sedating antihistamines improve her symptoms but she is concerned that the symptoms persist and she does not want to take a medication long term. The initial physical exam was unremarkable, and spirometry, a measure of breathing capacity, was normal. A slightly elevated erythrocyte sedimentation rate, a marker of inflammation, was the only laboratory abnormality.

She worked on an upper floor of a multi-storied office building that has had recurrent water incursion from windows, sliding doors, and the roof. The floors have been covered with carpet tiles that had become repeatedly wetted at the perimeter of the building, then cleaned and moved to other parts of the floor. Stained ceiling tiles have been replaced at various intervals over the past four years. Her symptoms exacerbated after renovation to replace the rugs and repair the water damage. *Stachybotrys chartarum* was isolated from wall samples. Her provider acknowledged her concerns and discussed health risks and possible environmental interventions-supporting her effort to be allowed to move her office to a location on another floor that had not suffered water damage. Initially, she did well. But then she had marked flare-ups of symptoms even though her work location was changed a number of times in efforts to reduce her exposures. She cleared completely during a vacation week. But over time her symptoms became increasingly severe and did not resolve as quickly after leaving work as before. She developed nasal congestion, sinusitis, cough and some wheezing at the end of each work day that at times persisted through the night and resolved only on weekends. Spirometry reflected a defect that improved with a bronchodilator, consistent with asthma. With the increased severity of symptoms, the health provider discussed whether she should continue to work in this building. The risk discussion now included consideration of potential consequences to her employment. With a concern over developing chronic illness, she transferred to a different building that had not experienced any water problems and her symptoms gradually improved. Sometime later, her employer asked her to move back to the building. The building owner had addressed some environmental concerns, but technical assessments fell short of concluding that all water sources had been addressed. Should this patient move back to the building? What should the discussion of the health risks and environmental interventions focus on now? Throughout the time that the provider cared for this patient, the provider acknowledged the difficulty in assigning building relatedness. However the actions of moving her office to various locations and monitoring her illness served to give stronger evidence of a building-related illness than her initial symptoms and pathology. Together the patient and provider developed a solution where work materials would be copied to allow her to continue work in the remote location without further exposure to suspected environmental agents.

CONCLUSIONS:

Risk communication tools serve health directors and health providers in managing the complexity of environmental exposure to moisture and mold. As our understanding of the health risks associated with chronic moisture in buildings deepens and concerns for patients and the public increase, physicians and health officials will be increasingly tasked to determine the scope of the public health problem related to mold and moisture in indoor environments and to communicate information about the risk to individuals and groups. Detailing examples where risk communication had been effective to encourage appropriate environmental responses supports efforts to include environmental change in homes, schools and offices as part of patient treatment and community response.

REFERENCES:

- Brandt M et al. Mold prevention strategies and possible health effects in the aftermath of hurricanes and major floods. *MMWR Recomm Rep*, 2006:55(RR-8): p. 1-27.
- Brown CM, Redd SC and Damon SA, Acute idiopathic pulmonary hemorrhage among infants. Recommendations from the Working Group for Investigation and Surveillance. *MMWR Recomm Rep*, 2004:53(RR-2): p. 1-12.
- Bush RK et al. The medical effects of mold exposure. *J Allergy Clin Immunol*, 2006:117(2): p. 326-33.
- Craner J. A critique of the ACOEM statement on mold: undisclosed conflicts of interest in the creation of an "evidence-based" statement. *Int J Occup Environ Health*, 2008:14(4): p. 283-98.
- EPA, U.S.E.P.A. Molds and Moisture. Available from: http://www.epa.gov/mold/.

- Fisk WJ, Lei-Gomez Q, and Mendell MJ. Meta-analyses of the associations of respiratory health effects with dampness and mold in homes. *Indoor Air*, 2007:17(4): p. 284-96.
- Heseltine ER., Jerome, ed. WHO guidelines for indoor air quality : dampness and mould World Health Organization: Copenhagen. 2009. 248.
- IOM, Climate Change, the Indoor Environment, and Health. Institute of Medicine of the National Academies, Washington, DC. 2011. p. 320.
- IOM, Damp Indoor Spaces and Health. Institute of Medicine of the National Academies: Washington, DC. 2004. p. 356.
- Lin I and Petersen D. Risk Communication in Action: The Tools of Message Mapping. 2007.
- Moorman JE et al. National surveillance for asthma--United States, 1980-2004. *MMWR Surveill Summ*, 2007:56(8): p. 1-54.
- Mudarri D and Fisk WJ. Public health and economic impact of dampness and mold. *Indoor Air*, 2007:17(3): p. 226-35.
- Mudarri D, ed. The Indoor Air Quality Guide: Best Practices for Design, Construction and Commissioning Available through http://www.ashrae.org/publications/page/IAQGuide. American Society of Heating, Refrigerating and Air-Conditioning Engineers, Inc. Atlanta, GA. 2009.
- Navy and Marine Corps Public Health Center, Industrial Hygiene Field Operations Manual Chapter 13.5 Risk Communication and Mold. 2011.
- Prezant BW, Miller DM; David J, ed. Recognition, Evaluation, and Control of Indoor Mold. 2008.
- USHSA, Communicating in a Crisis: Risk Communication Guidelines for Public Officials. 2002 November 15 2011; Available from: http://www.hhs.gov/od/documents/RiskCommunication.pdf. 2011.

CRITICAL ISSUES IN ART CONSERVATION AFTER A WATER INTRUSION EVENT: PITFALLS OF THE EMERGENCY RESPONSE

Karen H. Kahn

ABSTRACT

The procedure for responding to disasters involving water intrusion into a building containing cultural and heritage materials has become an area of increased interest among art preservation and conservation groups. Many articles have been written on the subject which influence, if not determine, how cultural institutions organize, prepare and conduct salvage and recovery after a disaster (Hawks et al, 2011). Little, if any, review has been made of the methods advocated for use by emergency response teams and how the many risks related to their work are managed.

The author of this article reviewed literature related to salvage and recovery from museums and other cultural institutions, and compared it to the industry standards followed by building water restoration experts. The author concludes, based on this review, that to the extent the conservationist literature recommends procedures for stabilizing the environment by removal of building materials and drying, its guidelines markedly depart from the well-established industry standards for water restoration. This article clarifies how, in the process of salvage and recovery, building contents are subject to increased damage from improper handling of hazardous building materials.

This paper discusses: 1) The Field Guide to Emergency Response (Heritage Preservation, 2006), a primary reference guide which cultural institutions rely upon during the emergency phase of salvage and recovery; 2) The methods which the Field Guide to Emergency Response proposes be utilized by cultural institutions to salvage and recover cultural and heritage materials [hereinafter "contents"] from a building which has sustained water damage; 3) How the methods utilized by cultural institutions compare with the published industry standards for responding to emergencies involving water damage and why the former methods lead to even greater damage; and 4) How managing the risk in the emergency response phase can achieve the goal of properly restoring a building and its contents when decisions are not controlled by the economics of the response.

INTRODUCTION

The critical relationship between building and contents restoration after a significant water intrusion event is poorly understood by cultural institutions, and their methods for salvage and recovery are unique to the building water restoration community. The methods utilized by cultural institutions appear to be largely based on the Field Guide to Emergency Response, whose recommended protocols are highly problematic, rather than published industry standards which provide guidance for responding to emergencies in water-damaged buildings. As discussed in this paper, self-help methods recommended by the Field Guide to Emergency Response have the potential to cause greater damage by the emergency responder's unknowing release of contaminants and the resulting spread of contamination beyond affected areas. While cost effective in the short term, self-help methods increase the cost of cleanup exponentially, and in the worst circumstances, allow contaminants to persist in the building environment and on contents.

These effects can be avoided if art conservators, owners of buildings, curators, collectors of art and insurers engage experts in the field of water restoration such as industrial hygienists, certified asbestos and lead consultants, water mitigation and remediation experts before deciding on a salvage and recovery strategy. Such communications are necessary to achieve the return of properly restored contents to a properly restored building. Emergency responders must also understand, well in advance of disaster, the laws and regulations which must be followed for the protection of workers, building occupants and the public. The authors of the Field Guide to Emergency Response assume that this level of knowledge exists among responders and based on this assumption, the Field Guide to Emergency Response dispenses with important admonitions in regard to the existence of hazardous materials. The integrity of the building and contents, as well as the health of the emergency responders and others is thereby placed at risk. Art conservators, building restorers, their experts and insurance adjusters should keep open lines of communication so that information is shared regarding any hazardous exposures which the building and its contents have sustained.

I. "The Field Guide to Emergency Response"-An Authoritative Reference Guide for Cultural Institutions

A primary reference guide upon which cultural institutions rely during the emergency phase of salvage and recovery is the "Field Guide to Emergency Response"(Heritage Preservation 2006). This reference guide for emergency salvage describes itself as "A vital tool for cultural institutions" and comprises a Handbook, Salvage Wheel (Heritage Preservation 2011) and DVD. The Guide, a FEMA sponsored publication of the National Heritage Foundation, is the result of a collaborative effort among conservators and others referred to as "The Field Guide to Emergency Response Team", "Field Guide to Emergency Response Advisers" and "DVD Content Matter Experts".

Although the Guide is widely cited by authors publishing on the subject in the United States, its' recommendations for salvage and recovery are, in important respects, misguided, misinformed and inconsistent with accepted industry standards for responding to an emergency involving water damage as described by the

Institute of Inspection, Cleaning and Restoration Certifications (IICRC)in its publication, "S500, Standard and Reference Guide for Professional Water Damage Restoration" (hereinafter "IICRC" (2006)).

II. Field Guide to Emergency Response Protocols Lead to Cross-Contamination through the Physical Disturbance of Building Materials and Air Circulation

The Field Guide to Emergency Response recommends removal of flood soaked insulation, wallboard and carpets (Heritage Preservation 2006:II:19;40); it recommends use of the HVAC system to prevent mold growth (Heritage Preservation. 2006:III;39-40) and to lower humidity and temperature (Heritage Preservation 2006:II;19). The Field Guide to Emergency Response recommends the use of fans to circulate air (Heritage Preservation 2006:III;42).

These recommendations are made without regard to whether hazardous building materials have been disturbed and whether contaminants such as mold, asbestos and lead have been released during the recovery process recommended by the Field Guide to Emergency Response. If hazardous materials have been released by the physical removal of building materials which contain flood water, mold, asbestos, lead and other hazardous substances, these contaminants will spread through the use of HVAC systems and fans. The Field Guide to Emergency Response does not address the spread of contaminants and the authors appear to be misinformed in this regard.

A. The Failure to Appreciate that Hazards Exist Leads to the Unknowing Spread of Contaminants

Mold

The Field Guide to Emergency Response educates responders to avoid working in areas where mold is present if they have respiratory or immune-related problems; however, no guidance is provided with regard to the specific types of respiratory problems mold can cause or aggravate, nor why those who are immune compromised should avoid exposure. Removal of moldy building materials, without the implementation of environmental controls, is likely to release spores and cross-contaminate a building and its contents. The removal work can also create significant exposures and impact the health of workers and building occupants. Even if responders avoid contact with moldy items, their mere presence in any area where there is disturbance of mold, and potentially in any areas of cross-contamination, may place them at a risk for adverse health effects.

Although not regulated by Federal or State law, as of the date of publication of the Field Guide to Emergency Response Handbook, accepted guidelines exist with regard to handling of mold in water-damaged buildings. (ANSI/IICRC S500, 2006)

These guidelines are not referenced in the Field Guide to Emergency Response's bibliography.

Water

The Field Guide to Emergency Response's description of contaminated water is limited and ambiguous. It describes contaminated water as water from a "sewage back-up" or water containing "hazardous materials". (Heritage Preservation 2006:III;35). There is no reference to Category Three water or "black" water as described by the IICRC S500. (ANSI/IICRC S500, 2006:14). Moreover, as water evaporates and aerosolizes, contaminants from water can be pulled into HVAC systems and circulate through the use of fans.

Other Hazardous Exposures

Other hazardous exposures are only minimally referenced in the Field Guide to Emergency Response (Heritage Preservation, 2006:III;50). For example, while asbestos is discussed in the last section of the Field Guide to Emergency Response's "Top Ten Problems to Expect", it is relegated to the end of the ten steps, and follows discussions about "Water", "Extreme Environmental Conditions", "Mold" "Mud" "Bleeding Dyes" "Corrosion" "Soot and Ash" "Broken Objects" and "Pests". None of the health hazards associated with exposure to asbestos are mentioned, although asbestos is known to cause lung disease and cancer. There is no discussion of the OSHA regulations which, under circumstances clearly defined in the regulations, require personal protective equipment including specially fitted respirators, personal exposure monitors, negatively pressurized containment chambers, air sampling inside and outside of containment chambers and disposal of asbestos waste at specially designated disposal sites (29 CFR 1910.132, 40 CFR 61.150 et seq. and see 29 CFR 1926).

The Field Guide to Emergency Response warns "if asbestos has been exposed during an emergency, it must be handled by professionals" (Heritage Preservation 2006:111;51). Since responders are not told that they should presume asbestos is present in a building constructed prior to 1980 and are not told what materials are likely to contain asbestos, responders are not in a position to know that asbestos has been "exposed", nor when it is necessary to follow OSHA mandates for worker protection.

The Field Guide to Emergency Response's failure to provide sufficient information to emergency responders removing flood soaked insulation and wallboard in a pre-1980 building results in unknowing contamination of the building, its contents and exposures to those performing the work. Additionally, if the HVAC system or fans have been used to circulate air, widespread contamination can result. This is also true where other hazardous materials are present but not identified.

The Field Guide to Emergency Response does not reference the IICRC S500 Standard which provides information about products containing asbestos, health hazards and OSHA mandates.

III. How Methods Utilized by Cultural Institutions Following the Field Guide to Emergency Response Compare with Industry Standards and Cause More Extensive Damage

The Organization which developed the S500 Standard

The IICRC is an organization, comprised of industry experts, trade associations, educational institutions, training schools and other organizations, which published an industry Standard for building water restoration. The S500 Standard is a consensus document based on the IICRC Standards Committee's review of scientific and industry literature, information from sources in the scientific community, international, national and regional trade associations with expertise and experience in professional disaster restoration, restoration service companies, the insurance industry, and many others (ANSI/IICRC S500, 2006:10). The S500 has been approved by the American National Standards Institute [hereinafter "ANSI"].

IICRC standards recommend inspection and documentation of hazardous conditions in a building before starting work. The standards warn against physical disturbance of materials until a building is assessed for hazards to avoid the spread of contamination and unsafe exposures (ANSI/IICRC S500, 2006:9.3.2;32). The IICRC also warns that hazardous materials such as asbestos and lead must be handled by experts with certifications in their field. (ANSI/IICRC S500, 2006: 10.5.2;44). The IICRC references OSHA regulations regarding the handling of hazardous materials such as asbestos (ANSI/IICRC S500, 2006:7.6;24), and states that "The presence of a hazardous or regulated material shall be carefully evaluated to determine if the restorer and its employees are qualified to work in that environment". (ANSI/IICRC S500, 2006:10.5.1;44).

The S500 Standard describes five principles of water damage restoration: the health and safety of workers and building occupants; documentation of conditions and work procedures; mitigation (controlling the spread of contaminants and controlling moisture); principles of drying; and cleaning and repair.(ANSI/IICRC S500, 2006:1;93-95) It calls for the identification of hazardous materials before they are physically disturbed so that the appropriate experts can be engaged and the environment controlled to prevent cross-contamination (ANSI/IICRC S500, 2006:9.3.2;32).

Knowledge that hazards exist leads to the ability to control them.

HVAC Use and Air Circulation

The IICRC Standard, as distinct from the Field Guide to Emergency Response, recommends limits on the use of HVAC systems and air movers to reduce humidity, temperature and moisture. The Standard warns against their use where there is concern that HVAC ducts have been contaminated with hazardous materials. The Standard further warns against the use of airflow in an environment known or suspected to have been "severely" contaminated and where there is a "significant possibility" of cross-contamination or health and safety issues. In such situations, restorers are advised to minimize the use of airflow for the purpose of drying(ANSI/IICRC S500, 2006:12.1.19;50-51 and 12.1.20;51).

Mold

The S500 Standard recommends that if mold is discovered in the course of restoration work, the work should be stopped and contained. "Further drying and mold remediation" should be performed by trained remediators following the IICRC S520" (ANSI/IICRC S500, 2006:7.5;25).The S500 dedicates an entire Chapter to the health effects of microbial contamination and provides a bibliography of peer-reviewed medical articles which have been published on the subject (ANSI/IICRC S500, 2006:3;100-104). The Field Guide to Emergency Response provides no similar references, nor even reference to the S500 or S520 Standards.

Contaminated Water

The S500 Standard defines three categories of water: clean, gray and black. The category of water which is of most concern in a water damaged building is Category Three, also known as "black water". Black water includes sewage water, ground surface water, rising water from rivers or streams, or water that has been in contact with pesticides, heavy metals, toxic organic substances or other hazardous materials outside or inside a building (ANSI/IICRC S500, 2006:9.6;33-34). Category Three water contains microorganisms which when aerosolized have the potential to cross-contaminate unaffected areas in a building and its contents (ANSI/IICRC S500, Appendix D, 2006:310-311). Professional judgment must be used to decide how to dry an environment which has Category Three water (ANSI/IICRC S500, 2006:316).

Other Hazardous Materials

The S500 reference guide, Chapter 9, outlines the Federal regulations which are at play when work involves contact with asbestos and lead. The Standard provides that if asbestos is known or presumed, work should stop and a licensed asbestos abatement contractor engaged (ANSI/IICRC S500, 2006:7.7;24). The Standard further alerts responders to engage an abatement contractor where building products have been identified to contain lead. The Standard advises emergency responders to identify hazardous materials during the initial inspection.

The Field Guide to Emergency Response is not directed to professionals with expertise in water restoration but to the staff of cultural institutions who may not have any education or experience dealing with emergencies involving hazardous materials. Because those following Field Guide to Emergency Response recommendations unknowingly create hazardous conditions, the hazards may never be investigated and discovered. The failure to appreciate that hazards exist results in the persistence of contaminants in the building and on contents.

Further, the Field Guide to Emergency Response assumes that because the law requires institutions to identify where asbestos and lead containing products are located in buildings (Heritage Preservation, 2006:51), it is therefore unnecessary to provide adequate warnings. The authors therefore assume a certain level of knowledge which in all likelihood does not exist among cultural institutions. And as the authors of the Field Guide to Emergency Response acknowledge: "After an emergency, there is little time for salvage training" (Heritage Preservation, 2006:II;32).

IV. Proper Risk Management Involves Reliance on Industry Standards

Proper risk management is accomplished by following accepted industry standards. During an emergency, risk is properly managed when experts are engaged to inspect, test, evaluate and perform the work. Risk is managed when the health and safety of workers and building occupants is protected, and laws and regulations are followed. There should be continuity in the communications between building and content restorers so that each understands the nature and scope of the hazards which have to be addressed during the emergency phase of their work and in the process of restoration.

As illustrated by the differences in approach among proponents of the Field Guide to Emergency Response and those of the S500 Standards Committee, risk management decisions often conflict with the best interests of cultural institutions due to the significant cost involved in proper restoration. Risk management decisions must be carefully evaluated to determine whether the economics of restoration are controlling the restoration methods beeing used. However, without an understanding of the principles of water restoration and industry standards, cultural institutions do not have the knowledge to demand that the emergency phase of addressing a disaster through restoration of its building and contents, be approached differently.

CONCLUSION

Successful restoration of a water-damaged building, and the contents it contains, depends on early identification of hazards and the avoidance of activities which create greater harm, or risks of harm to the structure and contents. Art conservators appear to have little education regarding industry standards for building water restoration. If emergency responders are unaware that they are working with hazardous materials, contamination may never be discovered and remediated. Contaminants may therefore be introduced into the art restorers' studio, art storage facilities or any building to which the contaminated objects are moved. If a building is properly restored but its contents are not, the unrestored property will re-introduce contaminants. Conversely, a building which is improperly restored, will re-contaminate properly restored contents.

All of these unfortunate results can be avoided if emergency responders are properly educated to follow well-established methods, and to engage professionals, when necessary, to inspect, test, evaluate and develop a strategy for stabilizing the environment and restoring the building and its contents.

REFERENCES

- ANSI/IICRC S500. Standard and Reference Guide for Professional Water Damage Restoration. Institute of Inspection Cleaning and Restoration Certification (IICRC) Vancouver, WA. 3rd edition. 2006. In [http://iicrc.org/standards/iicrc-s500/] 9/2012
- ANSI/IICRC S500. Standard and Reference Guide for Professional Water Damage Restoration. Appendix D, IICRC Technical Advisory on "In-Place" Drying. Vancouver, WA, 3rd edition. 2006:310-311.
- CFR; 29 CFR 1910.132, 40 CFR 61.150 et seq. and see 29 CFR 1926.
- Hawks C et al. Health & Safety of Museum Professionals, Society for the Preservation of National History Collections. Managing editor Butts SH. New York. 2011.
- Heritage Preservation. Emergency Response and Salvage Wheel. Washington, DC. 2011. In [https://www.heritagepreservation.org/catalog/Wheel1.htm] 9/2012
- Heritage Preservation. Field Guide to Emergency Response. Supra, Washington, DC. 2006
- Heritage Preservation. Field Guide to Emergency Response. Washington, DC. 2006.
- Heritage Preservation. Field Guide to Emergency Response. Section II, The Response Team. Washington, DC. 2006.

INDEX

AUTHORS AND CONTACT INFORMATION (PAGE)

- Harriet M. **Ammann** Ph.D., D.A.B.T.; Affiliate and Contact Information Associate Professor, Environmental Health, School of Public Health and Community Medicine, University of Washington, Seattle Washington; Ammann Toxicology Consulting LLC. 14, 311
- Dr. Ojan **Assadian**, Institute of Hygiene and Environmental Medicine, University Medicine Greifswald, Walther-Rathenau-Str. 49a, Germany 17489 Greifswald. 174
- Daniel **Aubin**, National Research Council's Construction Portfolio, Ottawa, Canada. 148, 149, 154
- Christiane **Baschien**, Federal Environment Agency, Germany. 328
- Kerstin **Becker**, Federal Environment Agency, Germany. 328
- Dr. Harald **Below**; Institute of Hygiene and Environmental Medicine, University Medicine Greifswald, Walther-Rathenau-Str. 49a,D- 17489 Greifswald, Germany . 24, 174, 175, 180
- Doris **Betancourt**, National Risk Management Research Laboratory, US Environmental Protection Agency, Research Triangle Park, NC. 125, 234
- Thomas **Brüning**, Institute for Prevention and Occupational Medicine of the German Social Accident Insurance Institute of the Ruhr University Bochum (IPA); Bürkle de la Camp Platz 1, 44789 Bochum, Germany. 46
- Jürgen **Bünger**, Institute for Prevention and Occupational Medicine of the German Social Accident Insurance Institute of the Ruhr University Bochum (IPA); Bürkle de la Camp Platz 1, 44789 Bochum, Germany, phone: +49 234 302; 4556, fax: +49 234 302 4505, buenger@ipadguv.de . 15, 46, 47, 50, 53,
- David A. **Butler**, Ph.D.; Senior Program Officer; Institut of Medicine, Washington D.C., USA. 17, 21
- Grace **Byfield**, Center for Microbial Communities and Health Research, RTI International, 3040 Cornwallis Rd, Research Triangle Park, NC. 125
- V **Campos**; Microbiology Department, Biological Sciences Faculty, Universidad de Concepción. Chile. 259
- A.R. **Cavaliere**, Professor Emeritus, Department of Biology, Gettysburg College. 293
- Denis **Charpin**, Department of Pulmonology and Allergy, Marseille University Hospitals, France. Phone: +33-491-968631, Fax: +33-491-968902, Email: denis-andre.charpin@ap-hm.fr . 15, 68,
- Yong Joo **Chung**, US EPA, ORD, NHEERL, RTP, NC, USA. 222
- Eugene C. **Cole**, Dr PH, Brigham Young University, UT, USA . 359
- Lisa B **Copeland**, US EPA, ORD, NHEERL, RTP, NC, USA. 222
- Jean M. **Cox-Ganser**, Ph.D. Centers for Disease Control and Prevention, National Institute for Occupational Safety and Health, Morgantown, WV, USA. 15, 23, 25, 27
- Timothy **Dean**, National Risk Management Research Laboratory, US Environmental Protection Agency, Research Triangle Park, NC. 125

- Robert **DeBernardo**, MD MBA MPH Center for Indoor Environments and Health, University of Connecticut Health Center. 363
- Anja **Deckert**, Institute for Prevention and Occupational Medicine of the German Social Accident Insurance Institute of the Ruhr University Bochum (IPA); Bürkle de la Camp Platz 1, 44789 Bochum, Germany, 46
- Leylâ **Deger**; Institut national de santé publique du Québec, Montréal, Canada. 73
- Anthony **Devine**, Center for Microbial Communities and Health Research, RTI International, 3040 Cornwallis Rd, Research Triangle Park, NC. 125
- Don **Doerfler**, US EPA, ORD, NHEERL, RTP, NC, USA. 222
- G. **Escalante**, Microbiology Department, Biological Sciences Faculty, Universidad de Concepción. EULA-CHILE Center. CHILE. 259
- Karin **Foarde** , Center for Microbial Communities and Health Research, RTI International, 3040 Cornwallis Rd, Research Triangle Park, NC. 125
- Michel **Fournier**; Direction de santé publique, Agence de la santé et des services sociaux de Montréal. CANADA. 73
- Yves **Frenette**, Hygiéniste de l'environnement DSP Montreal, QC , CANADA. yfrenett@santepub-mtl.qc.ca; 207
- Laurie M. **Freyder**, MPH; School of Public Health & Tropical Medicine and School of Medicine, Tulane University Health Science Center, New Orleans, LA 70112. 61
- Manfred **Gareis**; Max Rubner Institute, Institute for Microbiology and Biotechnology Kulmbach, Germany. 185
- Denis **Gauvin**, Institut national de santé publique du Québec, Québec City, Canada. 148
- Sophie **Goudreau**; Direction de santé publique, Agence de la santé et des services sociaux de Montréal, Canada. 73
- N **Gqaleni**; Department of Occupational and Environmental Health, Howard College Campus, University of KwaZulu-Natal, Durban 4041, South Africa. Telephone: +27 31 260 4280; Facsimile: +27 31 260 4650, email: gqalenin@ukzn.ac.za; 280
- Jan **Grajewski**, Kazimierz Wielki University , Institute of Experimental Biology, Division of Physiology and Toxicology, Chodkiwicza Street 30, Bydgoszcz, Poland. 15, 185
- Sirkku **Häkkilä**, University of Turku, Aerobiology Unit, CERUT, Finland, sirhak@utu.fi; 201
- Ulla **Haverinen-Shaughnessy**, National Institute for Health and Welfare, Environmental Health Department, P.O. Box 95, 70701 Kuopio, Finland; Email ulla.haverinen(at)thl.fi; 215
- Frank **Hoffmeyer**, Institute for Prevention and Occupational Medicine of the German Social Accident Insurance Institute of the Ruhr University Bochum (IPA); Bürkle de la Camp Platz 1, 44789 Bochum, Germany; 46
- Anne **Hyvärinen** Ph.D.; Kuopio, Finnland; 30
- Louis, **Jacques**, Direction de santé publique, Agence de la santé et des services sociaux de Montréal; Département de médecine sociale et préventive, Faculté de médecine, Université de Montréal; Département de santé environnementale et santé au travail,

Faculté de médecine, Université de Montréal; Clinique interuniversitaire de santé au travail et de santé environnementale, Institut thoracique de Montréal; Department of family medicine, McGill University, CANADA; 73

- N **Jafta**; Department of Occupational and Environmental Health, Howard College Campus, University of KwaZulu-Natal, Durban 4041, South Africa. Telephone: +27 31 260 4280; Facsimile: +27 31 260 4650, email: gqalenin@ukzn.ac.za; 280
- Joseph Q. **Jarvis**, M.D., MSPH, JQJCI, Salt Lake City, UT, USA; 118
- Eckardt **Johanning**, M.D., M.Sc., PhD, Fungal Research Group Foundation, Inc., 4 Executive Park Drive, Albany, N.Y., USA; johanningmd.com 195
- Robert N. **Jones**, MD, School of Public Health & Tropical Medicine and School of Medicine, Tulane University Health Science Center, New Orleans, LA 70112; 61
- Heinz-Jörn **Moriske**, Federal Environment Agency, Germany; 328
- Karen H. **Kahn**, Esq., Consultant, Berkeley, Calinfornia, USA; 372
- E B **Kern**, MD, Gromo Foundation for Medical Education and Research; Department of Otorhinolaryngology, State University of New York at Buffalo, Buffalo, New York. 89
- Kaye H. **Kilburn**, M.D. Ralph Edgington Professor of Internal Medicine University of Southern California Keck School of Medicine (ret.); President of Neuro-Test, Inc. 3250 Mesaloa Lane - Pasadena, California 91107 Telephone: (626) 798-4299 - Fax: (626) 798-3859; E-mail: kayekilburn@neuro-test.com; 130
- Jean **Kim**, Center for Microbial Communities and Health Research, RTI International, 3040 Cornwallis Rd, Research Triangle Park, NC; 125
- H. **Kita**, MD; Department of Immunology Mayo Clinic, Rochester, Minnesota; 89, 93
- Marike **Kolossa-Gehring**, Federal Environment Agency, Germany; 328
- Prof. Dr. Axel **Kramer**; Institute of Hygiene and Environmental Medicine, University Medicine Greifswald, Walther-Rathenau-Str. 49a,D- 17489 Greifswald, fon +49-3834-515542, fax +49-3834-515541, Email: kramer@uni-greifswald.de; 174
- Pierre **Lajoie**, Institut national de santé publique du Québec, Québec City, Canada; 148
- John J. **Lefante**, PhD; School of Public Health & Tropical Medicine and School of Medicine, Tulane University Health Science Center, New Orleans, LA 70112; 61
- C **León**; Microbiology Department, Biological Sciences Faculty, Universidad de Concepción. Chile: 9
- De-Wei **Li**; The Connecticut Agricultural Experiment Station Valley Laboratory, Windsor, CT (dewei.li@ct.gov); 236
- Wolfgang **Lorenz**, Ph.D. Institut für Innenraumdiagnostik, Marconistr. 23, 40589 Düsseldorf, GERMANY, Tel.: +49-211-99958160; Wolfgang.Lorenz@infid.de; 336
- Erwin **Maertlbauer**, D.V.M., Dr., Dr. Prof., Milk Hygiene, University of Munich, Munich, Germany; 163
- Stefan **Mayer**, Institution for statutory accident insurance and prevention in the trade and goods distribution (BGHW), M5, 7, D-68161 Mannheim, Germany; 52
- Marc **Menetrez**; National Risk Management Research Laboratory, US Environmental Protection Agency, Research Triangle Park, NC; 125

- M.A, **Mondaca**. Microbiology Department, Biological Sciences Faculty, Universidad de Concepción. EULA-CHILE Center, Universidad de Concepción. Chile; 259
- Philip R **Morey** PhD CIH; ENVIRON International Corporation, Gettysburg, PA, USA, pmorey@environcorp.com; 15, 263,
- BSc. hon. Judith **Mueller**; Labor Urbanus GmbH, Wagnerstrasse 15, DE-40212 Düsseldorf, Germany. Tel: +49 211378070 Fax: +49 211378071; email mueller@labor-urbanus.de; 246, 253
- Aino **Nevalainen** Ph.D., Professor, Kuopio, Finnland; aino.nevalainen@thl.fi; 15, 30
- Hélène **Niculita-Hirzel**, Institute for Work and Health, University of Lausanne and Geneva, Rue du Bugnon 21, 1011 Lausanne, Switzerland, Tel. +41 (0)21 314 71 47; Fax : +41 (0)21 314 74 20; Email: Helene.Hirzel@hospvd.ch; 270
- BA **Nkala**; Department of Occupational and Environmental Health, Howard College Campus, University of KwaZulu-Natal, Durban 4041, South Africa. Telephone: +27 31 260 4280; Facsimile: +27 31 260 4650, email: gqalenin@ukzn.ac.za; 280
- Anne **Oppliger**; Institute for Work and Health, University of Lausanne and Geneva, Rue du Bugnon 21, 1011 Lausanne, Switzerland; Tel. +41 (0)21 314 74 16; Fax : +41 (0)21 314 74 30: Email: Anne.Oppliger@hospvd.ch; 270
- Dr. rer. nat. Urban **Palmgren**, Labor Urbanus GmbH, Wagnerstrasse 15, DE-40212 Düsseldorf, Germany.; Tel: +49 211378070 Fax: +49 211378071; email urban.palmgren@t-online.de; 108, 246, 253
- Ju-Hyeong **Park**, Sc.D., M.P.H., C.I.H.; National Institute for Occupational Safety and Health, Division of Respiratory Disease Studies, Morgantown, West Virginia; 15, 101
- Stéphane **Perron**, Direction de santé publique, Agence de la santé et des services sociaux de Montréal; Département de médecine sociale et préventive, Faculté de médecine, Université de Montréal; Department of epidemiology and biostatistics, McGill University. CANADA; 73
- Céline **Plante**, Direction de santé publique, Agence de la santé et des services sociaux de Montréal: CANADA; 73
- J U **Ponikau**, MD; Gromo Foundation for Medical Education and Research; Department of Otorhinolaryngology, State University of New York at Buffalo, Buffalo, New York; 89
- Thomas G. **Rand**, 1Saint Mary's University, Halifax, NS, Canada. thomas.rand@smu.ca; 156
- Roy J. **Rando**, ScD; School of Public Health & Tropical Medicine and School of Medicine, Tulane University Health Science Center, New Orleans, LA 70112; 61
- Monika **Raulf-Heimsoth**, Institute for Prevention and Occupational Medicine of the German Social Accident Insurance Institute of the Ruhr University Bochum (IPA); Bürkle de la Camp Platz 1, 44789 Bochum, Germany; 46
- Helena **Rintala** Ph.D.; Kuopio, Finnland; 30
- E. Neil **Schachter**, MD; Maurice Hexter, Professor of Medicine and Community Medicine; Mount Sinai School of Medicine; NY, USA; 37

- Paula **Schenck**, MPH Center for Indoor Environments and Health, University of Connecticut, Health Center. 363
- Hans **Schleibinger**, National Research Council's Construction Portfolio, Ottawa, Canada; email: Hans.Schleibinger@nrc-cnrc.gc.ca; 148
- Richard **Shaughnessy**, Ph.D., University of Tulsa, OK, USA; 359
- D A **Sherris**, MD; Gromo Foundation for Medical Education and Research; Department of Otorhinolaryngology, State University of New York at Buffalo, Buffalo, New York; 89
- Audrey **Smargiassi;** Département de santé environnementale et santé au travail, Faculté de médecine, Université de Montréal; Institut national de santé publique du Québec, Montréal, Canada; 73
- Michael **Sulyok**, University of Natural Resources and Life Sciences Vienna, Center for Analytical Chemistry (CAC), IFA-Tulln, Konrad Lorenz Strasse 20, A-3430 Tulln, Austria; 52
- Hanna **Szczepanowska**; Research Conservator, MCI, Smithsonian Institution, Washington DC; E-mail: szczepanowskah@si.edu; tel: 301-238-1232; www.si.edu/mci; 293
- Regine **Szewzyk**, Federal Environment Agency, Germany; 328
- Dirk **Taeger**, Institute for Prevention and Occupational Medicine of the German Social Accident Insurance Institute of the Ruhr University Bochum (IPA); Bürkle de la Camp Platz 1, 44789 Bochum, Germany; 46
- Martin **Täubel** Ph.D., Kuopio, Finnland; 30
- Robert L. **Thivierge**; Centre hospitalier universitaire Sainte-Justine, Département de pédiatrie, Montréal, Canada; Université de Montréal, Développement professionnel continu, Centre de pédagogie appliquée aux sciences de la santé, Montréal, Canada; 73
- Magdalena **Twarużek**, Kazimierz Wielki University , Institute of Experimental Biology, Division of Physiology and Toxicology, Chodkiwicza Street 30, Bydgoszcz, Poland; 185
- R **Urrutia**; Microbiology Department, Biological Sciences Faculty, Universidad de Concepción. EULA-CHILE Center, Universidad de Concepción, Chile; 259
- Vera **van Kampen**, Institute for Prevention and Occupational Medicine of the German Social Accident Insurance Institute of the Ruhr University Bochum (IPA); Bürkle de la Camp Platz 1, 44789 Bochum, Germany; 46
- Stephen **Vesper**, US EPA, ORD, NERL, Cincinnati, OH, USA; 222
- Vinay Vishwanath, University of Natural Resources and Life Sciences Vienna, Center for Analytical Chemistry (CAC), IFA-Tulln, Konrad Lorenz Strasse 20, A-3430 Tulln, Austria; 52
- Marsha D. W. **Ward**, US EPA, ORD, NHEERL, Cardiopulmonary and Immunotoxicology Branch ,NC, USA; Tel #: (919) 541-1193; FAX: (919) 541-0026; e-mail: ward.marsha@epamail.epa.gov; 222
- Donald M. **Weekes**, CIH, CSP, InAIR Environmental, Ltd., Ottawa, ON, CANADA; 347

- Doyun **Won**, National Research Council's Construction Portfolio, Ottawa, CANADA; 148
- Chin S. **Yang**, Prestige EnviroMicrobiology, Inc., Voorhees, New Jersey (chins.yang@prestige-em.com); 236
- Wenping **Yang**, National Research Council's Construction Portfolio, Ottawa, CANADA; 148

INDEX

Index: Fungi and Bacteria